Midterm No. 1.
after 2nd section II
textile fibers

Introductory

Textile Science

Introductory

Textile Science

SECOND EDITION

Marjory L. Joseph

San Fernando Valley State College

HOLT, RINEHART AND WINSTON, INC.
New York   Chicago   San Francisco   Atlanta
Dallas   Montreal   Toronto   London   Sydney

TO RUTH W. AYRES

The following photographs were taken by John King, New Canaan, Conn.: 5.18, 8.10, 16.6, 18.6, 19.8, 22.25, 22.30, 23.3, 26.3, 26.5, 29.2, 30.4, 30.6; Pls. 7, 9, 12.

Library of Congress Catalog Card Number: 70-171525
ISBN: 0-03-086679-0
Printed in the United States of America
2 3 4 5   071   9 8 7 6 5 4 3 2 1

Preface

In writing *Introductory Textile Science* I have wanted to provide for university students taking their first course in the subject matter a text that could accomplish the following objectives:

- Stimulate within each reader a desire to recognize and appreciate textile fabrics.
- Offer sound scientific theory concerning fibers, including their processing and treatment, so that the student will have a solid foundation for a clear understanding of textiles, their selection, use, and care.
- Establish guides that might help the student use scientific data in the selection, use, and care of textiles.
- Cause the reader to seek an understanding of the interrelationships among fibers, yarn structure, fabric structure, finishing, and coloring; and the importance of these interrelationships in the selection, use, and care of textiles.
- Convey information that can help the student become aware of the complexity of the textile industry and understand existing legal controls and policies.

There is no place in today's complex society where man is completely isolated from textile fabrics. They clothe him, protect him, decorate him, and provide him with many hours of enjoyment. Fabrics cover the furniture and floors of his home; they drape or curtain his windows. Textiles serve in kitchens, bathrooms, bedrooms, and living rooms. Fabrics make public buildings attractive, sometimes sound-proof, and specialized fabrics are used in industry for a variety of purposes.

Since we all have some experience with textiles virtually every day of our lives, each of us can profit from greater knowledge of textile fibers and fabrics and the ways they can be used. If the prospective consumer knows something about the textile he wishes to purchase, and if the merchant and sales staff know something about the products they offer, sales of textile products can become relatively intelligent transactions.

The five years since the publication of the first edition of *Introductory Textile Science* have been filled with important developments in textiles. Man has walked on the moon wearing and using special textile products that eventually will prove valuable to many consumers on earth. Chemists have perfected new fibers, changed existing fibers, found new ways to make fabrics, and modified existing techniques while develop-ing new ones for finishing textiles—all in the interest of giving con-sumers more satisfying textile products. Thus, in addition to the goals initially established for *Introductory Textile Science,* I have been eager for the new edition to achieve the following:

■ Report the most recent data on new and modified fibers, on changes in yarn manufacture, on developments in fabric structure, and on innovations in textile finishing or wet processing.
■ Introduce current economic data indicating changes and trends in fiber use and consumption.
■ Integrate with the content of chapters on textile fibers data on second- and third-generation fibers, fiber modification, and current usage in generic terms and trade names; and bring greater clarity to the material on fiber morphology, at the same time adding new concepts to it.
■ Cause the material on yarn manufacture to reflect the newest methods, including texturizing.
■ Place increased emphasis on knitting and consider new fabric construction processes, such as stitch-knit.
■ Expand the unit on finishing to include new developments in resin finishes, current understanding of durable press and mini-mum-care techniques, and recent developments in color appli-cation.
■ Enrich and enlarge the illustration program so as to provide greater clarity in visual demonstrations.

The text has been organized to help the teacher rearrange the units to suit the needs of individual groups of students. It is my belief that each class of students is unique and may very well require a modified approach to subject matter in order for instruction to become fully effective. One group could make a more successful start with the study of fibers, while another could respond commendably if completed fabrics serve as the original stimulus, to be followed by fiber investigation.

While realizing full well that many introductory textile courses do not claim chemistry as a prerequisite, I have felt it necessary to acknowledge that fibers are chemical in nature and that, for this reason, the relevant chemistry be included as integral to the study of textiles. However, in view of known difficulties I have made every attempt to use only the most elementary chemistry and to explain it clearly and succinctly within the context of its use in *Introductory Textile Science.* The introduction to the chemistry of fibers now included in the text will surely simplify, rather than complicate, our understanding of fiber behavior and will help provide an explanation for various reactions related to finishing, dyeing, and maintenance of fabrics.

As the book's title would suggest, the emphasis I believe should be made in modern education is on theoretical information, but I also hold that theory must be interpreted for the benefit of both students and consumers. Thus, the text should prove usable for home economics majors as well as for students in design and merchandising. I also hope that it will appeal to anyone wishing to know more about textile fabrics and their selection, use, and care.

No single text can be sufficient unto itself. The suggested references should help the reader become responsive to other interpretations and concepts and advance the student toward knowledge of ever-developing trends in textile fibers, fabrics, finishes, and care techniques. Informed awareness should lead to an enthusiastic desire to keep pace with current innovations by reading periodical publications, technical bulletins, and news items.

Textiles is a dynamic field. It combines natural science, social science, and art in a total subject that is new every day. A large segment of the population derives economic security from various manufacturing and marketing processes associated with textiles. I should feel gratified indeed if *Introductory Textile Science* could enable the reader to develop some sense of the magnitude of the textile industry and to acquire interest in and respect for fabrics and the purposes their selective use and care can serve.

Many persons and organizations helped in the preparation of the manuscript for the original edition of *Introductory Textile Science,* and I am pleased once again to express my gratitude to Dr. Ruth L. Galbraith, Mrs. Elsie Crouthamel Bassett, and Mrs. Lorraine Trebilcock for their critical scrutiny of my writing, its content, sources, and teachability; to the corporations that granted permission for the use of their illustrations; to Jack Walter for the sketches that have now become the line drawings illustrating the text; to Mr. P. B. Cole of Montgomery Ward for a sample of tufted carpeting used as the subject of a photographic illustration; to Mr. William Huling for help in processing photographs that I was bold enough to take; to Miss Christine Hamilton and Mrs. Dorothy Blackman for their expert typing of the manuscript; and to Mrs. Blackman for tracking down research data and for her valuable editorial assistance.

In preparing the new edition I incurred still further debts that I am happy to acknowledge. Once again, Mrs. Bassett was good enough to make a searching critique of the manuscript. In addition, I enjoyed the encouragement, wisdom, and science of Dr. J. W. Weaver, who generously pored over these statements to ensure as much accuracy, completeness, and currency as human endeavor could attain. And on this occasion I had the good services of Miss Doris Overocker for the preparation of the typescript.

In citing the help that friends and colleagues have extended to me, I wish at the same time to absolve them of and assume full responsibility for any errors or omissions that the reader may discover.

As always, I must admit my dependence upon the support and tolerance of my family during the months and years that I have worked on *Introductory Textile Science.*

Granada Hills, Calif.   M.L.J.
January 1972

Contents

Introductory

Textile Science

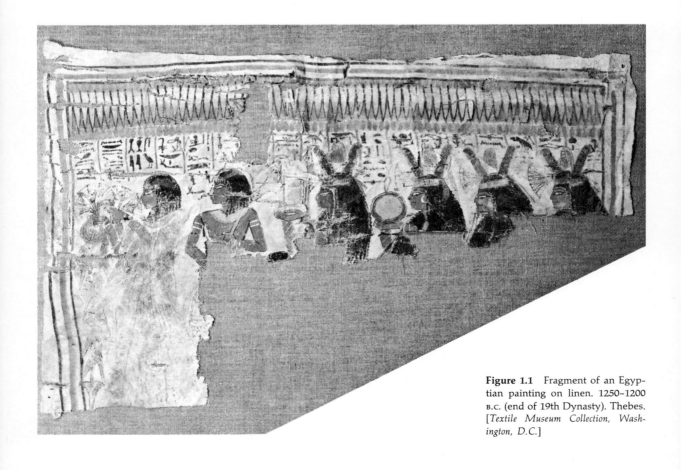

Figure 1.1 Fragment of an Egyptian painting on linen. 1250–1200 B.C. (end of 19th Dynasty). Thebes. [*Textile Museum Collection, Washington, D.C.*]

1

Introduction
to Textiles

Textile fibers and their use predate recorded history. Archeological evidence indicates that textiles of fine quality were made thousands of years before the oldest preserved accounts that refer to them. The early history of textile fibers and fabrics has been determined by such archeological finds as spinning whorls, distaff and loom weights, and fragments of fabrics found in such locations as the Swiss lake regions and Egyptian tombs (Fig. 1.1). Tales told by parents to children from generation to generation, designs and evidence of fabrics used in planning or constructing pottery items, and, eventually, written records yield further information.[1]

All early fibers were composed of natural plant or animal products. Wool, flax, cotton, and silk were most important and were employed most frequently. Later, mineral matter in the form of rock asbestos was used. History indicates that, at first, plant and animal fibers were subjected to only a minimum of processing, and most early fabrics were probably made by a simple plain weave interlacing of groups of fibers and yarns or by the knotting or plaiting fibers, grasses, or

[1]The reader interested in pursuing the early history of fibers and fabrics is referred to the books by Baity, Leggett, and Walton listed in the Bibliography.

3

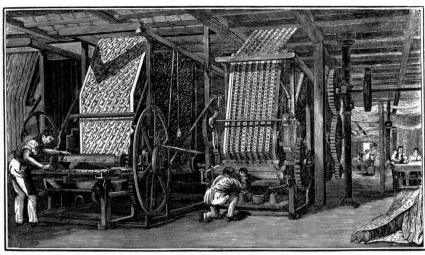

CALICO PRINTING.

Figure 1.2 American wood engraving showing interior of a textile factory. 1836. [*Library of Congress, Washington, D.C.*]

other raw materials (see Fig. IV.1). Man's ingenuity and desire to enhance his appearance led, over the centuries, to the development of complicated fabrics and, within the past hundred years, to technological expansion.

Spinning and weaving probably began during the Stone Ages. Early Stone-Age man made a covering for his body by wrapping animal furs around himself. As he learned how to fasten these skins together, clothing was created. It seems likely that weaving was developed as part of the process of interlacing branches and leaves to construct shelters. Eventually, man learned to make yarns and threads, and, finally, to manufacture cloth.

The Industrial Revolution of the eighteenth and nineteenth centuries transferred the processing of fibers and the manufacture of fabrics from the home or small cottage shop to the factory (Fig. 1.2). Mechanization gained importance, and gradually the textile industry expanded. As processing equipment and techniques were developed and as machine power replaced manpower, the use of cotton, wool, linen, and silk increased. Cotton and wool were especially affected, and the growth and production of these became the concern of governments throughout the world. Tariffs were levied, wars fought, and regimes toppled due to political, social, and economic pressures that accompanied industrial advances in textiles.

For centuries man depended upon nature for textile fibers. However, early in the twentieth century the first man-made fiber, rayon, became a practical reality, followed in the 1920s by cellulose acetate. Since the late 1930s scientists have produced literally dozens of new fibers. In fact, greater advances have been made in fibers, fabrics, finishes, and textile processing techniques since 1900 than in the four to five thousand years of recorded history prior to the twentieth century.

Since 1960 major changes have occurred in the manufacture of yarns and fabrics. Second-, third-, and even fourth-generation developments in man-made fibers have affected the behavior, use, and care of these products. Chemical finishing of fabrics and developments in dye chemistry are providing the consumer with goods that fulfill his expanding requirements. Continued research in textiles will, it is hoped, meet the challenges of the 1970s.

Textile fibers have almost limitless uses, and man is continually finding new applications for these essential materials. Fibers are inherent in the creation of clothing, home furnishings, and household textiles. They are used in the building trades; chosen for insulation in appliances; selected by industry for such products as filter cloths, pulley belts, and conveyor belts; and found in all forms of transportation. In sum, fibers play a role in nearly every type of activity or situation conceivable, including man's conquest of space.

The word *textile* comes from the Latin *textilis* and from the verb *texere*, which means "to weave." Today, a *textile* is freely defined as *any product made from fibers,* and the name is applied to nonwoven fabrics, knitted fabrics, and special fabric constructions in addition to woven goods. When the word *textiles* is used with the term *fiber,* it refers to *any product capable of being woven or otherwise made into fabrics.* The broad definition of the term is accepted and used today by nearly everyone who works with fibers or fabrics at any stage of manufacture or processing.

The textile industry is one of the largest in the world. If all facets of this vast economic giant are considered, it is quite possible that it involves more people and more money than any other industry. Even when growth, production, manufacturing, and processing of fibers to fabrics are the only areas considered, the textile industry still ranks at the top of the economic structure in terms of both manpower and dollar value.

In 1961 over 6.5 billion pounds of fiber were consumed in the United States alone. In 1965, 8.5 billion pounds were used and in 1969, 9.8 billion pounds. Figures for the per-capita consumption of textile fibers in the United States are given in Table 1.1.

Per-capita consumption of fibers dropped slightly between 1950 and 1960. An increase occurred after 1960 and continued to 1969. This increase may be partially attributed to economic conditions. The slight drop in 1969 and again in 1970 can be explained by several factors. An increase in the use of man-made fibers provides more yardage at less weight, so total yardage has probably remained nearly the same or increased slightly over preceding years. Furthermore, the fashions of these two years required less material. The longer styles of the 1970s

Table 1.1   United States Civilian Per-capita Consumption of Fibers in Pounds[a]

Year	Man-made Fibers	Cotton	Wool	Total
1950	9.5	29.3	4.5	43.3
1955	11.0	25.4	3.3	39.7
1960	10.0	23.3	3.4	36.7
1962	12.7	22.9	3.4	39.0
1964	16.2	22.5	3.0	41.7
1966	20.2	25.1	2.9	48.2
1968	26.6	22.0	2.6	51.2
1969	27.8	20.6	2.3	50.7
1970	27.7	19.8	1.7	49.2

[a] Adapted from *Textile Organon.*

may increase yardage consumption considerably over that of the immediate past. Another factor that might account, in part, for the apparently lower per-capita consumption was the economic recession that occurred during 1970.

Total fiber consumed in the United States followed much the same pattern as per-capita consumption. It increased steadily through 1969, partly because of the general population increase and the affluent economy of the 1960s. A slight drop in 1970 may be due to a leveling off in consumer needs and to the factors cited for the decrease in per-capita consumption. It will be interesting to observe fiber consumption patterns in the 1970s.

World production of textile fibers has increased steadily since 1950. Table 1.2 shows the pattern. Note the changes in percent of production

Table 1.2   World Production of Man-made Fibers, Cotton, Wool, and Silk[a]
Data in millions of pounds. Percentage of total in parentheses.

Year	Man-made Fibers		Cotton	Raw Wool	Silk	Grand Total
	Rayon and Acetate	Noncellulosic				
1950	3553 (17)	153 (*)	14,654 (71)	2330 (11)	42 (*)	20,732
1955	5030 (17)	587 (02)	20,926 (71)	2789 (10)	64 (*)	29,389
1960	5749 (17)	1548 (05)	22,295 (68)	3225 (10)	68 (*)	32,885
1962	6315 (18)	2381 (07)	23,052 (66)	3257 (9)	73 (*)	35,078
1964	7245 (18)	3728 (10)	24,930 (64)	3263 (8)	72 (*)	39,238
1966	7370 (19)	5473 (14)	23,274 (59)	3387 (9)	78 (*)	39,580
1968	7776 (17)	8336 (18)	25,629 (56)	3537 (8)	83 (*)	45,372
1969	7837 (17)	9683 (21)	24,820 (54)	3548 (8)	84 (*)	45,972
1970	7573 (16)	10871 (23)	24,947 (53)	3499 (8)	83 (*)	46,973

* Less than 1%
[a] Adapted from *Textile Organon*

of noncellulosic fibers as compared with other fibers, especially rayon, acetate, and wool. These figures should indicate clearly the importance of textile fibers to both the national and the world economy.

In addition to the economic aspects, textile fibers and fabrics are important because of their esthetic properties. Man's ever-present desire to enhance his surroundings is apparent in his constant search for new and different fashion fabrics. Developments in fabric and yarn structure, as well as in finishes and methods of color application, occur continually.

Still another factor that exercises considerable effect on the textile industry is the prevalent quest for apparel and domestic fabrics that are not only attractive but comfortable and durable. The modern consumer also wants fabrics that require a minimum of care. This desire has been satisfied to some degree by tremendous scientific advances in finishing technology, which have created fibers and fabrics in the form of "durable press," "no-iron knits," and other products that need little care, give good service, and at the same time are beautiful to look at.

The question is frequently asked, "Why is scientific knowledge of textiles valuable and important?" There is no short answer to such a question, but this book should provide key ideas that will help inform the reader. Every individual uses textile products in some form and should be concerned with the selection and maintenance of these products. This concern can best be realized through intelligent awareness of available technical data. Knowledge of proper maintenance techniques for the many fibers, fabrics, and finishes will help ensure satisfaction. Understanding of chemical, physical, and microbiological properties of fibers, and of the behavior that results from techniques used in creating fabrics, will guide the consumer in making wise selections when purchasing textile products.

Fabrics are characterized by certain properties, components, or definitive parts. Each individual fabric does not possess every possible component; however, many fabrics are produced by combining all definitive parts, and every fabric exhibits certain properties.

Fabrics are composed of textile fibers. These fibers may be short or long, fine or coarse, soft or stiff, smooth or rough. Fibers are the basic building blocks used in manufacturing a fabric. In some fabrics—such as felts and other nonwoven cloths—fibers are arranged in a predetermined pattern and processed directly into a fabric. Frequently, however, the fibers are initially formed, by one of several possible techniques, into yarns. These yarns may be simple or complex, single or ply, smooth or rough, highly twisted or loosely twisted. The yarns are then made into fabrics by one of the various methods used for fabric construction, such as knitting, weaving, knotting, or braiding.

After manufacture of the actual fabric, selected finishing or "wet-processing" procedures are applied in order to create the desired end product. Nearly every fabric receives at least one finish, and most receive several. By comparing "nonfinished" with "finished" fabric, one can easily see why many textile authorities say that modern fabrics are the result of modern finishes.

A final definitive part or property of a fabric is color. Some authors include color as a subdivision of wet processing or finishing, while others list color as an independent item. This book will discuss finish and color separately, for it is easier to understand both techniques if they are discussed individually.

In summary, definitive or component parts of a textile product include the fiber or fibers used (fiber content), the way fibers are arranged or made into yarns (yarn structure) for processing into fabrics, methods used to form arranged fibers or yarns into fabrics (fabric structure), finishing procedures and processes (finish), and coloring methods (color application). All fabrics have fiber content, some form of fiber arrangement, and fabric structure. The other properties may not be involved in every fabric; however, most fabrics utilize yarn structures and nearly all have received finishing treatments and color in some form.

Part I of this text discusses general fiber theory and classification. The remainder of the book has been arranged in the order in which the definitive parts normally evolve in the manufacture of textile fabrics. Part II is devoted to a study of the chemical, physical, microscopic, and biological properties of fibers. It contains a summary of the history of natural fibers and a brief look at the background of man-made fibers. Processes used in production and guides for selection and care are given for each fiber. Part III is devoted to yarn structure; Part IV to fabric structure. Part V deals with finishing procedures and color application. Part VI is aimed at the end-use of fabrics and includes information that should prove helpful to the consumer.

The study of textiles is based upon scientific principles, and these will be stressed throughout this text.

Fiber Theory
and Classification

As the textile industry has grown, a science has evolved, which, like any other discipline, has its own language, terminology, and methods of categorization. The scope of textile knowledge and technology has expanded at a fantastic rate in this century, so it has been necessary to refine the definitions and classes that have been established—to sharpen the tools of the trade. The next three chapters explain many of the descriptive terms that are used in textile science and outline a basic system of classification that will help the student to cope with the myriad textile fibers available today.

Fiber
Theory

All fibers available on today's market come from natural vegetable, animal, or mineral matter, or from manufacturing processes that utilize natural fibrous materials or synthesize fibers from other chemicals. Textile fibers are found in natural sources such as seed pods (cotton) or animal hair (wool). They can be manufactured from natural fibrous materials (rayon) or synthesized from chemicals that bear no resemblance to fibrous forms (nylon). Despite the source, however, it is possible to identify certain qualities common to all fibers.

FIBER PROPERTIES

In order to qualify as a suitable substance for use as a textile fiber, the product must possess certain essential properties or characteristics. These primary properties include high length-to-breadth (width) ratio, tenacity or adequate strength, flexibility or pliability, cohesiveness or spinning quality, and uniformity. Secondary properties are those that are desirable but not essential. They may improve consumer satisfaction with the fiber. Characteristics in this group include physical shape, specific gravity, luster, moisture regain, elastic recovery, elongation, resilience, thermal behavior, resistance to biological organisms, and resistance to chemicals and other environmental conditions. The following discussion includes definitions and descriptions of these properties as they apply to textile fibers.

Primary Properties

High Length-to-Width Ratio   Fibrous materials must possess adequate staple or fiber length, and the length must be considerably greater than the diameter. This is referred to as the length-to-width or length-to-breadth ratio. A minimum ratio of 100 is usually considered essential, and most fibers have ratios that are much higher.

Fibers shorter than $\frac{1}{2}$ inch are seldom used in yarn manufacturing. Short fibers such as cotton measure from less than $\frac{1}{2}$ inch to more than 2 inches in length. A cotton fiber 1 inch in length may have a diameter of 0.0007 inch, which would give a length-to-diameter ratio of 1400. Typical ratios for several natural fibers are[1]

cotton	1400
wool	3000
flax	170
ramie	3000
silk	33×10^6

The fiber molecules, as well as the fiber itself, are usually long and extremely thin. The monomers that form the fiber molecules or fiber polymer produce linear chains, and these are all characterized by minute diameter compared to molecular or polymer length.

Fibers that are measured in inches or fractions thereof are called *staple fibers*. Those measured in meters or yards and, in the case of manufactured fibers, in kilometers or miles, are called *filament fibers*.

Tenacity   A second primary property of textile fibers is adequate strength. While strength varies considerably among the different fibers, it is important that the substance possess sufficient strength to be worked and processed by machinery, as well as to provide adequate durability in the end-use to which it is allocated. *Tenacity* is the term usually applied to strength of individual fibers. The American Society for Testing and Materials (ASTM) defines tenacity as

> the tensile stress when expressed as force per unit linear density of the unstrained specimen; for example, grams-force per tex, or grams-force per denier.[2]

Denier is a unit of yarn number and is equal to the weight in grams of 9000 meters.

[1]R. W. Moncrieff, *Man-made Fibers*, 5th ed. (New York: John Wiley & Sons, Inc., 1971), p. 20.

[2]American Society for Testing and Materials, *Yearbook*, Vol. 24 (1970), p. 42.

Table 2.1   Fiber Tenacities[a]
Data obtained at 20°C (70°F), 65 percent relative humidity

	Grams per Denier		*Grams per Denier*
Asbestos	2.5–3.1	Glass	6.3–6.9
Cotton, raw	3.0–4.9	Modacrylic	2.5–3.0
Flax	2.6–7.7	Nylon, regular	2.5–6.7
Hemp	5.8–6.8	Nylon, HT	7.5–8.3
Jute	3.0–5.8	Polypropylene olefin	3.0–7.0
Ramie	5.5	Polyester	4.6–7.0
Silk	2.4–5.1	Saran	1.1–2.3
Wool	1.0–1.7	Spandex	0.75–0.9
Acrylic	2.0–4.0	Rayon, regular	0.7–3.2
Acetate	1.2–1.5	Rayon, HT	3.0–5.0
Triacetate	1.2–1.4	Rayon, HWM	2.5–5.0

[a] Adapted from E. R. Kaswell, *Textile Fibers, Yarns and Fabrics* (New York: Reinhold Publishing Corporation, 1953); and 1970 Fiber Properties Chart, *Textile Industries.*

Tenacity is determined by mechanical devices and proper mathematical conversion formulas. Tenacities of selected fibers are given in Table 2.1.

Fiber strength can also be measured in pounds per square inch. This is called *tensile strength* and is obtained by determining the force in pounds required to break a fiber cross-sectional mass equivalent to 1 square inch. This measure is seldom used except for cotton, since it is easier and somewhat more accurate to test a mass of cotton fibers than a single fiber.

It is significant that fiber strength does not always indicate comparable yarn or fabric strength. Relatively weak fibers like wool can be made into durable fabrics because of other important fiber properties, such as high elongation, superior elastic recovery, and resilience. For most fabrics a minimum fiber tenacity of approximately 2.5 grams per denier may be desirable. Fibers with higher strength are particularly useful in sheer and lightweight fabrics. They are also of importance where product end-use requires a high degree of strength, as in work clothes and industrial fabrics.

Yarn construction that permits fiber slippage does not make optimum use of fiber tenacity, and the resulting yarn may be weak. This is an important factor for consideration in the manufacture of yarns composed of staple fibers.

Flexibility   *Flexibility or pliability*—the property of bending without breaking—is a third necessary characteristic of textile fibers. To create yarns and fabrics that can be creased, that have the quality of drapability and the ability to move with the body, that give when walked or sat upon, and, in general, that permit freedom of movement, fibers used must be bendable, pliable, or flexible. Many substances in nature resemble fibrous forms, but because they are stiff or brittle, they do not make practical textile fibers.

It is further accepted that a fiber must flex or bend repeatedly in order to be classified pliable or flexible. As with the other properties, fibers of different types vary in their degree of pliability. The degree of flexibility determines the ease with which fibers, yarns, and fabrics will bend and is important in fabric durability.

Spinning Quality or Cohesiveness   A fourth property required of fibers is spinning quality, or *cohesiveness*. This characteristic can best be described as the ability of fibers to stick together in yarn manufacturing processes. The cohesiveness of fibers may be due to the longitudinal contour or the cross-section shape that enables them to fit together and entangle sufficiently to adhere to each other, or it may be the result of the surface or skin structure of the fiber, which causes fibers to stick together. When fiber shape and surface do not contribute to the cohesive quality, the same end result is attained by using fibers of filament length that are easily twisted into yarn form. In the latter situation, fiber length is considered to be the equivalent of cohesiveness. This supports the use of the term *spinning quality* as a substitute for cohesiveness.

The introduction of texture in filament fibers in the form of coiling, zigzagging, and other shaping also contributes to cohesiveness.

Uniformity   Many textile authorities consider uniformity to be an essential property of fibers. To make yarns it is important that fibers be similar in length and width, in spinning quality, and in flexibility. Man-made fibers can be controlled during production so that a relatively high degree of uniformity is maintained and irregularities are held to a minimum. Furthermore, any manufactured fiber can be made to conform to other existing fibers, natural or otherwise. Yarns that are composed of generally uniform fibers are preferred, because they are regular, they appear smooth, and they accept dyestuffs more evenly. Natural fibers are not so uniform as man-made fibers, for it is difficult to control all the aspects of nature. Therefore, it is essential to blend fibers from many different batches in order to produce quality yarns and fabrics. (See the discussion of yarn manufacturing on page 190 for further information about blending natural fibers.)

Secondary Properties

Additional characteristics possessed by fibers vary, and not all fibers possess all properties. Some of the secondary characteristics are used as a basis for describing and classifying fibers and may produce end products with choice qualities.

Table 2.2   Specific Gravities of Selected Fibers

Fiber	Specific Gravity	Fiber	Specific Gravity
Asbestos	2.10–2.80	Acetate	1.32
Cotton	1.54–1.56	Triacetate	1.30
Flax	1.50	Glass	2.54
Hemp	1.48	Modacrylic	1.30–1.37
Jute	1.48	Nylon	1.14
Ramie	1.51	Polypropylene olefin	0.90–0.91
Silk	1.34	Polyester	1.22–1.38
Wool	1.30–1.32	Saran	1.72
Acrylic	1.17–1.19	Spandex	1.00–1.21
		Rayon	1.46–1.54

Physical Shape   The shape of a fiber includes, in addition to the high length-to-breadth ratio previously discussed, average length, surface contour, surface irregularities, and cross section. These characteristics are the basis for the description of both the macroscopic and the microscopic appearance of a fiber. In turn, they are responsible for certain differences in yarn and fabric properties that will be discussed in later chapters.

Specific Gravity   The *specific gravity* of a fiber indicates the density relative to that of water at 4°C (39°F). The density of water at that temperature is 1. Density indicates the mass-per-unit volume expressed as grams per cubic centimeter or pounds per cubic foot.[3] Because density is commonly determined on balances or scales, it is frequently, but incorrectly, called weight. In this case, specific gravity can be referred to as relative weight compared with water. The correct expression for density is *mass-per-unit volume*, while *relative mass-per-unit volume* can be used to describe specific gravity.

Fibers with a high specific gravity such as glass (2.54) are compact and have a higher density than nylon, with a specific gravity of 1.14. Fabrics composed of fibers with comparatively high specific gravity are heavier than fabrics made of fibers with low specific gravity, provided other factors in the fabric construction are identical.

It is interesting to note that fibers with a specific gravity of less than 1 will float on water. For these products special consideration may be required in the choice of laundering techniques. Table 2.2 gives the specific gravity of many fibers available on today's market.

Luster   *Luster* refers to the gloss, sheen, or shine that a fiber possesses. It is the result of the amount of light reflected by a fiber, and

[3] ASTM *Yearbook* (1970), p. 24.

it determines the fiber's natural brightness or dullness. Silk, a natural fiber, has a high luster, while cotton, also a natural fiber, has a low luster. Man-made fibers are manufactured with luster controlled. These fibers may have a very high luster, or luster may be reduced through the addition of pigments, such as titanium dioxide. High luster is not always a desirable property. Thus it is advantageous to be able to choose fibers with different degrees of luster.

Moisture Regain and Moisture Absorption   Textile fibers, in general, have a certain amount of water as an integral part of their structure. ASTM defines *moisture regain* as

> the moisture in a material determined under prescribed conditions and expressed as a percentage of the weight of the moisture-free specimen.[4]

Moisture regain is determined by ASTM according to the following procedure.[5] Fibers are conditioned thoroughly to air dry weight at standard conditions (65 percent relative humidity and 70°F ± 2°F). The sample is weighed, dried to bone dry state, reweighed, redried, and reweighed to constant weight. The regain is calculated according to the following formula:

$$\text{Percent regain} = \frac{\text{conditioned weight} - \text{dry weight}}{\text{dry weight}} \times 100$$

Moisture content, while similar to regain, is determined under certain prescribed conditions and is usually expressed as the percent of moisture based on moist or conditioned weight. The procedure differs from the one for regain in that the sample is returned to a constant weight at standard or other prescribed conditions. The formula is:

$$\text{Percent moisture} = \frac{\text{conditioned weight} - \text{dry weight}}{\text{conditioned weight}} \times 100$$

When moisture content of fibers at relative humidities other than the standard 65 percent is desired, the latter method is used with the conditioned weight obtained at the prescribed relative humidity.

Moisture regain and moisture absorption are sometimes used as synonymous terms. However, moisture absorption is frequently used to indicate the moisture percent at either 95 percent relative humidity or 100 percent relative humidity. This is more accurately referred to as *saturation regain* (Table 2.3).

---

[4] ASTM *Yearbook* (1970), p. 35.
[5] American Society for Testing and Materials, Test D2495-66T.

Table 2.3   Moisture Regain

Fiber	(Commercial) Regain %; Standard Conditions	% Regain at 20°C (70°F); 95% Relative Humidity
Asbestos	1.0	3.0
Cotton, raw	8.5	15.0 (approx.)
Cotton, mercerized	8.5–10.3	
Flax	12.0	
Hemp	12.0	
Jute	13.75	
Ramie	6.0	
Silk	11.0	25.0+
Wool	13.6–16.0	29.0±
Acrylic	1.5 (1.3–2.5)	2.5–5.0
Acetate	6.5	14.0
Triacetate	3.2–3.5	8.8
Glass	0.0	0.3
Modacrylic	0.4–3.0	1.0–3.0
Nylon	4.5	6.5–8.5
Polypropylene olefin	0.0	0.0
Polyester	0.4–0.8	0.4–2.0
Saran	0.0	0.1
Spandex	0.3–1.3	1.0±–2.0
Rayon	11.0 (10.7–13.0)	20.0–27.0

Fibers with good moisture regain will accept dyes and finishes more readily than fibers with low regain. A few fibers have no regain, and this creates problems in processing. The relation of fiber strength to moisture content is an important consideration in evaluating fiber behavior. Some fibers are stronger when wet than dry, others are weaker when wet, and still others exhibit no change. Therefore, the care or maintenance of textile products is influenced by the strength regain relationship. For example, a fiber with low wet strength, such as rayon, requires careful handling during laundering to prevent undue stress on the wet fiber.

Cotton, which has greater strength when wet than dry, can be laundered with ease. A fiber with little or no regain will wash and dry quickly. Moisture regain also influences comfort. For a discussion of this factor, see Chapter 32.

Elastic Recovery and Elongation   The amount of stretch or extension that a fiber will accept is referred to as *elongation*. *Breaking elongation,* therefore, is the amount of stretch that occurs to the point where the fiber breaks.

The term *elastic recovery* designates the percent of return from elongation toward the original length. If a fiber returns to its original

Table 2.4   **Breaking Elongation and Elastic Recovery**

Fiber	Standard Conditions *% Dry Elongation*	Average Range *% Immediate Recovery at x% Elongation*
Cotton	3–10	75 @ 2%*
Flax	2.7–3.3	65 @ 2%
Jute	1.7–1.9	74 @ 1%
Ramie	3–7	52 @ 2%*
Silk	10–25	92 @ 2%*
Wool	20–40	99 @ 2%*
Acrylic	25–46	92–99 @ 2%*
Acrilan	36	99 @ 2%*
Creslan	40	90 @ 1%
Orlon	15–22	97 @ 2%*
Zefran	33	99 @ 2%
Acetate	23–45	94 @ 2%*
Triacetate	25–40	90–92 @ 2%*
Glass	3–4	100 @ 2%*
Modacrylic	33–39	79–97 @ 2%*
Dynel	38	94–100 @ 2%*
Verel	35	79 @ 2%*
Nylon, regular	26–40	100 @ 8%*
high-tenacity, filament	16–20	100 @ 4%*
high-tenacity, staple	23–58	100 @ 2%
Olefin: propylene	25–75	100 @ 2%*
Polyester, regular	19–23	97 @ 2%*
high-tenacity	11–28	100 @ 2%*
Spandex	440–700	100 @ 2%*
Rayon, regular	15–30	82 @ 2%*
high-tenacity	9–26	70–100 @ 2%*
high-wet-modulus	9–18	95+ @ 2%*

*Results were obtained on electronic tensile apparatus using a load rate of 10 grams per denier per minute and a thirty-second duration of load and one-minute recovery.

length from a specified amount of attenuation, it is said to have 100-percent elastic recovery at *x*-percent elongation (Table 2.4).[6]

The elastic recovery of a fiber can be determined by several accepted procedures. Recovery can be measured after applying stress for a short time, such as thirty seconds, then releasing the stress and determining the immediate elastic recovery, usually measured after a one-minute recovery period. Table 2.4 includes data on immediate recovery after stress to elongate the fiber the specified degree, usually 2 percent for the more commonly encountered fibers.

[6]J. Gordon Cook, *Handbook of Textile Fibers* (London: Merrow Publishing Co., 1968).

Delayed recovery is also determined by measuring fiber length a specified amount of time after removal of stress. The time of recovery may be minutes or hours. Recovery is sometimes determined following extended application of stress, that is, a stress application of minutes or hours. This is followed by either immediate or delayed recovery measurements. The longer the time during which stress is applied, the greater the tendency to have some permanent deformation or set that prevents complete elastic recovery.

The amount of elongation is an important factor in evaluating elastic recovery. Some fibers with low elongation have excellent elastic recovery; however, this property is of little value because of insignificant elongation. Thus, it becomes obvious that elongation and elasticity must be considered together in fiber evaluation. A fiber with extremely high elongation but medium to low elastic recovery might be undesirable, because the product could not return to size after extension.

Resiliency   *Resiliency* is the ability of a fiber to return to shape following compression, bending, or similar deformation. It is important in determining the crease recovery of a fiber or fabric, and it plays a significant role in the rapidity with which flattened carpet pile will regain its shape and restore its appearance.

This property is evaluated on a comparative basis from excellent to poor. Elastic recovery is a significant factor in the resiliency of a fiber, and usually good elastic recovery indicates good resiliency.

Flammability and Other Thermal Reactions   In the chapters concerning specific fibers, there are paragraphs devoted to thermal reactions. These sections are primarily descriptive and indicate the behavior of the fiber at various temperatures. Burning characteristics of the fibers are important in determining care and use, and they serve as helpful guides in fiber identification. Thermal characteristics, such as reactions to wet and dry heat, must also be considered in the treatment of fibers.

Additional Properties

In the chapters that follow, additional properties that affect fiber performance will be considered in terms of each specific fiber. These characteristics are treated descriptively in this book with some fiber comparisons provided. The emphasis is placed upon fiber groups and points out the importance of the characteristic and its influence upon selection, use, comfort, appearance, durability, and maintenance.

These properties include the way the fiber reacts to selected chemicals, to environmental conditions such as sunlight and other climatic variables, to microorganisms such as bacteria and fungi, and to insects such as moths and carpet beetles.

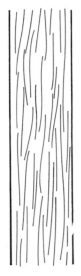

Figure 2.1 Schematic diagram illustrating highly oriented molecules within a fiber.

Figure 2.2 Schematic diagram showing molecules in random or amorphous arrangement, with low orientation.

An important factor that usually must be considered in determining the success or failure of a fiber is cost. The production or growth and processing of a fiber must be sufficiently economical so that the final price of goods does not exceed what the consumer is willing to pay at a given time. This does not rule out fibers that require vast sums of money for their research and development, but it would eliminate fibers that were extremely costly to produce in quantity. It also rules out natural fibrous materials that require exorbitant sums for growth and processing and would, thus, make the price of the final product too high for adequate sales.

FIBER MORPHOLOGY

Morphology refers to form and structure of a substance. In relation to fibers it applies to the biological structure of natural fibers, to the shape and cross section or appearance of all fibers, and to microscopic characteristics. The chapters on fibers consider morphology under the heading of microscopic appearance, shape, and structure.

MOLECULAR ARRANGEMENT

Fiber molecules are *polymers*—large molecules produced by linking together many monomers—and have a high length-to-width ratio just as does the actual fiber. A recent article described a fiber molecule in comparative terms by stating that if a typical fiber molecule is $\frac{1}{8}$ inch in diameter, it would be 40 feet long.[7] Actually, of course, the fiber molecule is ultramicroscopic, but this example suggests the relative length and width.

The arrangement of molecules within the fiber varies widely. They may be *highly oriented,* which means that they are parallel to each other and to the longitudinal axis of the fiber (Fig. 2.1); or they may exhibit a *low degree of orientation* and be at various angles to and crisscross one another (Fig. 2.2). High orientation is associated with good fiber strength and low elongation, while low orientation tends to produce the reverse properties. Other characteristics that may be affected by the degree of orientation include moisture regain or absorbency and flexibility.

Molecular orientation and crystallinity are terms used to help define molecular arrangement, and they are not entirely synonymous. *Crystallinity* indicates that fiber molecules are parallel to each other, but they need not be parallel to the longitudinal fiber axis. Orientation of molecules, as stated previously, describes molecules that are ar-

---

[7]"The Solid State of Polyethylene," *Scientific American,* Vol. 211, 5 (November 1964), p. 81.

ranged parallel to each other (crystallinity) *and* to the longitudinal axis of the fiber. Figure 2.1 is a graphic representation of fiber molecules that are highly oriented and, therefore, exhibit crystallinity, while Figure 2.3 shows fiber molecules that are somewhat crystalline in arrangement but do not exhibit orientation to the fiber axis.

When fiber molecules are arranged in random or amorphous groupings (Fig. 2.2), they may lie far apart, crisscross, or fall in other irregular ways. The extent to which amorphous areas occur is thought to vary widely from slight disorder to extremely disoriented arrangements, depending upon the fiber and specific areas within the fiber.

Figure 2.4 illustrates a molecular arrangement in which both oriented and amorphous areas exist in the same fiber. This duality is quite common.

J. W. S. Hearle has added a new concept to our knowledge of the arrangement of molecules within a fiber. The *fringed fibril theory* suggests that each macromolecule has the opportunity to pass through both highly oriented areas and amorphous regions.[8] This concept has considerable merit, since it could provide a sound basis for fiber behavior in processing, use, and care.[9]

Fiber molecules may slip back and forth within the fiber, just as some fibers may slip within yarns. Molecular slippage is likely to be at a maximum in amorphous areas and lowest in crystalline areas. When a high degree of molecular slippage occurs, the fiber will have high elongation, but the molecules may not return to the optimum position, with the result that the fiber will be weak and have poor elastic recovery. In some fibers, molecules are joined to parallel molecules by cross-linkage, which is brought about by actual chemical reaction (Fig. 2.5). On the other hand, in some types of fibers, molecules are held in place by associative forces. These forces can be broken easily, but they can be reestablished at the same or new locations.

An important type of associative force in textiles is called *hydrogen bonding*. According to chemists, hydrogen bonding occurs between hydrogen atoms and small electronegative atoms such as fluorine, oxygen, and nitrogen. The hydrogen bond is present when a proton from the hydrogen atom is shared between two negative electrons from two different atoms.[10] Hydrogen bonding is found in cellulose and

Figure 2.3 Schematic diagram illustrating molecules in a crystalline arrangement but not oriented.

Figure 2.4 Oriented and amorphous areas existing within the same fiber. [Redrawn from *Encyclopedia of Polymer Science and Technology*, Vol. 4, p. 456. By permission of Wiley-Interscience.]

Oriented   Amorphous

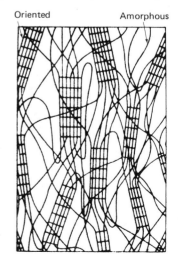

[8] J. W. S. Hearle and R. H. Peters, *Fiber Structure* (London: Butterworth & Co., 1963), p. 209.

[9] For more discussion on this and other related ideas, see Hearle and Peters, pp. 209–304; and H. F. Mark, N. G. Gaylord, and N. M. Bikales, eds., *Encyclopedia of Polymer Science and Technology*, Vol. 4 (New York: Wiley-Interscience, 1966), p. 449 ff.

[10] R. Q. Brewster and W. E. McEwen, *Organic Chemistry* (Englewood Cliffs, N.J.: Prentice-Hall, Inc., 1962), p. 148; and M. J. Sienko and R. A. Plane, *Chemistry*, 3d ed. (New York: McGraw-Hill, Inc., 1966), p. 327.

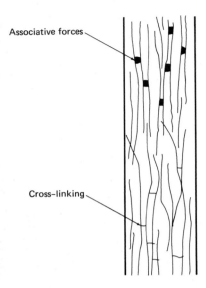

Associative forces

Cross-linking

Figure 2.5 Schematic diagram illustrating associative forces and crosslinking between molecules.

protein fibers. It may occur in manufactured fibers that include hydrogen atoms and any one or more of the electronegative atoms mentioned. It is important in fiber structure, because it may aid in producing fibers that have desirable properties such as crease resistance and dimensional stability.

When molecules do not contain the polar groups necessary for hydrogen bonding, weaker associative forces can be established between molecules, provided the molecular segments lie very close together (4–5Å). These forces depend upon proximity and upon positive and negative charges; they are often called *van der Waal's forces.*

3

Fiber
Classification

Systems of classification have been in use for hundreds, perhaps thousands, of years. Man, as a scientist, learned in the early stages of scientific development that placing like items together facilitated his understanding of these items.

When the study of textile fibers was new, a simple type of fiber classification was used, which was based on a systematic arrangement of fibers into the categories of animal, vegetable, and mineral matter. With the development of man-made fibers this early classification was made obsolete, and new systems had to be devised. It was apparent to those studying textiles that the day would come when trying to remember all the properties and characteristics of each individual fiber used by man would be practically impossible. Thus, the scientist reasoned that if fibers could be arranged or classified into like groups, man could become acquainted with general properties of each group, and then only the additional knowledge concerning special qualities of individual fibers would be required for intelligent selection, use, and care of textile products.

Over the years, many systems of classification have been recommended. Some are no longer sufficiently discerning for the innumerable fibers on the market at the present time. Other systems formerly used are still helpful for major divisions, but, because of the complexity of the current fiber situation, it is now advisable to add selected subclassifications.

In 1960 the Textile Fiber Products Identification Act (frequently referred to as the TFPIA) became effective. This legislation requires most textile products sold at retail to have labels stating the textile fiber content. To reduce consumer confusion the legislation established sixteen "generic" or family names for the grouping of all manufactured or man-made fibers. Since that time two more generic classifications have been added. In general, the fibers within a specific group possess the same essential properties and require the same basic care. Therefore, any system of fiber classification advocated since the passage of this legislation should include these generic terms at least as a partial basis for fiber categorization. The classification system proposed in this text (Tables 3.1 and 3.2) indicates generic terms with an asterisk and with a lower-case initial letter.

The use of the terms *animal, vegetable,* and *mineral* in fiber classification has been challenged, because there are differences between the behavior of vegetable cellulose used in the man-made cellulosic fibers and vegetable protein used in some of the man-made protein fibers. In fact, there is much greater similarity between vegetable protein fibers and animal fibers. Furthermore, there has been criticism directed at using the fairly common division of natural as opposed to man-made or manufactured fibers, and these categories are somewhat deceiving.

Most generic groups include trademarked fibers—those bearing a word, letter, device, or symbol that points distinctly to the origin or ownership of the particular fiber. These trademark names advertise the company as well as indicating a particular product. Trademark names listed in Table 3.1 are capitalized.

The use of the terms *thermoplastic* and *nonthermoplastic*—meaning the ability or lack of ability of a substance to become soft upon application of heat—has some merit in fiber classification, but since these terms fundamentally describe the properties of certain fibers, this distinction, too, may cause confusion and misinterpretation.

The system for classification adopted in this text is based upon the following:

1. the principal origin of the fiber. Fibers either occur as fibrous forms in nature or they are manufactured; thus, a major breakdown to indicate scientifically the origin of the fiber produces two groups: namely, the natural fibers and the man-made fibers.
2. the general chemical type. Fibers can be protein, cellulosic, mineral, or synthesized.
3. the generic term. The generic term or family name as specified

in the Textile Fiber Products Identification Act provides scientifically cogent bases for grouping like fibers.

4. the inclusion of common names of fibers. Many people are familiar with common names and trade names for fibers; therefore, examples of these should be included in any classification as a means of clarifying the group name to the student.

The examples below will clarify the classification method used and the way it meets the previously stated concepts related to fiber grouping:

1. the major origin of the fiber
   example: natural   man-made
2. the general chemical type
   example: cellulosic   man-made cellulosic
3. the general type or generic grouping
   example: seed hair   rayon
4. specific fiber name
   example: cotton   Bemberg rayon

In using Tables 3.1 and 3.2, it is important to remember that the degree of specificity used may vary according to the needs of a particular group of students. Further subdivision and discussion of fibers, including mention of additional fiber trade names, will be found in the chapters devoted to the various fiber groups.

The list of fibers under each category is not complete. For example, there are many more bast fibers than those listed. Because of the tremendous number of trademarked fibers available, Table 3.1 could not be complete without overstepping the boundaries of this book. Where the generic name is familiar, such as "rayon," the list of trademark names is sketchy.[1] Specific fibers included are those that are generally considered to be the commercially important ones in the group and those with which consumers are more likely to come into contact. The student who wishes to study some of the less familiar fibers in more detail will find interesting data in the references listed in the Bibliography.

It should be noted at this time that the legislation referred to and used in this fiber classification does not provide new names for the natural fibers. Except for hair fibers, natural fibers are to be listed by using their common name in the labeling of textile products. Legislation, as cited in the Wool Products Labeling Act, provides that all hair fibers may be labeled as "wool." However, the name of the actual

---

[1]For a complete list of all man-made fibers sold in the United States and abroad, see Adeline A. Dembeck, *Guidebook to Man-made Textile Fibers and Textured Yarns of the World,* 3d ed. (New York: United Piece Dye Works, 1969).

Table 3.1   Man-made or Manufactured Fibers

A. Man-made Cellulosic Fibers
   1. rayon*
      a. cuprammonium (discontinued in U.S., 1971)
         Bemberg
         Cupioni
         Sunspun
         Parfé
      b. saponified cellulose rayon
         Fortisan
      c. viscose rayon
         1. regular and high-tenacity rayon
            Avisco
            Coloray
            Enka
            Fibro
            Jetspun
            Narco
         2. high-wet-modulus rayon
            Avril (Fiber 40)
            Nupron (Fiber 24)
            Vincel
            Xena
            Zantrel 700
            Zantrel polynosic (Fiber HM)
B. Man-made Modified Cellulosic Fibers
   1. acetate*
      a. secondary acetate
         Acele
         Celafil
         Celanese
         Celaperm (solution dyed)
         Chromspun (solution dyed)
         Crystal
         DuPont
         Estron
         Chavacete
      b. triacetate**
         Arnel
C. Man-made Protein Fibers
   1. azlon*
      a. casein protein
         Merinova
      b. zein protein
         Vicara (discontinued)
      c. peanut protein
         Ardil (discontinued)
      d. soybean protein

      e. miscellaneous
         chicken feathers
         cottonseed protein
D. Man-made Synthesized Fibers
   1. condensation polymer fibers
      a. nylon*
         1. nylon, Type 6,6
            Antron
            Astroturf
            Blue "C"
            Cadon
            Cumuloft
            DuPont
            Monsanto
            Phillips 66
            501
         2. nylon, Type 6
            Ayrlyn
            Beaunit
            Caprolan
            Enka
            Nytelle
            Touch
            Unel
         3. nylon, Type 6T
            Nomex
         4. nylon, Type 11
            Rilsan
         5. nylon, Type 610
            Nylex
            Quill
         6. nylon (type not available)
            Qiana
            PACM
         7. nylon, bicomponent
            Cantrece
      b. polyester*
         Avlin
         Blue "C"
         Dacron
         Encron
         Fortrel
         Kodel
         Quintess
         Terylene
         Trevira
         Vycron

Table 3.1 Man-made or Manufactured Fibers—*Continued*

2. additional polymer fibers
   a. anidex*
      ANIM/8
   b. acrylic*
      Acrilan
      Courtelle
      Creslan
      Orlon
      Zefran
      Anywear
      Nomelle
      Wintuk
   c. modacrylic*
      Dynel
      Verel
      Kanekalon
   d. nytril*
   e. olefin*
      1. polyethylene
         Fortiflex
         Lus-Trus
         Tyvek
      2. polypropylene
         Durel
         Eastman
         Escon
         Herculon
         Marvess
         Meraklon
         Nypel
         Polyloom
         Typar
         Vectra
   f. saran*
      Enjay
      Saran
   g. vinal*
      Kuralon
      Vilon
      PVA
   h. vinyon*
      Vinyon HH
      Rhovyl
      Vogt
      Voplex
3. Elastomers
   a. spandex*

Duraspan
Glospan
Lycra
Numa
Unel
Vyrene
   b. rubber*
      Contro
      Lactron
      Lastex
   c. lastrile**
E. Man-made Mineral Fibers
   1. glass*
      Beta
      Ferro
      Fiberglass
      PPG
      Stranglas
      Uniglass
   2. metallic*
      Alistran
      Brunsmet
      Chromeflex
      Chromel-R
      Fairtex
      Lurex
      Malora
      Metlon
F. Other Man-made Fibers
   1. Alginate
   2. inorganic
      Avceram—carbon silica
      Fiberfax—alumina-silica
      Lexan—polycarbonate
      quartz fibers
      Thornel—graphite
      Boron
      Sapphire
   3. unclassified
      A-tell
      PBI—polybenzimidazole
      Raycelon—biconstituent (rayon + nylon)
      Source—biconstituent (nylon + polyester)
      Tricelon—biconstituent (nylon + acetate)
      Arnel-plus—biconstituent (nylon + triacetate)
      Teflon—tetrafluoroethylene
      Kynol—Phenolic flame resistant

*Generic terms as identified in the Textile Fiber Products Identification Act.
**Terms that may be used as generic names when the fiber meets special requirements listed by the TFPIA.

animal, such as vicuña, has sufficient selling power that products of this type usually are labeled according to the animal name.

This classification system may appear to be slightly more complex than those suggested by certain other authors. However, an additional breakdown at this level reduces the problems frequently encountered in discussion of the individual textile fibers. It has been found that this system is workable, and it is scientifically sound.

There is and always will be some disagreement between fiber manufacturers and the Federal Trade Commission (FTC) regarding the group to which a fiber is assigned. This is especially true as new fibers are developed that might not fit into existing categories or that have sufficient variations that the manufacturer would like a new generic group. A current dispute involves biconstituent fibers. It is undecided, at present, how these should be categorized.

The original TFPIA provided for adding new generic classes when necessary. Two new names—lastrile and anidex—have been added to accommodate new fibers developed in recent years.

While TFPIA requires that most textile products be labeled with

Table 3.2   Natural Fibers

A. Cellulosic Fibers	B. Protein Fibers
1. Seed hairs	1. Animal-hair fibers
a. cotton	a. wool (sheep)
b. milkweed	b. specialty hair fibers
c. kapok	1. alpaca
d. cattail	2. camel
2. Bast fibers	3. cashmere
a. flax	4. guanaco
b. ramie	5. llama
c. hemp	6. mohair (Angora goat)
d. jute	7. vicuña
e. sunn	c. fur fibers
f. kenaf	1. mink
g. urena	2. muskrat
3. Leaf fibers	3. Angora rabbit
a. abaca	2. Animal secretion
b. pineapple	a. silk (*Bombyx mori, et al.*)
c. agave	1. cultivated (mulberry)
1. sisal	2. dupioni
2. henequen	3. tussah (wild)
d. palm	b. spider silk
e. New Zealand flax	C. Mineral Fiber
f. yucca	1. Asbestos (rock source)
g. palma istle	D. Natural Rubber
4. Nut husk fibers	
a. coir (coconut)	

the generic name of the fiber, trademark names usually accompany the generic term as advertising for the manufacturer and as an attraction to consumers. The required legislation is helpful, but it does not take the place of sound scientific knowledge. The consumer who can recognize fiber names and identify, classify, and evaluate fibers is qualified to make intelligent selections of fibers and fabrics and wise decisions concerning care and handling of textile merchandise. This consumer has the technical information required for understanding why fibers react as they do to physical, chemical, and biological stimuli.

Fiber
Identification

Qualitative identification of a fiber is difficult and may require several tests. It is not the intent of this book to include all the possible tests; however, simple ones used in identification are described in this chapter with brief comments concerning their use and importance. In addition to providing methods of fiber identification, some tests contribute insight into problems of processing and care.

Much of the information presented here is in chart form. The discussion indicates general procedures and suggests interpretation. Reference is made to additional aids for fiber analysis, and bibliographical information is included. Further reactions and properties helpful in analyzing fibers and fiber behavior will be found in the chapters concerning specific fibers.

It is essential that textile teachers and researchers keep up to date concerning fiber data. Furthermore, a consumer with a certain amount of training in textiles will find the ability to tentatively identify fibers extremely valuable. He or she may wish to verify label information and may need to know what fibers or fiber groups are in a fabric in order to apply desirable care techniques.

TESTS FOR FIBER IDENTIFICATION

The Burning Test

The burning test is a good preliminary test. It does not identify fibers specifically, but it provides valuable data regarding appropriate care. In cases of yarns composed of two or more fibers the test will usually give the reaction of the fiber that burns most easily; however, if a heat-sensitive fiber is involved, it might melt or withdraw from the flame. The application of flame as a part of identification is a preliminary test and will indicate general groupings only. The burning test is simple, but the individual must exercise care to avoid harming himself and must guard against the possibility of fire. The procedure follows:

1. Select one or two yarns from the warp of the fabric (see p. 235).
2. Untwist so the fibers are in a loose mass.
3. Hold the yarns in forceps, and move them toward the flame from the side.
4. Observe the reaction as they approach the flame.
5. Move them into the flame, and then pull them out of the flame and observe the reaction.
6. Notice any odor given off by the fiber.
7. Observe ash or residue formed.
8. Repeat the procedure for the filling yarn (see p. 235).

If the fabric does not have yarn structure, a small sliver of the fabric can be used, but if more than one fiber is involved, the results may be misleading.

Typical reactions of fibers to flame are given in Table 4.1.

Several cautions should be cited regarding the burning test. Dyes and finishes may alter the flammability of fibers. Some finishes may reduce or prevent flaming, while others may increase flammability. Both dyes and finishes can affect the color of the residue and, in some instances, the shape of the residue. For comparative results it is important to follow the procedure carefully.

Microscopic Evaluation

It is possible to be quite definite in identification of some fibers through microscopic observations. Fibers are mounted to provide views of their lengthwise dimensions, or techniques are used to obtain cross sections, and these are mounted. Unfortunately, several of the man-made fibers are so similar that additional analysis is required before positive identification can be made.[1]

Reference to microscopic appearance is included in the discussion concerning individual fibers. Table 4.2 cites microscopic appearance of selected fibers.

---

[1] For excellent photos of microscopic characteristics of fibers, the reader is referred to American Society for Testing and Materials, *ASTM Standards on Textile Materials,* 1970; American Association of Textile Chemists and Colorists, *Technical Manual,* 1970; J. L. Stoves, *Fiber Microscopy* (Princeton, N.J.: D. Van Nostrand Company, Inc., 1958); A. N. J. Heyn, *Fiber Microscopy* (New York: Wiley-Interscience, 1954); and Milton Harris, ed., *Handbook of Textile Fibers* (New York: Textile Book Publishers, Inc., 1954).

Table 4.1 Chart of Burning Characteristics of Fibers

Fiber	Approaching Flame	In Flame	Removed from Flame	Odor	Residue
Natural Cellulose cotton and flax	does not shrink away; ignites upon contact	burns quickly	continues burning; afterglow	similar to burning paper	light, feathery; light to charcoal gray in color
Man-made Cellulose rayon	does not shrink away; ignites upon contact	burns quickly	continues burning; afterglow	similar to burning paper	light, fluffy residue; very small amount
Man-made Modified Cellulose acetate	fuses and melts away from flame; ignites quickly	burns quickly	continues to burn rapidly	acrid (hot vinegar)	irregular-shaped, hard, black bead
Natural Protein wool	curls away from flame	burns slowly	self-extinguishing	similar to burning hair	brittle, small black bead
silk	curls away from flame	burns slowly and sputters	usually self-extinguishing	similar to burning hair	beadlike, crushable, black
weighted silk	curls away from flame	burns slowly and sputters	usually self-extinguishing	similar to burning hair	the shape of fiber or fabric
Man-made Proteins azlons	curls away from flame	burns slowly	self-extinguishing	similar to burning hair	brittle, small black bead
Natural Mineral asbestos	does not melt (safe fiber)	glows red if heat is sufficient	returns to original form	none	same as original
Man-made Mineral glass	will not burn	softens, glows red to orange	hardens; may change shape	none	hard, white bead
metallic	*pure metal* no reaction	glows red	hardens	none	skeleton outline
	coated metal melts, fuses and shrinks	burns according to behavior of coating		none	hard, black bead

Table 4.1   Chart of Burning Characteristics of Fibers—*Continued*

Fiber	Approaching Flame	In Flame	Removed from Flame	Odor	Residue
Man-made Synthesized					
acrylic	fuses away from flame; melts; ignites readily	burns rapidly with hot flame and sputtering; melts	continues to burn and melt; hot molten polymer will drop off while burning	acrid	hard, black, irregular bead
modacrylic	fuses away from flame; melts (considered safe)	burns slowly if at all; does not feed a flame; melts	self-extinguishing	acrid, chemical odor	irregular, hard black bead
nylon	melts away from flame; shrinks, fuses	burns slowly with melting	self-extinguishing	celery	hard, tough, gray or tan bead
polyester	fuses; melts and shrinks away from flame	burns slowly and continues to melt	self-extinguishing	chemical odor	hard, tough, black or brown bead
olefin	fuses; shrinks, and curls away from flame	melts and burns	continues to burn and melt; gives off black sooty smoke	chemical odor	hard, tough, tan bead
saran	fuses, melts, and shrinks away from flame	melts; yellow flame	self-extinguishing	chemical odor	irregular, crisp black bead
vinal	fuses, shrinks, curls away from flame	burns with melting	continues to burn with melting	chemical odor	hard, tough, tan bead
vinyon	fuses and melts away from flame	burns slowly with melting	self-extinguishing	acrid	hard, black, irregular bead
Elastomeric					
spandex	fuses but does not shrink away from flame	burns with melting	continues to burn with melting	chemical odor	soft, sticky, and gummy
rubber	shrinks away from flame	burns rapidly and melts	continues burning	sulfur or chemical odor	tacky, soft black residue

Table 4.2   Microscopic Appearance of Textile Fibers

Fiber	Longitudinal Appearance	Cross-sectional Shape
Man-made		
acetate	distinct lengthwise striations; no cross markings	irregular shape with crenulated or serrated outline
Arnel		
acrylic		
Acrilan, Courtelle, Creslan, Zefran	rodlike with smooth surface and profile	nearly round or bean shape
Orlon	broad and often indistinct lengthwise striation; no cross markings	dog bone
bicomponent	lengthwise striations; no cross markings	irregular mushroom or acorn
Orlon		
modacrylic		
Dynel	lengthwise striations; no cross markings	irregular worm or ribbonlike
Verel	broad and often indistinct lengthwise striation; no cross markings	dog bone
nylon		
nylon 6, nylon 6,6 regular	rodlike with smooth surface and profile	round or nearly round
Antron, "501," and Cadon	broad, sometimes indistinct lengthwise striations; no cross markings	trilobal
olefin		
polyethylene, polypropylene	rodlike with smooth surface and profile	round or nearly round
polyester		
Dacron, Fortrel, Kodel, and Vycron	rodlike with smooth surface and profile	round or nearly round
Dacron type 62	broad, sometimes indistinct lengthwise striations; no cross markings	trilobal

Table 4.2 Microscopic Appearance of Textile Fibers—_Continued_

Fiber	Longitudinal Appearance	Cross-sectional Shape
rayon		
viscose, regular	distinct lengthwise striations; no cross markings	irregular shape with crenulated or serrated outline
high-tenacity viscose	rodlike with smooth surface; indistinct striations or none	slightly irregular shape with few serrations
high-wet-modulus viscose Avril, Lirelle, Zantrel	rodlike, smooth surface	round or oval shaped
Cuprammonium, Fabelta Z-54	rodlike with smooth surface and profile	round or nearly round
saran Saran	rodlike with smooth surface and profile	round or nearly round
spandex Lycra	broad, often indistinct lengthwise striation; no cross markings	dog bone
Natural		
cotton, mercerized and not mercerized	ribbonlike convolutions (twists) sometimes change direction, and are less frequent in mercerized fibers; no significant lengthwise striations, but lumen may appear as striations in some fibers	tubular shape with tubes usually collapsed, and irregular in size
flax, bleached	bamboolike, pronounced cross-marking nodes; no significant lengthwise striations	very irregular in size as well as shape; round and oval are most prevalent
silk, boiled off	smooth surface and profile, but may contain nodes; no significant lengthwise striations	mostly triangular with point of triangle usually rounded off; irregular in size and shape
wool, cashmere, mohair and regular (Merino)	rough surface, cross markings due to surface scales; medulla or central fiber core sometimes apparent in coarse grades	round or nearly round; medulla may appear shaded

Solubility

The solubility of a fiber in specific chemical reagents is frequently a definite means of identification. This is especially true when solubility data are combined with other test results. In addition, the advisability of using certain chemicals on fibers is determined by solubility testing. A few substances used in stain removal and cleaning may damage some fibers, and knowledge of these reactions will prevent serious harm to fabrics.

Table 4.3 provides an identification scheme that uses a system of elimination. It can be used for unknowns composed of one or more fibers. Table 4.4 indicates general fiber solubility in selected reagents, so it can support Table 4.3. The relation of solubility reagent to solutions encountered in the home is given.

Acetone, which dissolves acetate and Arnel, is found in many fingernail polish removers. Vinegar contains acetic acid; however, while glacial acetic acid is concentrated, vinegar has only 6 percent acid. It will not destroy fibers but will weaken those affected by the glacial acetic. Sodium hypochlorite (pH 11) with 5 percent available chlorine is standard undiluted bleach such as Clorox® and similar products. Cresol (meta) is a component of Lysol®. Cresol is diluted in Lysol and will weaken affected fibers noticeably but not destroy them. Other

Table 4.3   Scheme for Identification by Fiber Solubility
Use the same sample of yarns or fabric until the total substrate has been destroyed. Rinse the residue thoroughly after each test. Observe the behavior of the substance at each step. It may be of value to observe the residue under the microscope. The sample should remain in the solution for five minutes before moving to the next step.

Chemical	Removes
1. Glacial acetic acid, 25°C (75°F)	acetate, triacetate
2. Hydrochloric acid 1:1, 25°C (75°F)	nylon 6, and 6,6
3. Sodium hypochlorite, 25°C (75°F) 5 percent available chlorine	silk, wool
4. Dioxane, 100°C (212°F)	saran
5. meta Xylene, at boil	olefins
6. Ammonium thiocyanate, at boil, 70 percent by weight	acrylics
7. Butyrolactone, 25°C (75°F)	modacrylics and nytriles
8. Dimethyl formamide, 95°C (200°F)—not always effective	spandex
9. Sulfuric acid, 75 percent by weight, 25°C (75°F)	cellulosics
10. meta Cresol, 95°C (200°F)	polyesters

Table 4.4   Solubilities of Fibers

Solution	Concentration	Temperature	Fibers Affected
Acetone	100%	25°C (75°F)	acetate, dynel modacrylic, triacetate
Acetone	80%	25°C (75°F)	acetate
Acetone	100%	50°C (120°F)	acetate, triacetate, modacrylics
Acetic acid	glacial	25°C (75°F)	acetate, triacetate
Acetic acid	glacial	95°C (200°F)	acetate, triacetate, nylon 6, nylon 6,6
meta-Cresol		25°C (75°F)	acetate, triacetate, nylon 6, nylon 6,6, silk, modacrylics, spandex
meta-Cresol		95°C (200°F)	acetate, triacetate, nylon 6, nylon 6,6, spandex, modacrylics, polyesters
Phenol	90%	25°C (75°F)	acetate, triacetate, modacrylic, nylon 6, nylon 6,6, spandex
Phenol	90%	95°C (200°F)	acetate, triacetate, modacrylic, nylon 6, nylon 6,6, polyester, spandex, acrylics (damaged)
NaOH	5%	boiling	acetate, triacetate, silk, wool
Sodium hypochlorite	5% available Cl	25°C (75°F)	wool, silk
Formic Acid	90%	25°C (75°F)	acetate, triacetate, spandex, nylon 6, nylon 6,6, acrylics (weakened)
HCl	1:1	25°C (75°F)	nylon 6, nylon 6,6
HCl	concentrated	25°C (75°F)	acetate, triacetate, wool, silk, nylon 6, nylon 6,6, acrylics, spandex, rayon (with heat)
Sulfuric acid	75% w/w	25°C (75°F)	acetate, triacetate, nylon 6, nylon 6,6, cellulosics, silk
Dimethyl formamide		25°C (75°F)	acetate, triacetate, modacrylics, acrylics (Zefran requires heat), saran, spandex
Xylene		boiling	olefins, rubber, saran

chemical reagents used in fiber identification are not usually encountered in either laundering or dry cleaning procedures. They are included here for identification and scientific value.

SUGGESTED IDENTIFICATION PROCEDURES

The first step in identification should be the burning test, which will indicate broad groupings of fibers. This should be followed by the microscopic test to establish definitive groups. Solubilities will further classify the fibers, so relatively definite identifications can be made. When there is still doubt, confirmatory tests can be used by experts. These include testing for specific gravity, melting point, refractive index and index of birefringence, X-ray diffraction, infrared spectrophotometry, and chromotography.

The handling of chemicals requires certain safety measures. Laboratories should have normal precautionary procedures posted for quick reference, and emergency phone numbers should be located easily. Most of the chemicals listed in Tables 4.3 and 4.4 require ordinary precautions. One of the most critical of these is to protect eyes and skin. It is essential to prevent contact between chemicals and the human eye. However, if an accident should occur, the eye must be rinsed immediately with water or a soothing substance that will counteract the injuring chemical. Prompt action is a must. Many chemicals will irritate the skin or seriously burn it. It is best to protect the skin from contact with any chemicals, especially concentrated acids and bases.

All the chemicals cited in Tables 4.3 and 4.4 can be used without a fume hood, provided there is adequate ventilation. However, the fumes from cresol, phenol, and formic acid are unpleasant, and a hood makes their use less annoying. Dioxane and meta-xylene should not be used near a direct flame, for they are highly flammable.

Textile
Fibers

If Part II of this book had been written at the turn of this century, it might have contained only two, possibly three chapters. Today no fewer than twelve distinct categories of fibers can be isolated, and many authorities would place the total far higher. Even when the number is held at a manageable dozen, many subclassifications and variants must be accounted for. Volumes have been written about each of the textile fibers described in the following chapters; within the limits of this text it is not possible to examine them in meticulous detail. However, for each of the fibers discussed there is a brief outline of the history, production, and properties, as well as an indication of the end-uses to which the fiber is most often applied.

5

Natural Cellulosic Fibers

Fibrous materials are found in nearly all plant life, but some plants in particular have proved to be important sources of textile fibers used in the manufacture of yarns and fabrics. These plant fibers are composed largely of cellulose and, therefore, are classified as natural cellulosic or vegetable fibers. The term *natural cellulosic* is preferred, for it indicates the simple chemistry of the substances and provides a scientific method for comparing natural cellulose with man-made cellulose fibers.

Cellulose is a linear polymer or long-chain molecule built by combining several thousand anhydroglucose units. While glucose, a simple sugar, is soluble in water, cellulose is not because of the immense size of the molecule. Cellulose is a carbohydrate; it contains the elements carbon (44.4 percent), hydrogen (6.2 percent), and oxygen (49.4 percent). The repeating units can be pictured by the diagram in Figure 5.1.

Two glucose units combine first to form cellobiose, then many cellobiose units combine to form cellulose. The number of anhydroglucose units in the cellulose molecules is referred to as the *degree of polymerization* of the molecule. Each repeating unit equals 2n or 2 anhydroglucose units. The glucose units are flipped over as they join

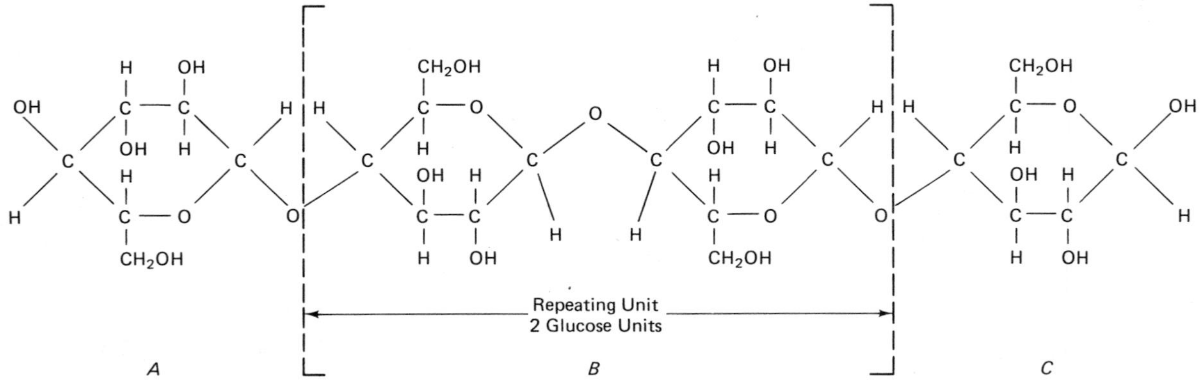

Figure 5.1 Diagram of a cellulose molecule. Repeating unit *B* is composed of two anhydroglucose units; *A* and *C* indicate terminal units of the molecule.

together, so in order to show the method of joining clearly, it is necessary to illustrate two units, as in Figure 5.1.

The cellulose molecules form into fibrils or bundles of molecular chains that combine in groups to form the cellulose fiber. Each fiber is composed of many cellulose molecules. These are not arranged in a completely parallel manner; rather, certain portions of the fiber may have the molecules lying parallel, while other areas are characterized by a somewhat random molecular arrangement. As discussed in Chapter 2, the parts of the fiber where molecules lie side-by-side and are held together by many associative forces are called crystalline (see p. 20). If the molecules, in addition to lying side-by-side, are parallel to the longitudinal axis, there is a high degree of molecular orientation. The cellulose fiber has areas where the molecules are not in complete crystalline arrangement. These areas of partial disorder approach a somewhat amorphous state, but evidence indicates there is no true amorphous area in the fiber.[1] The presence of high orientation and crystallinity is usually accompanied by great strength, low elongation, and low pliability. Areas of molecular disorder provide some pliability and slightly increase elongation.

The strength of the fiber is influenced, also, by the degree of polymerization of the fiber molecules. The higher the degree of polymerization, the stronger the fiber.

The molecules within the fiber tend to be held in place by hydrogen bonding (see p. 21). When cellulose fibers are bent, the hydrogen bonds are broken and new ones form, which results in creases or wrinkles that do not hang out.

The chemically reactive unit in cellulose is the OH, hydroxyl, group. This group may undergo substitution reactions in procedures to modify the cellulose fibers or in the application of some finishes

[1] W. E. Morton and J. W. S. Hearle, *Physical Properties of Textile Fibers* (London: Butterworth & Co., 1962), p. 24.

and dyestuffs. Substitution occurs when one or more hydroxyl units are removed and other ions or radicals, atoms, or groups of atoms, attach themselves to the carbon atoms. Modification can occur when the hydrogen of the hydroxyl group is removed by chemical action and other elements or compounds hook to the oxygen remaining.

Cellulose fibers have several properties in common. They burn easily and quickly with a yellow flame; they give off a smell like that of burning paper or leaves; they deposit a light, fluffy, grayish residue or ash. Cellulose is damaged by acid solutions, especially strong mineral acids, but it possesses excellent resistance to alkaline solutions.

In general, cellulose is low in elasticity and resilience; consequently, it wrinkles badly. Cellulose fibers are soft and absorbent, so they usually make comfortable products. The fibers launder readily and can withstand strong detergents, high temperatures, and bleaches (if properly used). This group of fibers is seldom damaged by insects, but fungi, such as mildew, will destroy cellulose or stain it severely. The specific properties of the major natural cellulose fibers will be treated in detail in the following discussion.

COTTON

HISTORICAL REVIEW

The origin of cotton is unknown. Archeologists have contributed valuable information concerning the fiber's early use, but there is a dearth of evidence to indicate when or where cotton first grew. Two authors have published a summary of their findings regarding the history of cotton in which they cite inconclusive evidence that cotton was grown and used in Egypt about 12,000 B.C. They further refer to verified findings that cotton was grown in India about 3000 B.C. Most authorities agree that India was the principal country in which cotton was widely accepted and used before 2500 B.C.

For centuries man believed that cotton was a product of the Old World and that it was brought to the shores of the Americas by early explorers. There is considerable speculation that cotton spread to both the Eastern and Western Hemispheres from a long-lost land that once existed in the Pacific. Another theory holds that cotton was carried from the Old World to the New World by way of a land route across the Bering Straits.

Today modern science has provided reliable data to indicate that cotton was indigenous to the land that is now North and South America as well as to Asia and Africa. Carbon 14 tests have produced evidence that cotton was grown and made into fabrics in Peru as early as 2500 B.C. Thus, it would seem that cotton culture occurred simultaneously in several areas of the world. This supposition is further upheld by the fact that botanical differences are evident between Eastern and Western cotton,[2] though it must be conceded that these differences could have evolved in adaptation to different environments.

One of the early written comments concerning cotton was made by Herodotus (484–425 B.C.), who told of a plant found in India that "instead of fruit, produces wool of a finer and better quality than that of sheep." The word *cotton* is derived from the Arabic *quoton* or *qutun*, which means a plant found in conquered lands. *Muslin* is also taken from the Arabic language and was applied to cottons woven in Mosel. Ancient writers described this cloth as being "so sheer that it was invisible when spread over the ground and saturated by dew."

Despite all the speculation, there is definite knowledge that cotton has been an important crop in India since 3000 B.C. and was well known in Egypt as early as 2500 B.C. Since that time cotton has become the major fiber throughout the entire world.

When Columbus landed in the Bahamas in 1492, he was welcomed by natives who brought him cotton fibers, yarns, and threads. He found cotton growing on the islands (Sea Island cotton) and garments, nets, and hammocks (*hamacas*) made of cotton. It would be logical to believe that if the natives presented cotton to Columbus, it must have been grown in the area for some time prior to his arrival. There is evidence that cotton culture in what is now the United States dates back about 2500 years. The area in which cotton was first grown includes parts of the present Utah, Texas, and Arizona.

Fragments of cotton fabrics have been found in dry caves and burial sites of the American Indians who inhabited the Southwest centuries ago, and anthropologists interpret this fact as indicating the importance of fabrics to early Indian cultures. These fragments have been analyzed to determine their age, and scientists believe they antedate the Christian era by five hundred years. Pueblo Indians such as the Hopi and Zuni have used cotton since the first century A.D.

It is not feasible to include a detailed history of the early development of cotton in the North American continent, but the interested reader will find numerous references for further investigation. A brief look at the recent history of cotton culture will provide a foundation for the study of the current cotton industry in the United States.

The first recorded planting of cotton on the East Coast occurred in Florida in 1536. However, the primary purpose of this early crop was enjoyment of the blossom, not fiber use. Cultivation of the plant for fiber and as a profit-making venture occurred in Virginia, where cotton was abundant, between 1607 and 1620. Records show that cotton was cultivated throughout the Carolinas about 1665; by 1700 the cotton grown there furnished clothing to one-fifth of the population.

[2]M. D. C. Crawford, *The Heritage of Cotton* (New York: G. P. Putnam's Sons, 1924); and H. B. Brown and J. O. Ware, *Cotton* (New York: McGraw-Hill, Inc., 1958).

Most early cotton was of the Sea Island variety, because the Churka or roller gin imported from India could separate seeds and fiber in this type only. However, as cotton culture moved inland, it was found that Sea Island cotton would not thrive, and upland varieties were adopted. These were impossible to gin on the Churka, so hand separation of seed and fiber was required. After the invention of the saw-type gin by Eli Whitney in 1793 production of upland cotton increased rapidly. The cash value of the cotton crop jumped from $150,000 in 1793 to over $8 million in a ten-year period.

Plants for manufacturing and processing cotton into yarns and fabrics were originally located in the New England states by virtue of the presence of adequate waterpower and manpower. After the Civil War, the manufacturing plants began to move south to be closer to the crop. At the present time, the production of cotton fabrics is concentrated in the Southeast and East, but cotton cultivation has spread across the southern states from the Atlantic to the Pacific, with a large percentage of cotton now obtained from the Southwestern states.

Cotton is still the major fiber in use in the world today. However, during the last three decades world consumption of cotton has dropped from approximately three-fourths of all fibers to less than 60 percent. In the United States the popularity of cotton has fallen behind that of man-made fibers; nonetheless, as an individual fiber it still accounts for more use than any other.

GROWTH AND PRODUCTION

The cotton plant is a member of the *Malvacae* or *Mallow* family (Fig. 5.2). There are several species that are of the genus or class *gossypium*. Cotton is cultivated most satisfactorily in warm, humid climates or in warm climates with adequate irrigation. A primary growing requirement is a long, frost-free period of from six to seven months with mild temperatures and about twelve hours of sunlight each day. During actual growth either an average of 3 to 5 inches of rain a month or the equivalent through irrigation is required, followed by a dry season as the fibers are maturing.

The seeds are sown by machine in parallel rows 3 to 4 feet apart. The blossom, which appears approximately one hundred days after planting, is exquisite. When it opens, it is creamy white or light yellow in color; by the second morning the blossom has changed to pink, lavender, or red, and by the end of the second day the flower falls, leaving the young boll or seed pod in which the fibers form (Fig. 5.3). Fifty to eighty days later the pod bursts open, and the fleecy cotton fibers are ready for picking. The flowers appear over a long period of time, and thus the harvest period of mature cotton is of similar duration.

left: **Figure 5.2**   In this photograph of a cotton plant, the bolls are quite visible.

right: **Figure 5.3**   The single cotton boll has loose, fluffy fibers.

The crop is cultivated eight to ten times during the growing season to keep the weeds under control. Frequent spraying helps to combat the numerous insects that damage cotton. The best known of these pests is the boll weevil, which destroys the actual fibers, but other insects—such as the cotton leaf worm, the cotton aphid, the spider mite, and the pink bollworm—attack the plant itself.

Before picking, especially if mechanical pickers are used, the plants are sprayed with defoliants, which cause the leaves to shrivel and fall off. As the cotton bolls mature and open, the fleecy fibers cascade out of the boll in the form of "locks" of fiber. These are ready for picking.

Cotton can be picked by hand, by mechanical picking machines, or by stripping devices (Fig. 5.4). Hand picking results in more uniform and better-quality cotton, because the picker can select the mature fibers only, and he can repick the fields as many times as it is profitable. However, labor for this type of picking is scarce, and wages are increasing, so it is becoming economically unprofitable to use hand pickers except in small fields.

Picking machines are of two types: the picker and the stripper. The picker pulls the fibers from the open bolls, while the stripper pulls the entire boll from the plant. Both machines are important. The picker works best on fields with lush growth and high fiber yields; strippers are more effective on fields of low yield and low-growing plants. Strippers always pull from two rows simultaneously in a "once-over" operation. Pickers are designed to pick one or two rows and can go over the fields several times each season.

PROCESSING

After the cotton is picked, it is taken to the ginnery, where the fiber, called *cotton lint* by the trade, is separated from the seed. The gin used today is much the same as the first saw-gin designed by Eli Whitney. In addition to separating lint and seed, the modern gin will remove some foreign matter, such as dirt, twigs, leaves, and parts of bolls. The roller gin is still used for some long staple fibers, but most cotton is ginned on the saw-gin. The seeds are a valuable by-product of the cotton industry and produce cattle feed and cottonseed oil. The fibers or cotton lint are packed into large bales at the ginnery. Each bale weighs about 500 pounds gross.

Samples of fibers are removed from the bales and used for determining the class. Factors in this classification include the staple length, the grade, and the character of the cotton. *Staple length* refers to the length of the lint and is determined to some degree by the variety of cotton. The cotton market recognizes five classes of fiber length:

1. very short staple cotton, less than $\frac{3}{4}$ inch
2. short staple cotton, $\frac{13}{16}$ inch to $\frac{15}{16}$ inch
3. medium staple cotton, $\frac{15}{16}$ inch to $1\frac{1}{8}$ inch
4. ordinary long staple cotton, $1\frac{1}{8}$ inch to $1\frac{3}{8}$ inch
5. extra long staple cotton, $1\frac{3}{8}$ inch and longer

Fiber grade is based on color, amount of foreign matter present, and ginning preparation. The color can vary from white to gray or yellow. Cotton may be spotted or tinged, and it may be bright or dull. The spotted and tinged cotton is a frequent result of "one-time"

Figure 5.4 A cotton-picking machine.

Figure 5.5 Photomicrograph of regular cotton, cross sections. [*E. I. DuPont de Nemours & Company*]

picking, in which bolls that have been opened for some time are mixed with newly opened bolls.

Foreign matter with the fiber includes leaves, twigs, broken bracts (parts of the boll or seed pod), dust, dirt, and sand. Hand-picked cotton usually has a low amount of foreign matter, while mechanically picked fibers can have much or little, depending on picking conditions. Fibers pulled off the ground have additional dirt and soil embedded in them.

The quality of the ginning influences the grade of the cotton. If the ginning is poorly done, irregular cotton results, which reduces the grade.

The system for grading cotton is controlled by the U. S. Department of Agriculture (USDA). Originally nine grades were defined, with "middling" considered the medium or average grade. The original nine grades were

1. middling fair (best)
2. strict good middling
3. good middling
4. strict middling
5. middling (average)
6. strict low middling
7. low middling
8. strict good ordinary
9. good ordinary (lowest)

The grading or classification of cotton by quality is done, today, in several ways. One method uses a grade classification dependent on a modification of seven of the nine grades cited above. Grades one and two are no longer used; the remaining seven have 24 intermediary steps. A second method classifies cotton by *micronaire fineness*; that is, it determines the weight in micrograms per inch of fiber. This is a more sophisticated technique than the previous one. Many mills determine the micronaire per bale, then blend to obtain a consistent average fineness for input to the pickers.

In addition to classification on the basis of staple length, micronaire fineness, color, and foreign matter present, cotton is also rated in terms of its character. This includes such qualities as fiber strength, uniformity, cohesiveness, pliability, elastic recovery, fineness, and resiliency. These properties are determined generally on a "bundle" of fibers rather than on individual ones. When all this information has been tabulated, the final quality of the cotton is assessed and used in establishing fiber price.

After ginning and classification are complete, the cotton bales are shipped to manufacturers, where yarns and fabrics are made. These construction procedures are discussed in Parts III and IV. Methods of coloring, and use and application of finishes are discussed in Part V.

FIBER PROPERTIES

Microscopic Properties

Cotton fibers are composed of an outer cuticle (skin) and primary wall, a secondary wall, and a central core or *lumen*. Immature fibers exhibit thin wall structures and large lumen, while mature fibers have thick walls and small lumen that may not be continuous, because the wall closes the lumen in some sections (Fig. 5.5).

The longitudinal view of the fiber (Fig. 5.6) shows a ribbonlike shape with twist (convolutions) at irregular intervals. The diameter of the fiber narrows at the tip. The lumen may appear as a shaded area or as striations; this is more obvious in immature fibers.

Fibers that have been swollen, as in mercerization, do not show the twist as clearly as does untreated fiber. Swollen fibers appear smooth and round when compared with the untreated. Immature fibers also have few convolutions.

The cross section of the fiber (Fig. 5.7) usually has three areas: the outer skin, the secondary wall, and the lumen. The contour varies considerably: some fibers are nearly circular, some are elliptical, and some are kidney-shaped. Immature fibers are generally more irregular in contour than mature fibers.

Figure 5.6 Photomicrograph of regular cotton, longitudinal view. [*E. I. DuPont de Nemours & Company*]

Physical Properties

Shape   Cotton fibers are fairly uniform in width—more so than most other natural fibers. The width varies between 12 and 20 microns, and the central portion of the fiber is wider than either end. Depending upon variety and growing conditions, the length of the cotton fiber used in yarn and fabric manufacture ranges from $\frac{1}{2}$ inch to $2\frac{1}{2}$ inches, with most fibers in the $\frac{7}{8}$-inch to $1\frac{1}{4}$-inch category.

The length, fineness, and uniformity of fibers aid in distinguishing the common varieties. American Upland cotton has a diameter of approximately 18 microns, a length of less than $1\frac{1}{8}$ inches, and smooth, regular convolutions. The great preponderance of cotton produced in the United States is Upland and includes such types as Acala, Deltapine, Crocker, Delfos, Empire, and Stoneville.

Sea Island and American-Egyptian cotton are long-staple varieties. They are fine in width, usually 15 microns or less, over $1\frac{1}{8}$ inches in length, and possess regular convolutions. Sea Island is grown on the islands off the Georgia coast, while American-Egyptian cotton is raised in Arizona, New Mexico, and California.

Long-staple American-Egyptian cotton is sold under such names as Supima and Pima. The amount of long staple cotton produced is small in comparison to shorter fibers of the Upland varieties. Long-staple cotton can be made into fine-quality and beautiful fabrics. However, it is more costly to process and more difficult to cultivate.

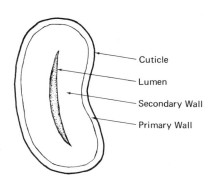

Cuticle

Lumen

Secondary Wall

Primary Wall

Figure 5.7 Diagram of the cross section of a cotton fiber.

Since the 1950s production of Sea Island has dropped markedly, while American-Egyptian, raised in the West and Southwest, has shown steady increase. However, long staple cotton represents a small amount of the total cotton crop.

Egyptian cotton, which is imported in small amounts, is long, less uniform than American-Egyptian or Sea Island, and slightly wider. It is usually yellower in color than comparable United States varieties.

Luster The luster of cotton is low unless finishes have been added.

Strength Cotton has a tenacity of 3.0 to 5.0 grams per denier. This produces a fiber of moderate to above-average strength. When wet, cotton increases in strength, so it may have a wet strength equal to 110 to 120 percent of its dry strength. This means that fiber care and wet processing techniques do not require modification to compensate for reduced fiber wet strength.

Elastic Recovery and Elongation Cotton has an elongation of 3 to 7 percent (and may be as high as 10 percent under certain conditions) at the break point. It is relatively inelastic, with a recovery of only 74 percent at 2-percent elongation. At 5-percent elongation it exhibits less than 50-percent elastic recovery.

Resiliency The resiliency of cotton is low.

Specific Gravity Cotton is one of the denser fibers. The specific gravity is 1.54.

Moisture Absorption The moisture regain for cotton at standard conditions is 8.5 percent. Regain at 95-percent relative humidity is approximately 15 percent, and at 100-percent relative humidity it may absorb 25 to 27 percent moisture.

In addition to an increase in strength, wet cotton is more pliable and less rigid than dry cotton.

Dimensional Stability Cotton fibers are relatively stable and do not stretch or shrink. Cotton *fabrics*, however, do tend to shrink as a result of tensions encountered during yarn and fabric construction. Consequently, the fabrics require treatment to render them less susceptible to shrinkage.

Thermal Properties

Cotton burns readily and quickly with the smell of burning paper. It leaves a small amount of fluffy gray ash. Long exposure to dry heat above 149°C (300°F) will cause the fiber to decompose gradually, and temperatures greater than 246°C (475°F) will cause rapid deterioration. Normal exposure to heat encountered in routine care and processing will not damage cotton, but it will scorch if ironed with too-high temperatures. Finishes, such as starch, increase the tendency to scorch.

Chemical Properties

Effect of Alkalies   Cotton is highly resistant to alkalies. In fact, they are used in finishing and processing the fiber. Most detergents and laundry aids are alkaline, so cotton can be laundered in these solutions with no fiber damage.

Effect of Acids   Strong acids destroy cotton, and hot dilute acids will cause disintegration. Cold dilute acids cause gradual fiber degradation, but the process is slow and may not be immediately evident.

Effect of Organic Solvents   Cotton is highly resistant to most organic solvents and to all those used in normal care and stain removal. It is, however, soluble in such compounds as cuprammonium hydroxide and cupriethylene diamine, and these are used for chemical analysis of cotton.

Effect of Sunlight, Age, and Miscellaneous Factors   Prolonged exposure to sunlight will cause the cotton fiber to become yellow and will gradually cause degradation. This damage is accentuated in the presence of moisture, some vat dyes, and some sulfur dyes.

If properly stored, cotton will retain most of its strength and appearance over a long period of time. It should be stored in dark, dry areas.

Biological Properties

Microorganisms   Cotton is damaged by fungi such as mildew and bacteria. Mildew will produce a disagreeable odor and will result in rotting and degradation of cotton. Certain bacteria encountered in hot, moist, and dirty conditions will cause decay of cotton.

Insects   Moths and beetles will not attack or damage cotton. Silverfish will eat cotton cellulose, especially if heavily starched, but they do more damage to paper and other products composed of relatively short cellulose molecules.

COTTON IN USE

Cotton is the most widely used fiber, and it is excellent for a multitude of purposes (Fig. 5.8). Fabrics of cotton are usually inexpensive, since the fiber is comparatively low in price, and further economies result

Figure 5.8   Cotton fabric (100 percent) decorated with hand-screened print. [*Marimekko of Finland*]

from easy-care properties. Cotton has virtually universal consumer acceptance. Fabrics of many different weights and many different constructions can be made from cotton. It is used for apparel fabrics, for household or domestic goods, and for industrial applications.

Cotton fibers produce fabrics that are characterized by good wearing qualities, excellent launderability, high absorbency, good color fastness if proper dyes are used, easy dyeability, good pliability and flexibility, and good heat resistance. Adequate finishing processes must be applied to produce cotton fabrics that will not shrink, as well as fabrics with crease resistance and wash-and-wear properties. Cotton accepts finishes that produce resistance to shrinkage, water repellency, flame retardation, wrinkle and crease resistance, minimum care properties such as durable press, and fabric stretch characteristics.

Comfort is the outstanding characteristic of fabrics made from cotton. This is the result of adequate fiber strength and cohesiveness, so fine yarns can be constructed and made into relatively thin and lightweight fabrics. A low degree of fiber loft—that is springiness or fluffiness—makes the fibers pack compactly into yarns and results in a high ratio of fabric thinness to weight. Furthermore, the fiber has good wickability—moisture will move along the fiber surface and through the fabric quickly[3]—which adds to comfort. Additional discussion of fabric characteristics related to comfort, maintenance, durability, and appearance will be found in Chapter 33.

FLAX

HISTORICAL REVIEW

Flax is considered by many to be the oldest fiber used in the Western world. Fragments of flax (linen) fabrics have been found in excavations at the prehistoric Swiss lake regions of Switzerland, which date back to about 10,000 B.C.

The use of linen in Egypt between 3000 and 2500 B.C. has been verified. At that time and for many years after it was used for mummy wrappings. These early fabrics were of a fineness that has never been duplicated. Examples have been found that were spun so fine that more than 360 single threads joined together formed one warp thread. Other fabrics were made with more than 500 yarns per inch.[4]

The use of flax spread from the Mediterranean region to other parts of Europe. Belgium became one of the important centers for growing flax because of the chemicals in the water of the River Lys.[5] This water

---

[3] E. R. Kaswell, *Textile Fibers, Yarns and Fabrics* (New York: Reinhold Publishing Corporation, 1953), p. 101.
[4] *Ciba Review*, No. 49, p. 1766.
[5] *Ciba Review*, No. 49, p. 1775.

Figure 5.9 The final spinning of flax. [*Belgian Linen Association*]

was found to be exceptional in retting flax (see p. 54), and it produced high-quality fibers (Fig. 5.9). The town of Courtrai, located on the Lys, became, as it remains today, a major center for the flax industry.

Linen fabric was introduced in Great Britain from Egypt about 1000 B.C., but actual use of flax, growing wild in England, probably did not occur until the first century A.D. At that time, it is likely that Britons used flax fiber for coarse yarns and fabrics. However, at the same time, Ireland was beginning to process flax into fine linen fabrics, and by A.D. 500 Irish linen was held in high esteem by rulers throughout Europe.

During the seventeenth century the British government, in an attempt to maintain control of the wool industry in England, encouraged the development of the Irish linen industry by permitting its expansion while limiting the production of wool.

Flax seed was brought to America by early settlers. Many colonists grew their own flax, spun their own yarns, and made their own fabric. The Industrial Revolution and development of the factory system took this enterprise out of the home. The machinery most suitable for processing flax was developed in Europe; thus, flax production for fiber use came to a virtual halt in America.

Today, the Soviet Union grows most of the flax for fiber. Other producers include New Zealand, Belgium, Ireland, and the nations of eastern Europe. The United States grows flax for seed but imports almost all of the flax fiber for textiles, usually in the form of finished linen fabrics.

top: **Figure 5.10** Machines pull up flax by the roots. It is bundled, and dried; the seeds are removed. Stalks are sorted for size, rebundled, and taken to retting tanks. [*Belgian Linen Association*]

center: **Figure 5.11** Along Belgium's River Lys, flax retted in the concrete buildings dries in the sunlight. [*Belgian Government Information Center*]

above: **Figure 5.12** Bundling flax. [*Belgian Government Information Center*]

GROWTH AND PRODUCTION

Flax is a bast fiber—a woody fiber obtained from the phloem of plants. It derives from the stalk or stem of the *Linum usitatissimum.*

The flax plant requires a temperate climate with generally cloudy skies and adequate moisture. Bright sunlight and high temperatures are damaging unless alternated with abundant rainfall. The plant must obtain considerable nutritive value from the soil, and thus, best-quality flax requires a crop-rotation program of five years.

Flax seed is planted by hand in April or May. When the crop is to be used for the production of fiber, the seed is sown close together, so the plant will grow dense but fine. The flax plant reaches a height of 2 to 4 feet. Its blossoms are a delicate pale blue. If the flax is to be used for fiber, it is pulled before the seeds are ripe; if to be used for flax seed, the plant is allowed to mature completely before harvesting.

PROCESSING

Pulling and Rippling

Flax for fiber is pulled by hand or by mechanical pullers (Fig. 5.10) to keep the roots intact, for the fibers extend below ground surface. Harvesting occurs in late August when the plant is a rich brown color. After drying, the plant is *rippled;* that is, it is pulled through special threshing machines that remove the seed bolls or pods.

Retting

To obtain the fibers from the stalk, the outer woody portion must be rotted away. This process, known as *retting,* can be accomplished by any of several procedures (Fig. 5.11).

Dew retting involves the spreading of flax on the ground, where it is exposed to the action of dew and sunlight. This natural method of retting gives uneven results but provides the strongest and most durable linen. Unfortunately, it requires a comparatively long period of time; the average exposure is from four to six weeks.

Pool retting is a process whereby the flax is packed in sheaves and immersed in pools of stagnant water. Bacteria develop in the water and rot away the stalk covering. When retting is complete, the water is drained away, and the flax is dried preparatory to the next step. The time required for pool retting is from two to four weeks.

Tank retting, similar to pool retting, utilizes large tanks in which the flax is stacked. The tank is filled with warm water, which increases the speed of the bacterial action. Retting is then accomplished in a few days.

Both pool and tank retting give good-quality flax that is uniform in size and strength and light in color.

Stream retting is practiced in some areas. The flax is stacked along the banks of slow-moving streams. The constantly moving water slows down the retting procedure considerably, but it reduces the unpleasant smell associated with dew and pool retting, and it produces good-quality flax. A long time is needed for this retting technique; it frequently takes as long as dew retting.

Chemical retting is accomplished by stacking the flax in tanks, filling the tanks with water, and adding chemicals such as sodium hydroxide, sodium carbonate, or dilute sulphuric acid. Chemical retting can be accomplished in a matter of hours instead of days or weeks. However, it must be carefully controlled in order to prevent damage to the fiber.

Breaking and Scutching

After retting is complete, the stalk is bundled together (Fig. 5.12) and passed between fluted rollers that break the outer woody covering into small particles. It is then subjected to the *scutching* process, which separates the outer covering from the usable fiber. Since the early nineteenth century this has been done by mechanical devices; before that it was a manual operation.

Hackling

After scutching, the flax fibers are *hackled* or combed (Fig. 5.13). This separates the short fibers (*tow*) from the long fibers (*line*). It is accomplished by drawing the fibers between several sets of pins, each successive set finer than the preceding. This process is similar to the carding and combing operation used for cotton (see Chap. 17) and prepares the fibers for the final step in yarn manufacturing (Fig. 5.14).

Spinning

The flax fibers are drawn out into yarn, and twist is imparted. Flax fibers are spun either dry or wet, but wet spinning is considered to give the best-quality yarn. Basically, the final yarn processing is similar to that used in making cotton yarns.

FIBER PROPERTIES

Microscopic Properties

Flax fiber is composed of fibrils or bundles of fiber cells held together by a bonding or gummy substance (Fig. 5.15). Under the microscope the longitudinal view of the fiber shows the width to be quite irregular

Figure 5.13 The large machine shown performs the final step in the production of flax before spinning. The pile of flax is partly combed and ready to be fed into the large combing machine, where it will travel from right to left and emerge as a wide, continuous ribbon (*far left*). [*Belgian Linen Association*]

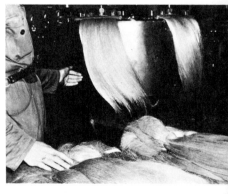

Figure 5.14 Sometimes the flax is removed from the combing machine before it is formed into a continuous ribbon. When this is done, the flax takes on the appearance of human hair. [*Belgian Linen Association*]

(Fig. 5.16). The central canal or lumen casts a shadow, giving a slightly darker effect down the center. There are no convolutions as in cotton, but longitudinal lines or striations can be seen. The points at which the fiber width changes are marked by swellings and irregular joint formations called *nodes*. These are similar to the joints on bamboo.

The cross section view (Fig. 5.17) clearly shows the lumen, the thick outer wall, and a somewhat polygonal shape. Immature flax may be oval in shape and usually has a larger lumen than mature fiber.

Physical Properties

Shape and Appearance   Flax fiber is not so fine as cotton; flax cells have an average diameter of 15 to 18 microns and vary in length from $\frac{1}{4}$ inch to $2\frac{1}{2}$ inches. Bundles of cells form the actual fiber as it is used in spinning, and these bundles may be anywhere from 5 to 20 inches long. Line fibers are usually more than 12 inches long, while tow fibers are less than 12 inches. The fiber is roughly cylindrical.

The natural color of flax varies from light ivory to dark tan or gray. The choice fibers from Belgium and the other Low Countries are a pale sandy color and require little bleaching. Flax possesses a high natural luster that produces attractive yarns and fabrics.

Strength   Flax is a strong fiber, which usually has a tenacity of 5.5 to 6.5 grams per denier. Occasionally, some fiber of inferior quality may have a tenacity as low as 2.6 grams per denier. Fabrics of flax are durable and easy to maintain because of the good strength. When wet the fiber is even stronger than when dry, with an increase in strength of about 20 percent.

Most linen fabrics used in apparel have received resin finishes to improve fabric performance. These finishes reduce the strength of the linen so that it may give poor durability. Linen, finished and unfinished, has poor flexing strength, which may result in reduced serviceability despite the natural strength based on pull measurements.

Elastic Recovery and Elongation   The amount of elongation that flax will undergo before breaking is very small. When dry the fiber will extend 2.7 to 3.3 percent. Within the limits of the possible elongation the fiber has little elasticity. At 2-percent extension it has an immediate elastic recovery of 65 percent.

Resiliency   Linen fabrics are prone to crease and wrinkle badly. They are somewhat stiff and possess little resiliency. However, finishes can be applied that help offset these disadvantages.

Specific Gravity   The density of linen is comparable to other cellulosic fibers used for apparel. Its specific gravity is 1.50.

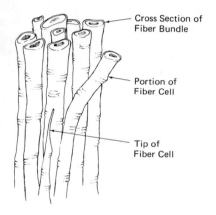

Cross Section of
Fiber Bundle

Portion of
Fiber Cell

Tip of
Fiber Cell

Figure 5.15 Schematic diagram of fibrils in the flax fiber.

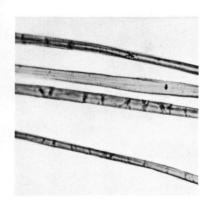

Figure 5.16 Photomicrograph of flax fiber, longitudinal view. [*E. I. DuPont de Nemours & Company*]

Moisture Absorption   Flax has a standard moisture regain of 12 percent. The saturation regain is comparable to that of other cellulosic fibers. Also, fabric structure can influence the moisture absorption.

Dimensional Stability   Flax fibers do not shrink or stretch to any marked degree. However, as in the case of cotton, yarns and fabrics are subject to some relaxation shrinkage unless preshrunk during finishing. This shrinkage is usually less than that experienced for cotton fabrics. Ironing linen while damp will reduce the apparent shrinkage by stretching the fabric back into position.

Thermal Properties

Like any other cellulosic fiber, flax burns. It is highly resistant to decomposition or degradation by dry heat and will withstand temperatures to 149°C (300°F) for long periods of time with little change. Above 149°C prolonged exposure will result in gradual discoloration. Safe ironing temperatures may go as high as 500°F as long as the fabric is not held at the high temperature for any length of time.

Chemical Properties

Flax is highly resistant to alkaline solutions; it is also resistant to cool dilute acids, but concentrated acids and hot dilute acids will cause deterioration. Flax has excellent resistance to dry cleaning solvents and other organic compounds encountered in maintenance.

There is a gradual loss of strength when linen fabrics are exposed to sunlight, but this is not serious. Flax makes a good choice for curtain and drapery fabrics.

If stored properly, linen will age remarkably well. Table coverings and sheets packed away for many years have proved to be as strong as comparable new fabrics.

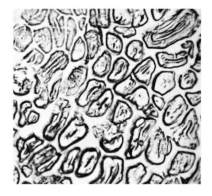

Figure 5.17 Photomicrograph of flax fiber, cross section. [*E. I. DuPont de Nemours & Company*]

Biological Properties

Dry linen has excellent resistance to mildew, but if it is very moist or is stored in a very humid atmosphere, mildew will grow rapidly and will damage the fiber. Other bacteria will not attack linen unless the fiber is exposed to severe conditions of dampness.

Flax is generally resistant to most household insects and pests.

LINEN IN USE

Flax has relatively high strength and can be made into fine yarns and sheer fabrics that are both strong and cool. Comfort is further enhanced by the wicking properties of linen. Fabrics composed of flax are popular

Figure 5.18   Table linens. [*Dansk Designs Ltd.*]

for wearing apparel because, in addition to being cool, they maintain their clean appearance and are easily laundered.

Linen is used for table coverings (Fig. 5.18) because it wears well, it looks attractive and elegant, and the yarns lie flat if a beetling finish has been applied. *Beetling* involves beating the cloth with large wooden blocks in order to flatten the yarns. This also gives the fabric a gloss. A high-speed beetling machine has been developed that utilizes metallic hammers, which can be controlled in various ways to give light or heavy blows to the fabric and produce a desirable product in less than half the time used by the older wooden faller types (Fig. 5.19).

Linen fabrics may be smooth and sheer, coarse, homespun in effect, or highly patterned by means of the Jacquard weave (see pp. 243–244). They are used for apparel and for a wide variety of domestic fabrics. There is considerable prestige associated with linen; it has a high heritage and sentiment value.

The resistance of flax to chemicals and laundering as well as to deterioration from age and sunlight, plus the strength of the fiber, combine to produce fabrics with long life. Linen fabrics respond satisfactorily to laundering, and it is the preferred method of cleaning. Because of the structure of the fiber, it does not soil quickly, and, unless stained, it does not require bleaching.

JUTE

Jute is also a bast fiber. Like flax, it has been used since the dawn of civilization; however, it attained economic importance only during the latter part of the eighteenth century. Today, jute is one of the most widely used fibers and is of special interest to countries of low economic standards, because it is quite inexpensive. Jute is grown and widely used in Brazil, India, and Pakistan. In fabric form it is frequently called *burlap*.

Recently, jute has enjoyed limited success as a fashion fabric for wearing apparel. It is a weak fiber, however, and its use is normally limited to products in which durability is not important.

The jute plant is cultivated in a manner similar to that of flax. The seeds are planted close together and grow to a height of 15 to 20 feet. The fibers are extracted from the stem in the same manner as for flax (by retting), followed by breaking and scutching. Yarn construction is also similar to flax.

Natural jute has a yellow to brown or gray color, with a silky luster. It consists of bundles of fiber held together by gummy substances that are pectinaceous in character. It is difficult to bleach completely, so many fabrics are bright, dark, or natural brown in color.

Jute has a tenacity of about 3 to 5 grams per denier. It has a low elongation of less than two percent and poor elastic recovery. The specific gravity is 1.5.

Under the microscope it resembles flax, except that the lumen, when viewed in cross section, is much less even than in flax.

Jute reacts to chemicals in the same way as do cotton and flax. It has good resistance to microorganisms and insects. Moisture increases the speed of deterioration, but dry jute will last for a very long time. Jute works well for bagging, because it does not extend and is somewhat rough and coarse. This tends to keep stacks of bags in position and resist slippage. It is widely used in the manufacture of linoleum and carpets for backing or base fabric.

RAMIE

Frequently called *China Grass*, ramie is a bast fiber that has been cultivated for hundreds of years in China and Formosa. There is some evidence that ramie was grown in the Mediterranean civilizations as early as flax; however, its use in China was of much greater importance. The history of the development of ramie is centered in the Orient, and only in comparatively recent centuries has it become important in Western civilizations. At the present time, the fiber is grown commercially in China, Japan, Egypt, France, Italy, Indonesia, Russia, and the United States.

Ramie cultivation in the United States was first attempted about 1855, but scientific management of ramie fields in the Florida Everglades was not begun until 1929. Since that time, the Belle Glade Experiment Station of the USDA has spent large sums of money on ramie growth and production.

The ramie plant (Fig. 5.20), a member of the nettle family, is a perennial shrub that can be cut several times a season after the neces-

left: **Figure 5.19**   A beetling machine for flax fabrics. [*The Irish Linen Guild*]

right: **Figure 5.20**   Ramie plant. [*United States Department of Agriculture*]

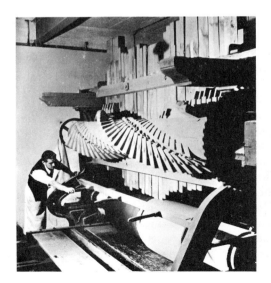

left: **Figure 5.21** Ramie plants during early growth. [*R. V. Allison, Belle Glade Experimental Station, Florida*]

left below: **Figure 5.22** Ramie stalks with decorticated and degummed fibers. [*R. V. Allison, Bell Glade Experimental Station, Florida*]

sary preliminary growth (Fig. 5.21). It can be started from seeds, which necessitates a developmental period of three years before fibers are formed; or it can be grown from root cuttings, which mature within two or three years.

After cutting, the ramie stalks are then decorticated by one of several machine or hand methods to remove the outer woody covering and reveal the fine fibers (Fig. 5.22). These are degummed to eliminate the pectins and waxes. Degumming is usually done by soaking the fibers in caustic soda for several hours, then bleaching and neutralizing in a dilute acid bath. Finally, the fiber is washed and dried.

Ramie fibers are long and very fine. They are white and lustrous and almost silklike in appearance. The strength of ramie is excellent and varies from 5.3 to 7.4 grams per denier. Elastic recovery is low, and elongation is poor. The fibers are somewhat stiff.

Ramie reacts chemically in the same manner as other cellulosic fibers, except that the fiber is not easily damaged by cold concentrated mineral acids. The high degree of molecular crystallinity and low molecular accessibility reduces the rate of acid penetration and thus the rate of damage. Of special interest is the fact that ramie is highly resistant to microorganisms, insects, and rotting. This is probably due to the presence of nonfibrous matter, which may contain material that is toxic to bacteria and fungi.

Ramie fabrics sometimes resemble a fine linen, or they can be heavy and coarse like canvas. The latter are of industrial importance, while the former are made into shirts, table coverings, napkins, and similar items. Because of its stiffness, ramie is frequently blended with other fibers, such as cotton or rayon. The ramie adds strength, while the cotton or rayon introduces softness.

HEMP

Hemp is a bast fiber that was probably used first in Asia. Records indicate that it was cultivated in China before 2300 B.C. Sometime during the Early Christian era, hemp was carried into Gaul and became an important fiber throughout Europe.

Today, hemp is grown on every continent and in nearly every country. It is a tough plant and will grow at altitudes to 8000 feet and in climates where temperatures are warm or hot. It can be replanted in the same fields more frequently than flax (Fig. 5.23).

The processing of hemp is very similar to that of flax. It requires retting to loosen the outer covering, followed by stripping or scutching to obtain the usable fibers (Figs. 5.24, 5.25). Hackling or drawing and spinning of the fibers into yarns is the final step before fabric manu-

Figure 5.23 Cutting hemp by hand. [*Ciba Review*]

Figure 5.24 Retting hemp. The fibers are laid in *rafts* and submerged by stone weights. [*Ciba Review*]

Figure 5.25 Stacking hemp to dry after retting. [*Ciba Review*]

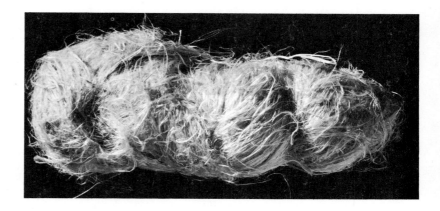

Figure 5.26   Hemp fiber. [*Ciba Review*]

facture. The fiber is dark tan or brown and is difficult to bleach, but it can be dyed bright and dark colors (Fig. 5.26).

The hemp fibers vary widely in length, depending upon their ultimate use. Industrial fibers may be several inches long, while fibers used for domestic textiles are about ¾ inch to 1 inch long. The specific gravity of hemp is 1.48. Its strength is 5.2 grams per denier. The elongation (1 to 6 percent) is low and its elasticity poor. Standard moisture regain is 12 percent, and it can absorb moisture up to 30 percent of its weight.

Hot concentrated alkalies will dissolve hemp, but hot or cold dilute or cold concentrated alkalies will not damage it. With the exception of cool weak acids, mineral acids will reduce the strength and eventually destroy the fiber completely. Organic solvents used in cleaning and bleaches, if handled properly, will not damage hemp.

The thermal reactions of hemp and the effect of sunlight are the same as for cotton. Hemp is moth resistant, but it is not impervious to mildew.

Coarse hemp fibers and yarns are woven into cordage, rope, sacking, and heavy-duty tarpaulins. In Italy, fine hemp fibers are used for interior design and apparel fabrics.

MISCELLANEOUS PLANT FIBERS

Other plant fibers discussed in this text include those that have limited use in the United States for specific products or those that have topical interest to the textile student.

Kenaf is a fiber somewhat similar to jute. During World War II many countries, including the United States, faced a severe shortage of certain fibers, and jute was one of these. Kenaf was found to grow and produce good fibers in both the United States and Mexico, and for some time was produced in quantity in both countries. However, while some is still grown here, most kenaf is raised in India and Pakistan.

The fiber is obtained from the stalk of the plant, and methods of cultivation and processing are similar to those used for jute. The plant grows about 10 feet tall, but the fibers are considerably shorter, because they break apart during processing.

Kenaf is a light-colored fiber that is frequently used in its natural state, but it can also be dyed if desired. It is not strong and has low elasticity. Water does not weaken kenaf, and thus, it finds widespread use in the manufacture of rope, cordage, and coarse fabrics, such as canvas, sacking, and carpeting.

Urena is another bast fiber that was developed as a result of the shortage of jute during the war. It has been grown in South America for centuries, probably since prehistoric eras, but present-day cultivation began in 1926, with the most rapid growth occurring since 1940.

At the present time urena is widely used in Africa. It grows to a height of over 10 feet, and the processing is like that required for jute and hemp. When properly processed, the fiber is nearly white; it is soft and has a natural luster. Urena is used for low-cost apparel, for coarse decorating fabrics, and for sacking.

Sisal is one of a group of fibers obtained from the leaves of plants. It is obtained from a plant that belongs to the *Agave* family and is raised in Mexico, especially in the Yucatán peninsula. The fiber is also cultivated in Africa, Java, and some areas of South America.

The leaves grow in a rosette form from a short trunk. They are cut when they are about four years old. The processing of the fibers involves separating them from the fleshy part of the leaf and removing pectins, chlorophyll, and other noncellulosic substances. Sisal can be dyed bright colors, using both cotton dyes and acid dyes normally used for wool. It is important in the manufacture of such items as matting, rough handbags, ropes and cordage, and native-style hats.

Henequen is a leaf fiber also from a member of the *Agave* family. Its area of production is much the same as that of sisal, and it is processed similarly. The major application of henequen at the present time is for agricultural twine.

Abaca is obtained from a plant that belongs to the banana family. In appearance it is often mistaken for the edible banana tree. The plant grows mainly in the Philippines, and most of the textile fiber is processed there. It is cultivated for appearance value in the southwestern United States, especially in California, in parts of Mexico, and, occasionally, in Florida. The leaf stalk from which the fiber is taken may reach a length of 25 feet. The fibers, therefore, are generally long; usable strands of abaca fiber may be up to 15 feet in length.

Good-quality abaca has a natural luster and is an off-white color. Poor grades are dark gray or brown.

The fiber is strong and flexible and is exceptionally good for making rope and cordage, place mats for outdoor or indoor use, and clothing. The fabric is delicate, lightweight, yet strong.

Piña fiber, from the leaves of the pineapple plant, is also produced in the Philippines. It is a white or light ivory colored fiber about 2 to 4 inches in length. The fiber is fine, lustrous, soft, flexible, and strong. It is highly resistant to water. Fabrics made of piña may be either soft and delicate or crisp. Considerable fiber is employed in making "peasant" type clothing and table coverings with elaborate embroidery. The fabrics are easily cleaned and will retain their appearance for long periods of time. Some rope and twine is also made from the pineapple fiber.

Coir, a fiber from a seed source, is used for matting and cordage. It is obtained from the coconut and is the fibrous mass between the outer shell and the actual nut.

The natural color of coir is a rich cinnamon brown, and the fiber is frequently left undyed for floor mats and outdoor carpets or patio coverings. Because of a high degree of stiffness, it is wrinkle resistant, and it is strong and impervious to abrasive wear. In addition, it can stand exposure to weather, especially water, so it proves to be a very practical fiber for mats and similar products. The fiber can be dyed dark colors, but it is difficult to bleach it sufficiently to produce pale hues.

Kapok is a seed hair fiber obtained from the Java kapok tree. The tree grows in tropical regions and rises to a height of 50 feet or more. The seed pods are similar to the cotton boll. Ginning is not necessary, since the fibers can be dried, after which the seeds will shake away easily.

Kapok is extremely light, buoyant, and soft. Its major use is for padding and stuffing, particularly in upholstered furniture and mattresses. Because it is nonallergenic, it makes an excellent filling for pillows. The best-known use for kapok is in life preservers. Kapok-filled preservers will support up to thirty times the weight of the preserver and will not become waterlogged. The fiber is difficult to spin into yarns, so it is primarily used for these other purposes.

Man-made
Cellulosic Fibers:
Rayon

The Textile Fiber Products Identification Act defines *rayon* as

> a manufactured fiber composed of regenerated cellulose, as well as manufactured fibers composed of regenerated cellulose in which substituents have replaced not more than 15 percent of the hydrogens of the hydroxyl groups.

Rayon received its name in 1924. Before that it had been called "artificial silk," and because of the general dislike of artificiality, many potential consumers avoided it. In addition early rayon was an inferior product. After the fiber was named *rayon,* and with improvement of its properties, it gradually gained public acceptance and today ranks next to cotton in amount used.

HISTORICAL REVIEW

The first written comment concerning the potential of creating man-made fibers is found in Robert Hooke's *Micrographia*, published in 1664. Hooke predicted that eventually there would be a way to duplicate the excrement of the silkworm. However, not until the nineteenth century did scientists actually make artificial fibers.

Figure 6.1 Spinnerettes used in the formation of man-made fibers.

In 1855 George Audemars made filaments from a solution of mulberry twigs in nitric acid. He pulled out a fine thread by inserting a needle into the solution and then removing it. The material adhering to the needle solidified in the air. E. J. Hughes, in 1857, created fibers from a solution of starch, glue, resins, and tannins.

The major breakthrough in the production of man-made fibers occurred in 1862, when Ozanam invented the spinning jet or spinnerette. This remarkable little device is the basis for all manufactured fiber production (Fig. 6.1; see also Fig. 6.6).

In 1883 J. W. Swan produced filaments by forcing a solution of cellulose nitrate in glacial acetic acid through a spinning jet. He exhibited his fibers to the Society of Chemical Industry in London in 1884, and the following year he had his yarns crocheted into nets. Swan used a denitration process to reduce the flammability of the fiber, and his product was used for the filament in the first electric light bulb.

Credit for the invention of rayon is generally given to Count Hilaire Chardonnet, who produced a cellulose nitrate that he dissolved in alcohol. This solution was forced through the spinnerette into water or warm air. The filaments hardened, were stretched to orient the molecules and introduce sufficient strength, and finally were denitrated and purified to reduce flammability. By 1889 Chardonnet had exhibited fabric samples, and by 1895 his company, the Societé Anonyme pour la Fabrication de la Soie de Chardonnet, was paying dividends.

The cellulose nitrate process for making rayon is of historical interest only. Other and better methods for producing regenerated cellulose have replaced Chardonnet's.

The viscose process was discovered in 1891 by the English scientists C. F. Cross and E. J. Bevan. Since that time the operation has been greatly modified and improved. The first American rayon plant, called the American Viscose Company, was opened in 1910. Today, this organization is a division of the FMC Corporation and is a major force in the rayon industry.

left: **Figure 6.2** Packing cellulose sheets into a steeping tank to form alkali cellulose. [*FMC Corporation*]

right: **Figure 6.3** Shredded alkali pulp. [*FMC Corporation*]

An important part of the mechanical development of rayon manufacturing was the invention of the spinning box by C. F. Topham in 1905. This box caught the fibers and imparted sufficient twist to hold the filaments in place.

Between 1916 and 1930 such companies as DuPont, Industrial Rayon, Celanese, American Enka, and North American joined the ranks of manufacturers of viscose rayon. In 1926, American Bemberg, now Beaunit Mills, introduced the cuprammonium process for manufacturing fibers.

Rayon was first employed in tire cord in 1937. To increase the practicability of rayon for automobile tires a group of manufacturers developed specifications for a unique high-strength rayon filament to be used in tire manufacturing. This fiber was identified by the term *Tyrex* and is still used for a large proportion of vehicle tires manufactured at the present time.

The decade of the fifties saw many improvements in existing rayon fibers, as well as the development of high-strength rayon, special carpet rayon, and high-wet-modulus rayon. However, during the same period several plants were closed because of increased use of other fibers and machinery obsolescence.

An important market for rayon today is the field of home-furnishing fabrics.

Figure 6.4 Crumbs treated with carbon disulfide (xanthate). [*FMC Corporation*]

MANUFACTURING PROCESSES

Viscose Process

The principal raw material for viscose rayon is wood pulp. *Cotton linters*—the cotton fibers that are too short for yarn or fabric manufacturing—are also used. The wood pulp and linters are processed, and pure cellulose is extracted and formed into thin sheets that are about 2 feet square (Fig. 6.2). Considered chemically, the raw material is identical to pure cotton cellulose or to the cellulose in any of the vegetable fibers.

The sheets of cellulose are steeped in an alkali solution (17.5 percent solution of NaOH) until the cellulose is converted to soda or alkali cellulose. After the steeping process the alkali pulp is shredded into alkali cellulose crumb, which is aged for a specific time (Fig. 6.3). After aging, the cellulose crumbs are treated with carbon disulfide (Fig. 6.4). To this point the soda cellulose has been white, but the carbon disulfide changes it in color to a bright orange and chemically to a product called sodium cellulose xanthate. These molecular changes are shown in Figure 6.5.

The xanthate crumb is dissolved in dilute sodium hydroxide and forms a honey-colored liquid. This solution is aged until it reaches

1. Cellulose   +   Caustic soda $\longrightarrow$ Soda cellulose + Water
   $(C_6H_{10}O_5)n$ + nNaOH $\longrightarrow$ $(C_6H_9O_5Na)n$ + nH$_2$O

   NOTE: The reaction takes place on the number 2 carbon.

2. Soda cellulose + Carbon disulfide $\longrightarrow$ Sodium cellulose xanthate
   $(C_6H_9O_5Na)n$ +   CS$_2$ $\longrightarrow$ $C_6H_9O_4OCS_2Na$

3. Structural diagram of sodium cellulose xanthate

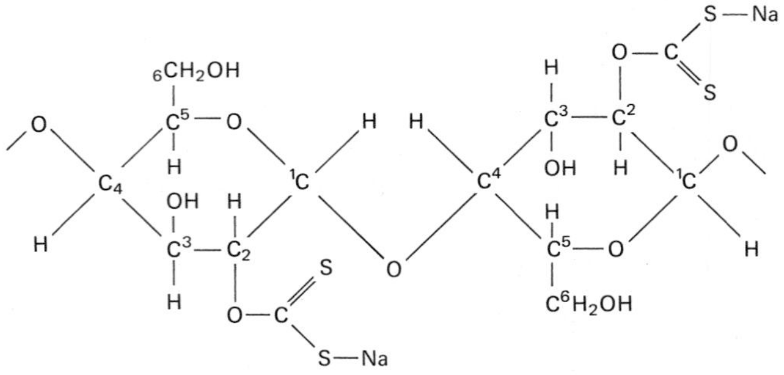

4. Sodium cellulose xanthate   + Sulfuric acid (dilute) $\longrightarrow$ Cellulose (viscose type rayon) +
   $[2(C_6H_9O_4OCS_2Na)]n$ + H_2SO_4 $\longrightarrow$ $(C_6H_{10}O_5)n$ +
   Carbon disulfide + Sodium sulfate
   2 CS$_2$ +   Na$_2$SO$_4$

Figure 6.5 Chemical reactions in viscose manufacture.

the correct viscosity (Fig. 6.6). The name *viscose* is taken from the thick liquid formed in this manufacturing step. After attaining proper viscosity, the solution is pumped to the spinning tanks, delivered to the spinning machines, and forced by pump through a spinnerette into an acid bath (Fig. 6.7). The acid bath reacts with the solution, causing pure cellulose to coagulate into filament fibers. The filaments are thoroughly washed after coagulation to remove any impurities that might adhere to the fibers, as well as any color remaining from the xanthate step. The final chemical reaction is shown in Figure 6.5.

Finally, the filaments are collected (Fig. 6.8) and either combined directly into yarns or cut into short lengths for spinning into yarns by one of the methods for cotton or wool fibers (see pp. 192, 194).

The orifices in the spinnerette vary in size and number (Fig. 6.1). Small orifices produce fine filaments, and the number of openings determines the number of filaments in the complete yarn. When staple fibers are cut for spun yarns, a large spinnerette is used with approximately three thousand openings. These fibers are collected into a group called *tow* and then cut into the desired length for yarn construction.

The spinning or coagulating bath contains several chemicals in addition to the sulfuric acid used in regenerating cellulose: sodium sulfate to hasten the precipitation of filaments; glucose to contribute softness and pliability to filaments; zinc sulfate to produce **strength** and decrease the serrated cross section; and water to **provide the** necessary volume.

If dull fibers are desired, a delustering agent, usually titanium dioxide, is added to the chemical solution before extrusion. The delustering agent breaks up the light rays and reduces the shine. The degree of dullness can be controlled by the amount of titanium dioxide added.

below left: **Figure 6.6**   Viscose solution—xanthate crumb dissolved in NaOH. [*FMC Corporation*]

right: **Figure 6.7**   Forcing liquid through a spinnerette into a coagulating bath. [*FMC Corporation*]

below right: **Figure 6.8**   Collecting filaments into yarn as they emerge from the coagulating bath. [*FMC Corporation*]

Since the early 1950s several viscose manufacturers have produced solution-dyed fibers. American Enka markets solution-dyed filaments under the name Kolorbon; Viscose calls their product Avicolor; Courtauld's trade name is Coloray; and IRC Fibers Co. identifies theirs by the term Dy-Lok. The introduction of pigment into the spinning solution increases the colorfastness of the viscose.

Mechanical improvements in the past few years have resulted in rapid production of rayon and uniform spinning in combining filaments. One of the important developments has been the continuous spinning machine. This mechanical "genius" takes the fiber directly from the spinning bath, through the wash baths, between rollers that stretch the filaments and orient the fiber molecules, to spinning devices that add the desired amount of twist.

To make high-quality spun yarns crimp is added to staple fibers. The crimp causes the fibers to hold together with ease and gives body to the yarn. Crimp is produced either by mechanical devices or by adjusting the acid in the coagulating bath. Smaller proportions of acid result in more crimp in the fiber.

Special Viscose Rayon Fibers

Most leading rayon manufacturers produce viscose rayon with different tenacities. Regular-, medium-, and high-tenacity viscose differ primarily in strength and to some degree in elongation properties. In addition to standard viscose fibers, certain manufacturers are producing modified fibers that have properties considered desirable for selected end-use requirements. The two important modifications include cross-linked viscose and high-wet-modulus viscose rayon.

Cross-linked viscose rayon is distinguished by its high resistance to wrinkling. One major fiber of this type was Corval, but production of this product was discontinued in the summer of 1964. The fiber was characterized by a firm "hand," permanent crimp, good dimensional stability, and good covering power, as well as its wrinkle resistance. Today, cross-linked rayon is achieved by chemical finishing processes. (see p. 293).

High-wet-modulus viscose rayon includes such trademarked fibers as Avril (Fiber 40), Nupron (Fiber 24), Zantrel (Fiber 700), and Xena. The manufacture of these fibers differs from that of standard viscose in that the aging and ripening steps are either omitted or reduced, a weaker solution of sodium hydroxide is used in mixing, and the fiber filaments are extruded into a weaker solution of sulfuric acid for coagulation. The resulting fibers are more like cotton than viscose rayon in their mechanical, physical, and chemical properties. They are stronger and have better dimensional stability than standard viscose fibers.

High-wet-modulus rayons are excellent in blends and produce good wash-and-wear properties, because they accept "minimum-care" finishes somewhat better and require less resin than do other viscose fibers. Further desirable properties of high-wet-modulus rayons include: good stability to laundering and easy care; ability to be mercerized; and a crisp, lofty hand. (*Hand* relates to the "feel" of a fabric—the qualities that can be ascertained by touching it.)

Some American companies have attempted to obtain a new generic term for their high-wet-modulus rayons on the basis that they are not true rayon fibers. However, their chemical composition conforms to the Federal Trade Commission's definition of rayon, and, as yet, no new generic term has been approved or seriously considered.

One other special viscose fiber is latent-crimp rayon. This fiber differs from other viscose products in that it is processed in such a way that it will develop a high degree of crimp when subjected to water. This property enables yarn and fabric manufacturers to produce interesting and unusual design effects. It is used in floor coverings, home-furnishing fabrics, and novelty yarns.

Cuprammonium Rayon

E. Schweitzer, in 1857, discovered that cellulose would dissolve in a solution of ammonia and copper oxide. However, this reaction was not applied to the manufacture of fibers at that time. In 1891 cupra rayon was made in Germany by Max Fremery and Johann Urban, but it was not successful. Edmund Thiele, in 1901, devised the stretch spinning process that made practical fibers of good strength and fineness. He was employed by J. P. Bemberg, the major producer of cuprammonium rayon.

Further improvement in the manufacture of cuprammonium fibers occurred in 1940 with the introduction of a continuous spinning process. Despite the fact that the fiber has many highly desirable properties, it is no longer produced in the United States. Beaunit Mills, formerly the American Bemberg Corporation, suspended operation of its cuprammonium rayon plant early in 1971.

In the cuprammonium process, cellulose from cotton linters or wood pulp is purified and bleached to a pure white, then dissolved in a solution of ammonia, copper sulfate, and caustic soda. The solution is carefully controlled to maintain about 4 percent copper, 29 percent ammonia, and 10 percent cellulose. A clear blue liquid is formed that requires no aging before spinning and that is not damaged if extended storage is required. (However, it is essential that storage be maintained under an atmosphere of nitrogen.) Any undissolved cellulose and other impurities are filtered out of the liquid mixture before fibers are formed.

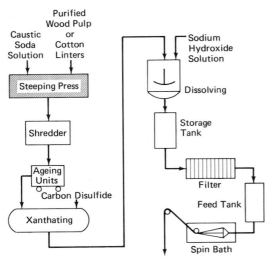

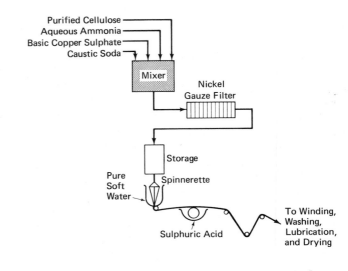

To Desulfurizing, Bleaching, Washing, Drying, Twisting, Skeining, and Coning

left: **Figure 6.9** Flow diagram of the viscose process.

above: **Figure 6.10** Flow diagram of the cuprammonium process.

The spinning solution is pumped through the spinnerette into a funnel through which soft water is running. The movement of the water stretches the newly formed filaments and introduces a small amount of molecular orientation. The fibers then move to the spinning machines, where they are washed, put through a mild acid bath to remove any adhering solution, rinsed, and then twisted the desired amount to form yarns. The viscose and cuprammonium processes are compared in Figures 6.9 and 6.10.

Saponified Cellulose Rayon

The Celanese Corporation manufactures rayon fiber, Fortisan, and Fortisan 36 by a process that involves the saponification of cellulose acetate. The fiber was made available commercially just before World War II, and during the war Fortisan found widespread military application. (In fact, it is still used for various products by the armed forces.) At the present time production is limited, but the fiber is desirable in home-furnishing fabrics and certain industrial applications. Compared to other fibers, it is rather expensive.

The first steps in manufacturing this product are identical to those involved in the production of acetate (see Chap. 7). The acetate is then treated with an alkali that reacts with the acetyl groups. The resulting sodium acetate splits off, and the acetyl radical is replaced by a hydroxyl (OH) group. Since it is primarily cellulose, the final product conforms to the definition of rayon. The reaction can be shown as follows:

Cellulose acetate + sodium hydroxide $\longrightarrow$

cellulose + sodium acetate

Figure 6.11 Photomicrograph of rayon fiber, regular viscose, longitudinal view. [*E. I. DuPont de Nemours & Company*]

Table 6.1   Average Degree of Polymerization for Rayon Fibers[a]

Fiber	Degree of Polymerization
Cotton	9,000–10,000
Viscose rayon	
regular	300–500
high-tenacity	400–600
high-wet-modulus	800–1,000
Cuprammonium rayon	400–500
Saponified cellulose rayon	500–700

[a] J. W. S. Hearle and R. H. Peters, *Fiber Structure* (London: Butterworth & Co., 1963), p. 12.

FIBER PROPERTIES

Molecular Structure

Rayon is composed of cellulose. Like cotton, it is a polymer of an-hydroglucose units (see Fig. 5.1). The number of glucose units—the degree of polymerization—in each molecule of rayon is considerably less than in natural cellulose fibers. The average degree of polymerization for rayons is cited in Table 6.1; that of cotton is included for comparison.

The high degree of polymerization for cotton is based upon recent research. The difference in degree of polymerization between cotton and rayon accounts for some of the physical variance in fiber properties, despite the fact that cotton and rayon are identical in chemical composition.

There is evidence that differences in molecular arrangement exist among the types of rayon as well as between rayon and natural cellulosic fibers. Such factors as molecular orientation, crystallinity, and presence of amorphous areas contribute to the behavior of the fibers.

Microscopic Properties

The length or longitudinal appearance of regular viscose rayon exhibits uniform diameter and interior parallel lines called *striations*. These striations are the result of light reflection by the irregular surface contour (Fig. 6.11). If the fiber has been delustered, it will have a grainy, pitted appearance; bright fiber is relatively transparent.

The cross section of the fiber shows highly irregular or serrated edges (Fig. 6.12). The presence of delusterants is indicated by a spotted effect, while bright fiber appears crystal clear.

High-tenacity viscose is similar to regular viscose when viewed under magnification, except that it may have less irregular contour in

Figure 6.12   Photomicrograph of rayon fiber, regular viscose, cross section. [*E. I. DuPont de Nemours & Company*]

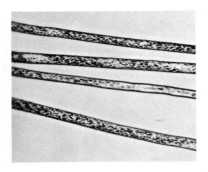

Figure 6.13 Photomicrograph of high-wet-modulus viscose rayon, longitudinal view. [*E. I. DuPont de Nemours & Company*]

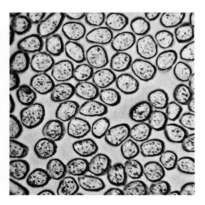

Figure 6.14 Photomicrograph of high-wet-modulus viscose rayon, cross section. [*E. I. DuPont de Nemours & Company*]

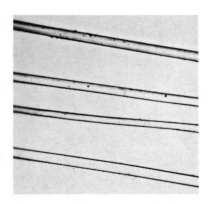

Figure 6.15 Photomicrograph of cuprammonium, longitudinal view. [*E. I. DuPont de Nemours & Company*]

cross section and, thus, show fewer striations in the longitudinal view. High-wet-modulus viscose rayon may appear nearly round in cross section and, therefore, unstriated in longitudinal view (Figs. 6.13, 6.14).

In the longitudinal view cuprammonium rayon is uniform in width; it is smooth surfaced, and there are no markings or striations (Fig. 6.15). The cross section is round or oval. It is relatively clear, and there are no irregularities in the contour (Fig. 6.16).

The longitudinal view of Fortisan exhibits a smooth, uniform diameter, with interior parallel lines or striations (Fig. 6.17). The cross section is irregular with a lobed contour (Fig. 6.18).

Physical Properties

Shape and Appearance   Rayon fibers can be produced in any length desired, and the width (diameter) of the fiber is controllable: it can vary from 12 microns or less to over 100 microns. Unless colored pigment is added to the solution before extrusion, the fibers are white.

Cuprammonium rayon and Fortisan are frequently produced in finer filaments than viscose rayons.

Strength   Tenacity in grams per denier is given in Table 6.2 for various types of rayon fibers.

Some rayons must be handled carefully during maintenance because of their low wet tenacity. This factor is responsible for many of the unfavorable opinions concerning rayon. Early rayon had such low wet strength that considerable consumer dissatisfaction resulted. It took many years to dispel this public image, and it has not entirely disappeared.

The new high-tenacity and high-wet-modulus viscose fibers are considerably stronger than old viscose and provide rayons with wet and dry strength that are equal to or better than cotton. Fortisan rayon is one of the strongest fibers on the market.

Table 6.2   Fiber Tenacity in Grams per Denier

Fiber	Dry Tenacity	Wet Tenacity
Viscose rayon		
regular	1.5–2.4	0.7–1.4
medium	2.4–3.2	1.2–1.9
high	3.0–5.0	1.9–4.3
high-wet-modulus	3.4–5.5	2.7–4.0
Cuprammonium rayon	1.7–2.3	0.95–1.35
Saponified cellulose		
rayon, Fortisan	6.0–7.0	5.1–6.0

Elastic Recovery and Elongation   The elastic recovery of regular viscose and cuprammonium rayon is low (Table 6.3). As tenacity increases, elastic recovery increases at low amounts of elongation; as elongation approaches the breaking point, the elastic recovery drops, and permanent deformation occurs. The strong rayon fibers have good elastic recovery at 2 percent extension; however, the possible elongation is much lower than for regular rayons, which have a medium amount of extension compared with other fibers. (See also Tables 2.1, 2.2, 2.3, and 2.4.)

Specific Gravity   The density of rayon fibers is similar to that of natural cellulosic fibers. Cited as specific gravity, it varies for the different types of rayon from 1.46 to 1.54. Fabrics made of rayon fibers will be similar to cotton in weight because of the similarity in fiber density, provided fabric construction and yarn size are identical.

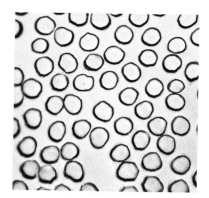

Figure 6.16 Photomicrograph of cuprammonium, cross section. [E. I. DuPont de Nemours & Company]

Resiliency   The resiliency of a fiber is important to the consumer, because it is a factor in determining the crease resistance of a fabric. Most rayon fibers have low resiliency, and, thus, the fabrics usually require finishes to produce adequate recovery from undesirable creasing and wrinkling.

High-wet-modulus rayon may exhibit slightly better resilience than regular rayon fibers. Fortisan filaments have better resiliency than other types of rayon, but this is due, primarily, to the high fiber stiffness.

Moisture Absorption   The moisture absorption of rayon fibers at two different relative humidities is cited in Table 6.4. The moisture regain and absorption of viscose and cuprammonium are higher than those of natural cellulosic fibers. In Fortisan rayon these factors are slightly higher than in natural cellulose fibers and slightly lower than in other rayons. Thanks to these moisture absorption properties, rayon fibers

Figure 6.17 Photomicrograph of saponified cellulose, longitudinal view. [E. I. DuPont de Nemours & Company]

Table 6.3   Elastic Recovery and Elongation of Rayon Fibers

Fiber	% Elongation		% Elastic Recovery at 2% Extension
	dry	wet	
Viscose rayon			
regular tenacity	15–30	20–40	82
medium tenacity	15–20	17–30	97
high tenacity	9–26	14–34	70–100
high-wet-modulus	6.5–18	7.0–33	95
Cuprammonium rayon	10–17	17–33	75
Saponified cellulose rayon	6	6	95–100

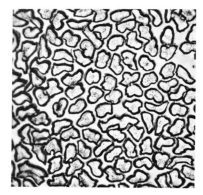

Figure 6.18 Photomicrograph of saponified cellulose, cross section. [E. I. DuPont de Nemours & Company]

Table 6.4   Moisture Regain, or Absorption, of Rayon Fibers

Fiber	Standard Regain at 20°C (70°F); 65% Relative Humidity	Saturation Regain at 20°C (70°F); 95% Relative Humidity
Viscose rayon	11.5–16	25–27
Cuprammonium rayon	12.5	27
Saponified cellulose rayon	10.7	20

accept color easily. The rearrangement of fiber molecules during manufacturing is considered partially responsible for the difference in moisture regain. Rayon fibers swell in water and lose strength.

Dimensional Stability   Regular rayons are subject to easy stretching during yarn and fabric manufacture, followed by relaxation shrinkage after laundering. However, the new high-wet-modulus viscose rayon fibers have been designed in such a way that they do not stretch easily and, thus, are less likely to suffer relaxation shrinkage. Fortisan rayon has good dimensional stability.

The degree of relaxation shrinkage that occurs in rayon fabrics is the result of fabric construction as well as fiber characteristics. Tightly woven fabrics will exhibit less size change because of the compact arrangement of yarns and fibers. Finishes are used on most rayon fabrics to control the dimensional stability (see Chap. 26).

Thermal Properties

Rayon fibers are cellulose, so they burn rapidly with a yellow flame. A small amount of light gray or off-white fluffy residue is left. When the flame is extinguished, there may be an *afterglow*. The fibers do not melt.

Hot water and iron temperatures from 300°F to 350°F can be used safely, but exposure to high temperatures for an extended period of time results in fiber degradation.

Conductivity

The electrical conductivity of rayon fibers is sufficient to prevent the buildup of static electric charges.

Chemical Properties

Effect of Alkalies   Strong alkali solutions cause rayon fibers to swell and eventually produce a loss of strength; weak alkalies do not damage rayon. Like cotton and other natural cellulose fibers, the high-wet-modulus fibers can be mercerized by the application of caustic soda solutions. Since the rayons are chemically identical to cotton, both respond in much the same manner to various chemical stimuli.

Effect of Acids   Hot and cold concentrated acids cause rayon fibers to disintegrate. Hot dilute acids result in fiber deterioration, but cold dilute acids have little or no effect.

Effect of Organic Solvents   Rayon fibers have good resistance to dry-cleaning solvents and stain-removal agents. Many other organic compounds appear to have little or no harmful effects on the fiber.

Effect of Sunlight, Age, and Miscellaneous Factors   Rayon fibers undergo deterioration when exposed to the ultraviolet rays of the sun. Conventional rayon, comparatively low in original strength, has a relatively shorter life in fabrics exposed to sun than the stronger cellulose fibers, such as cotton, flax, or high-modulus rayons. Rayons have good resistance to aging and, if properly stored, last for many years.

Strong oxidizing agents attack and damage most rayon fibers; however, cuprammonium has comparatively good resistance. Weak oxidizing chemicals, such as hypochlorite and peroxide bleaches, can be used safely on most rayons, but the manufacturers of Zantrel rayon advise consumers to avoid peroxide bleach. It is essential that any bleach be properly diluted and used according to directions.

Biological Properties

Figure 6.19   A leather-look fabric of 50 percent high-wet-modulus viscose rayon and 50 percent cotton [*FMC Corporation*]

Viscose and cuprammonium rayons are resistant to all insects except silverfish, which will destroy the fibers unless they are protected by special finishes. Fortisan rayon has good resistance to all insects.

Mildew will destroy rayons of all types, and soil increases the ease with which mildew forms on the fibers, thus accelerating the damage.

Rayon is subject to harm by rot-producing bacteria. However, high-wet-modulus viscose is fairly resistant to these bacteria, and special finishes increase the resistance of other rayon fibers.

RAYON FIBERS IN USE

Rayon fibers are used extensively in apparel (Fig. 6.19) and home-furnishing fabrics. High-tenacity rayons, including special fibers such as Tyrex and Fortisan, are used in automobile tires and various industrial applications. Because it can be produced in either filament or staple form, rayon offers more variety in fabric and yarn construction than do cotton or other natural cellulose fibers. Through control of fiber size, yarn number, fabric construction techniques, and finishes, fabrics can be produced that are sheer, heavy, soft, firm, stiff, or limp. Simple, complex, or textured yarns can be made from rayon fibers (see Parts III and IV).

Rayon staple is the nearest price competitor to cotton. In February 1971, rayon staple averaged $0.28 per pound, while market price for middling-grade cotton was approximately $0.27 per pound. High-wet-modulus rayon was slightly more expensive. Prices varied from $0.38 to $0.42, depending on denier, staple length, and type.

Blend or combination fabrics available to the consumer in 1971 included polyester and rayon, acrylic and rayon, nylon and rayon, silk and rayon, acetate and rayon, cotton and rayon, and flax (linen) and rayon. Rayon contributes absorbency and some softness when blended with polyesters, acrylics, and nylons. With cotton, rayon enhances the appearance of the fabric.

The high-wet-modulus rayon fibers have gained wide consumer acceptance. They compare favorably with cotton in strength, and they accept durable-press finishes easily.

Regular viscose rayon tends to stretch and may shrink during laundering. However, fabrics composed of improved rayon fibers can be laundered easily and safely. While the laundry treatment and handling is dependent upon yarn and fabric construction, finish and color application, and fiber content, rayon is not damaged by detergents and other laundry aids such as starches, fabric softeners, and water softeners. When essential for appearance, rayons can be bleached. Hypochlorite or peroxide bleaches can be used, but they must be properly diluted to prevent fiber damage.

Rayon can be ironed at medium to high temperatures although finishes used on the fiber may require medium to medium-low temperatures. In fact, since rayon is cellulose, it can tolerate the same general care given to cotton fabrics of the same type and with the same finishing and color application.

For many years the term rayon encompassed cellulose fibers made by viscose, cuprammonium, and other processes, as well as modified cellulose fibers such as cellulose acetate. In 1951 the Federal Trade Commission ruled that, if named on labels, *rayon* was to indicate man-made cellulosic fibers and *acetate* was to indicate fibers composed of cellulose acetate. The Labeling Law of 1960 made such labeling mandatory as part of the total fiber-labeling legislation. However, confusion still exists, and the phrases "acetate rayon" or "rayon acetate" are still heard. Both are incorrect and misleading. Blends or combinations of the two fibers should indicate the percent of each fiber present, and labels should inform the consumer that the fabric is composed of both rayon and acetate. Acetate is discussed in Chapter 7.

Manufacturers of several rayon fibers, such as Nupron, Xena, Avril (Fiber 40), and Zantrel, have established quality certification programs and permit the use of the trademark only when the fabric meets certain specifications established by the company producing the fiber. These programs assure the consumer that the fabric meets established performance characteristics as stated on attached labels.

Modified Cellulosic Fibers: Acetate and Triacetate

HISTORICAL REVIEW

Cellulose acetate, an ester of cellulose and acetic acid, was first made in 1869 by Paul Schutzenberger. However, it was not until 35 years later that a practical, safe, and relatively inexpensive technique for producing cellulose acetate was discovered by Henri and Camille Dreyfus. For several years the Dreyfus brothers manufactured their product in the form of lacquers, plastic film, and "dope" for use on early airplane fabrics. In 1913 they made filaments of the substance, but World War I interrupted their work, and successful commercial production of acetate fibers was postponed.

The Dreyfus firm produced acetate fibers in England in 1921, and in 1924 an allied organization started production of the fiber in the United States. Since that time several companies have established factories for the manufacture of acetate fibers. American companies now producing acetate include Celanese Corporation (the first manufacturer), DuPont, American Viscose Division of FMC Corporation, and Tennessee Eastman Company.

Triacetate fibers were developed along with regular secondary acetate. However, manufacture of triacetate was delayed until the middle of the twentieth century, when satisfactory and relatively safe solvents became available in sufficient quantity to make the production economically profitable. Further problems solved during the intervening years involved development of appropriate coloring processes for the triacetate.

Arnel, the triacetate fiber made in the United States by Celanese Corporation, was introduced in 1952, and large-scale commercial production began in 1955. Both acetate and triacetate are respected fibers for today's fabrics.

The confusion of nomenclature concerning rayon and acetate has been discussed (see p. 78). Despite the obvious differences in fiber properties, the public, as well as many apparel manufacturers, are so conditioned to using the word rayon to include cellulose acetate that acceptance of the proper terminology has been seriously delayed. It is the responsibility of well-informed teachers and consumers to recognize and insist upon the use of the correct generic term.

Acetate, as defined by the 1960 Labeling Law, is

> a manufactured fiber in which the fiber-forming substance is cellulose acetate. Where not less than 92 percent of the hydroxyl groups are acetylated, the term triacetate may be used as a generic description of the fiber.

MANUFACTURE OF ACETATE (SECONDARY OR REGULAR)

The steps in the manufacture of acetate fiber are illustrated in Figure 7.1. The raw materials for manufacturing acetate include cellulose, acetic acid, and acetic anhydride, plus sulfuric acid as a catalyst. Cellulose is obtained from either wood pulp or cotton linters. It is purified, bleached, and shredded. Shredded cellulose is fed into pretreatment tanks, where it is thoroughly mixed with glacial acetic acid—35 percent based on weight of cellulose—and held for a specified length of time.

Pretreated pulp is transferred to kneading machines called *acetylators*, where acetic anhydride is added. During this step the pretreated cellulose assumes liquid form as a new chemical compound, cellulose acetate. The chemical reaction is diagramed in Figure 7.2. At this point the cellulose acetate is a clear liquid called *acid dope*. It is emptied into special storage tanks, where it is aged or ripened. Water is added to the solution to reduce the acid concentration from 100 to 95 percent. During the ripening, hydrolysis occurs, which removes some of the acetyl groups from various parts of the cellulose molecule, replacing them with a hydroxyl radical, and producing acetate.

$$\begin{bmatrix} \text{acetyl radical} & = \text{OOCCH}_3 \\ \text{hydroxyl radical} = \text{OH} & \end{bmatrix}$$

Upon completion of ripening the secondary acetate is mixed with water, and the acetate precipitates out in the form of small flakes. The

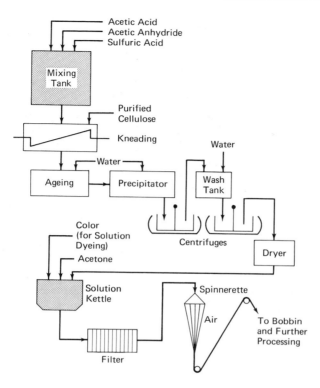

left: **Figure 7.1** Flow diagram illustrating the steps in the manufacture of acetate.

below: **Figure 7.2** Chemical reactions in the manufacture of acetate.

1. Cellulose + Acetic anhydride ⟶ Cellulose triacetate

2 anhydroglucose units _____ *n**

1 acetylated anhydroglucose unit

2. Cellulose triacetate + Water ⟶ Secondary acetate:

A ratio of 0.5 to 1.0 acetyl radicals ($OOCCH_3$) per anhydroglucose unit are replaced by hydroxyl (OH) radicals. Exact order is unknown.

*$2n$ = degree of polymerization

flakes are washed free of excess acid and dried. During the precipitation and washing process the excess acetic acid and the sulfuric acid are recovered for reuse, which helps reduce the cost of manufacture.

The cellulose acetate flake is dissolved in acetone to form the *spinning dope*. The spinning solution is forced through filters to remove any undissolved acetate and impurities, then forced through the spinnerette into a warm-air chamber. The acetone evaporates and is recovered; the acetate coagulates as it falls through the chamber, and as the filaments travel downward, they are twisted together to form yarn. The method of spinning acetate, is called *dry spinning* while that used in the manufacture of rayon is *wet spinning*.

MANUFACTURE OF TRIACETATE (ARNEL)

Triacetate is manufactured from the same raw materials as secondary acetate. The ripening stage, in which hydrolysis of the acetate occurs, is omitted in triacetate production. To produce the spinning solution, the dried triacetate flake is dissolved in methylene chloride and dry spun into a warm-air chamber.

FIBER PROPERTIES

Molecular Structure

Acetate is a polymer composed of units of cellulose acetate in which an average of 2 to 2.5 of the three OH units per glucose residue have been replaced by the $OOCCH_3$—acetyl radicals. The degree of polymerization of acetate is between 350 and 400. Triacetate differs from acetate only in that over 92 percent of the hydroxyl radicals have been replaced by acetyl groups. One authority states that complete acetylation occurs in triacetate manufacture.[1] The degree of polymerization is similar for both secondary acetate and triacetate.

Microscopic Properties

When viewed longitudinally, acetate is uniform in width with several lines parallel to the length (Fig. 7.3). These striations are farther apart than in viscose, and, logically, there are fewer striations. Bright acetate is clear, while dull or pigmented acetate appears speckled or pitted. The cross section of acetate is lobed in outline with irregular curves but no sharp serrations like those found in viscose (Fig. 7.4).

Triacetate is very similar in microscopic appearance to acetate (Fig. 7.5). It may have clearer striations, and the cross section is both lobed

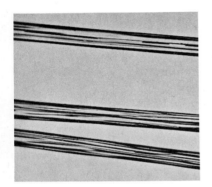

Figure 7.3 Photomicrograph of acetate, longitudinal view. [*E. I. DuPont de Nemours & Company*]

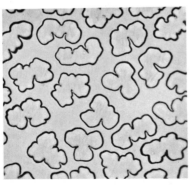

Figure 7.4 Photomicrograph of acetate, cross section. [*E. I. DuPont de Nemours & Company*]

[1]R. W. Moncrieff, *Man-made Fibers*, 5th ed. (New York: John Wiley & Sons, Inc., 1971), pp. 206, 225.

and somewhat serrated (Fig. 7.6). Positive separation of triacetate and secondary acetate by microscopy is not satisfactory.

Physical Properties

Shape and Appearance   Acetate and triacetate are man-made fibers, so they possess the potential of controlled length and diameter. The makers of spinning jets have developed jets with orifices of different shapes, which enables fiber manufacturers to produce modified fibers, such as flat filaments—*crystal acetate.*

Strength   Acetate has a dry tenacity of 1.2 to 1.5 grams per denier; when wet, the strength drops to approximately 0.9 gram per denier. Arnel possesses a dry tenacity of 1.2 to 1.4 grams per denier and a wet strength of 0.8 to 1.0 gram per denier. Although the strength of the various acetate fibers is low when compared with many other fibers, the product has desirable properties that make it attractive to the designer and consumer.

Elastic Recovery and Elongation   Acetate has the following elastic recovery and elongation:

    elastic recovery = 100 percent at 1-percent stretch
    elastic recovery = 48–65 percent at 4-percent stretch
    dry elongation = 23–45 percent
    wet elongation = 35–45 percent

If stretched more than 1 percent, acetate undergoes some permanent deformation and will never return to its original size.

Triacetate has the following elastic recovery and elongation:

    elastic recovery = 90–100 percent at 1–2 percent stretch
    elastic recovery = 80–84 percent at 4-percent stretch
    dry elongation = 25–40 percent
    wet elongation = 30–40 percent

Resiliency   Regular acetate does not have good recovery from crushing or wrinkling. Triacetate possesses resiliency, and wrinkles hang out quickly.

Density   The specific gravity of secondary acetate is 1.32 and of triacetate 1.30.

Moisture Absorption   Standard moisture regain for acetate is 6.5 percent and for Arnel 3.2 to 3.5 percent. At saturation acetate will absorb up to 14 percent moisture, while Arnel triacetate absorbs about

Figure 7.5 Photomicrograph of triacetate, longitudinal view. [*E. I. DuPont de Nemours & Company*]

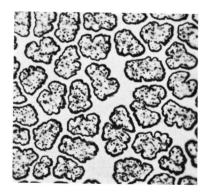

Figure 7.6 Photomicrograph of triacetate, cross section. [*E. I. DuPont de Nemours & Company*]

9 percent. When triacetate is specially processed, the standard moisture regain may be reduced to 2.5 to 3.0 percent.[2]

Dimensional Stability   Both acetate and triacetate fibers are comparatively resistant to stretch or shrinkage unless fabric relaxation occurs. The fibers may shrink if exposed to high temperatures. Arnel triacetate is especially resistant to stretch or shrinkage.

Thermal Properties

Acetate   Cellulose acetate (secondary) is a thermoplastic fiber and is easily softened by temperatures above 177°C (350°F). The fiber melts and burns evenly, forming a hard, black bead ash. It gives off a chemical odor or an odor similar to that of hot vinegar. Because of its sensitivity to high temperatures, the fiber should be ironed at low to medium settings with steam. If properly handled, permanent flattening of yarns and fibers for design effects—such as moiré taffeta—can be achieved.

Arnel   Triacetate melts and burns in the same manner as regular acetate. However, triacetate can be "heat treated" so it will withstand temperatures near 232°C (450°F) without damage. The thermoplasticity combined with heat treatment permits setting permanent pleats and creases in Arnel fabrics.

Chemical Properties

Effect of Alkalies   Dilute alkalies have little effect on acetate or triacetate. Concentrated alkalies cause saponification of both and eventually a loss in fiber weight and a reduction in the soft hand of fabrics.

Effect of Acids   Concentrated acids, both organic and inorganic, weaken the fibers and in most instances cause complete fiber disintegration. Hot acids, both dilute and concentrated, may cause decomposition or, at least, loss of strength. Cold dilute acids weaken the fiber if exposure is prolonged.

Effect of Organic Solvents   Petroleum products used in dry cleaning do not damage acetate fibers. However, such solvents as acetone, phenol, and chloroform will destroy the fibers. One should be cautious using fingernail polish remover, for it often includes acetone.

Effect of Sunlight, Age, and Miscellaneous Factors   Acetate loses strength and may develop splits after prolonged exposure to sunlight.

[2]Moncrieff, p. 229.

Over a period of time, acetate will become weaker from aging, though storage away from light and circulating air will delay this process. Arnel triacetate has excellent stability to aging and is more resistant to sunlight than silk, regular acetate, or nylon fibers.[3]

All acetates develop static charges, especially when dry, because they are poor conductors of electricity.

Biological Properties

Microorganisms   Fungi such as mildew and bacteria may discolor acetate and triacetate. Some weakening of acetate may occur, but triacetate retains its strength.

Insects   Moths and other household pests do not damage acetate fibers. However, silverfish may attack heavily sized fibers in order to eat the sizing. This is especially true of acetate fabrics that are delivered to the consumer as they come off the loom and, thus, contain the warp size used during fabric construction.

ACETATES IN USE

Regular acetate is preferred by many designers for its outstanding drapability and its desirable hand. Acetate can be made into fabrics of varying weight, thickness, and degree of softness or stiffness. Because of the thermoplastic property, it should be either dry cleaned or laundered and ironed at warm, not hot, temperatures. The relatively low moisture regain of acetate renders fibers resistant to damage by staining and to size change from shrinkage or stretch. Acetate fabrics should not be wrung or twisted when wet, for they retain creases. The fiber accepts special dyestuffs satisfactorily, and white acetate retains its whiteness, if properly laundered, with a minimum of bleaching. In many instances, no bleaching is required. In addition to use in a wide variety of apparel fabrics, acetate finds considerable use in household fabrics, such as drapery and upholstery materials.

Triacetate has the versatility and desirable properties of regular acetate, plus additional characteristics that influence its care and use. It is capable of receiving permanent pleats and creases that will withstand wear and maintenance (Fig. 7.7). It is dimensionally stable and can be processed and maintained at temperatures slightly higher than regular acetate. It has a somewhat crisper hand than acetate.

Arnel triacetate is found in blends, where it imparts stability and wrinkle resistance. Both acetate and triacetate are used in knitted fabric

[3]Moncrieff, p. 229; and C. Z. Carroll-Porczynski, *Manual of Man-made Fibers* (New York: Chemical Publishing Co., 1961), p. 40.

Figure 7.7   Tennis dress of Arnel triacetate jersey. [*Celanese Fibers*]

construction, but triacetate knits will hold their shape, retain permanent pleats, and pack beautifully. The manufacturers of Arnel maintain a quality certification program. Products labeled with the trademark Arnel have met certain specifications that are considered to be desirable by the fiber producer.

Arnel triacetate is more expensive than acetate, and both fibers are higher in price than viscose or cotton, but they cost less than many synthesized polymer fibers. Prices per pound for selected staple fibers of comparable size during 1971 were as follows: acetate, $0.40; triacetate, $0.48; viscose, $0.28; nylon, $0.98; acrylic, $0.90 to $0.98; polyester, $0.61.[4]

The trade names applied to acetate may indicate dope- or solution-dyed filaments. Chromspun and Celaperm are the most common solution-dyed acetate fibers (see p. 328). These fibers are more costly, and prices in 1971 ranged from $0.95 to $1.38 per pound. Rayon filaments, clear or solution dyed, ranged in price from $0.60 to $1.75, depending on denier. Comparable prices for selected white filament fibers of the same size were approximately acetate, $0.85; rayon, $1.09; triacetate, $1.09; nylon, $1.65; polyester, $1.63.

[4]*Modern Textile Magazine,* December 1970 and January 1971.

Natural
Protein Fibers

Natural protein fibers are obtained from animal sources. Most of the fibers in this group are the hair covering from selected animals; secretions from other animals constitute the remaining natural protein fibers.

Hair fibers include coverings from such animals as sheep, mohair goat, cashmere goat, camel, llama, alpaca, and vicuña; fur fibers are those obtained from animals more often used for their fur-producing value, such as mink, rabbit, beaver, and muskrat. Secretions are obtained from the larva or worm stage of the silkworm, *Bombyx mori*, from various wild silkworm species of the types *Antheraea* and *Attacus*, and from the spider that spins fine fibers in making its web. Hair and fur fibers have many properties in common with secretion fibers, but some characteristics are quite different.

All the protein fibers are composed of amino acids that have been formed into polypeptide chains with high molecular weight. The number and arrangement of alpha amino acid residues varies among the fibers, and there is considerable difference between hair and secretion fibers.

Fibers in this group have excellent moisture absorbency. Their standard moisture regain is high, and they absorb additional moisture at the saturation point. Protein fibers tend to be warmer than natural cellulose fibers. The low degree of electrical conductance contributes to a buildup of static charge, so the resulting yarns and fabrics tend to release much static electricity, although this problem is reduced in the presence of moisture.

Natural protein fibers have poor resistance to alkalies and can be dissolved in a 5-percent solution of sodium hydroxide at the boiling point. Most fibers in this group have good resistance to acids, the exception being silk which is damaged or completely destroyed by concentrated mineral acids. These fibers are harmed by many oxidizing agents, particularly chlorine-type oxidizing bleaches. Hydrogen peroxide, however, also an oxidizing agent, is used safely and successfully to bleach wool and silk. Sunlight causes white fabrics of the natural protein fibers to discolor slowly and turn yellow.

The density of fibers in this group is less than that of cellulosic fibers. As a group they have good resiliency and elasticity. All protein fibers except silk are comparatively weak and even weaker when wet, so care must be exercised in cleaning. The fibers burn slowly in flame, are considered self-extinguishing, leave a brittle, beadlike residue, and smell like burning hair or meat.

In recent years protein fibers have faced serious threats to their status by the increasing use of man-made fibers. However, they possess many outstanding properties, and these have resulted in their continued popularity. Further discussion will indicate reasons for this popularity and point out important characteristics of protein fibers.

WOOL

HISTORICAL REVIEW

The early history of wool is lost in antiquity. Sheepskin, including the hair, was probably used long before it was discovered that the fibers could be spun into yarns or even felted into fabric. There is no evidence to support the theory that wool was the first fiber to be processed into fabric, but it seems certain that, as a part of the skin, wool was used for covering and protection by prehistoric man.

Scientists have placed the first domestication of animals and the development of a primitive form of agriculture in the Mesolithic Age (10,000–6000 B.C.) or slightly earlier. The discovery of sheep bones in increasing abundance in the excavations of Mesolithic sites indicates use of these animals and their early domestication.

Jarmo, located in present-day Iraq, is thought to have been one of the first villages established by primitive man. The site, dated by Carbon 14 tests at approximately 4700 B.C., produced large numbers of animal bones, of which over 80 percent were of sheep and goats.

It has been interpreted that this quantity of animal bones indicates a high degree of domestication.

The earliest fragments of wool fabric have been found in Egypt, probably because of the preserving qualities of the climate, and have been dated about 4000 to 3500 B.C. There is indication, however, that wool fabrics were first made in Mesopotamia. Clay tablets from the Sumerian culture indicate that wool spinning was an important industry, and the trading of sheep was a part of the economic life of Sumeria.

Use of wool is cited frequently in the Old Testament. Wool centers flourished in Syria and Palestine, and woolen fabrics of excellent quality and beauty were produced. The Indo-European invasions were extremely important to the history of wool, for these aided in the development of wool trading and introduced the fiber to people who had not realized its value.

It is assumed that wool existed in Europe during the Late Stone Age, but there is no evidence of either fabrics of wool or domesticated herds of sheep until the Bronze Age (1500–800 B.C.). The earliest example of wool fabric found in Europe has been dated about 1500 B.C. and was unearthed in archeological digs in Germany. Danish sites have yielded excellent fragments of early woolen fabrics dated about 1300–100 B.C. These fabrics are rough and coarse and contain considerable wild sheep hair. Centuries later the wool industry flourished in Europe, and the breeding of sheep became a highly specialized activity.

Wool as a product for trade and barter dates back hundreds of years. Britain obtained sheep when the Celtic tribes moved into the area about the sixth century B.C. By the time of the Roman invasion of Britain, sheep raising and wool processing were important activities. Arrival of the Saxons in England in the fourth and fifth centuries A.D. brought the wool industry to a standstill until William the Conqueror reintroduced sheep into Great Britain.

Sheep were brought to the New World by Columbus on his second voyage in 1493, and Cortez carried sheepherding to the West and Mexico in 1521. The American Revolution emphasized the importance of wool clothing to the colonists in the temperate zones, for the supply from England was cut off. Wool of high quality was produced during the latter part of the eighteenth century, when George Washington encouraged the importation of fine Merino sheep from Spain.

Today, sheep are raised in every state on the contiguous United States and in most countries located in the temperate zones of both the Southern and Northern Hemispheres. The center for sheep raising in the United States is in the West.

Despite large areas in the United States suitable to wool production, much wool is imported. Clear evidence that a substantial amount of wool used in the United States must be imported is provided by the

Table 8.1   United States Consumption and Production of Wool[a]
Data in millions of pounds

Year	Total Wool Consumption	Total Mill Consumption	Total Wool Produced
1955	489.6	413.8	308.8
1960	538.5	411.0	298.9
1962	570.4	429.1	276.5
1964	490.8	356.7	237.4
1966	504.3	370.2	219.2
1968	466.3	329.7	198.1
1969	433.3	312.9	182.9

[a] Data from *Wool Situation*, published by the U.S. Department of Agriculture, February 1970, April 1970, and October 1970.

figures in Table 8.1, which presents comparative data regarding wool production and consumption. Tariffs and shipping charges increase the cost of wool products to the American consumer.

GROWTH AND PRODUCTION

To provide the finest-quality wool for the present-day consumer, production is carefully and scientifically controlled. Sheep are inoculated against disease, dipped in chemicals to protect them against insects, and fed diets designed to produce healthy animals. Sheep in small herds are kept in sanitary shelters, allowed to graze in clean pastures, and fenced in. Large herds in the western states are usually permitted to roam in open range land, but, even in these large areas there is always a sheepherder to watch the animals and keep them under control.

Several breeds of sheep are raised primarily for fiber. The finest quality wool is obtained from the Merino, a breed that originated in Spain and was brought to the United States in the late eighteenth and early nineteenth centuries. The animal is small-bodied, with many long loose folds of skin. The fiber is fine and of high quality.

The Delaine and the Rambouillet breeds were developed from the Merino and produce wool slightly coarser than Merino, but still of good fineness and quality. Coarse wools derive from the Navajo, Cotswold, Lincoln, and Romney breeds.

Medium fibers are obtained from such breeds as the Cheviot, Columbia (a breed developed in the United States from a cross of a Rambouillet ram and a Lincoln ewe), Corriedale, Montadale, Panama, and Southdown. In addition, there are several breeds that are raised for the value of their meat but may produce medium wool; these include the Hampshire, Shropshire, Suffolk, and Oxford. There are

Figure 8.1 Shearing sheep. [*United States Department of Agriculture*]

approximately two hundred breeds of sheep in the world, and about fifteen of these breeds are raised in the United States for their wool fiber.

Wool can be sheared from the living animal or pulled from the hide after the animal has been slaughtered for its meat. The sheared wool is called *fleece* or *clip wool*. Wool taken from the slaughtered animal hide is called *pulled wool* and is frequently inferior in quality to fleece or clip wool.

Shearing is done once a year in most states (Fig. 8.1); occasionally in the southern states wool may be sheared twice in order to keep the animal more comfortable and to eliminate extra long, coarse fibers. The sheep is sheared in early spring and the fleece removed in one piece by expert shearers. It is then rolled, packed into bags, and shipped to the nearest processing plant.

Pulled wool is removed from the hide by one of two methods. It may be treated with a dipilatory that loosens the fiber and permits it to be pulled away from the skin without damaging the hide, or it can be loosened by the action of bacteria on the root end of the fiber. Pulled wool is usually mixed with fleece by manufacturers.

Preliminary grading of wool fibers is done while the fibers are still in the fleece, because this step is important in determining cost. Fleeces are then shipped to the mill, where they are prepared for processing.

FIBER PROPERTIES

Molecular Structure

Wool is a protein substance called *keratin* and is composed of eighteen amino acid residues, of which seventeen are present in measurable amounts. The amino acid residues join together, and the molecules are arranged in such a manner as to give the fiber many of its desirable properties, such as resiliency and elasticity. The molecular structure frequently given for wool is reproduced in Figure 8.2. This figure shows the folded characteristic of the *alpha* keratin molecule. When extended, it is referred to as *beta* keratin. As soon as the tension used to form *beta* keratin is removed, the molecule attempts to return to the *alpha* or folded form. This is partly responsible for the ability of wool fiber to recover its shape after distortion.

Recent scientific analysis of wool provides evidence indicating a *helical* (or roughly spiral) rather than a folded form for the keratin.[1] Pauling and Corey have postulated the helical molecular arrangement

[1] J. W. S. Hearle and R. H. Peters, *Fiber Structure* (London: Butterworth & Co., 1963), p. 58; and Peter Alexander and R. Hudson, *Wool, Its Chemistry and Physics* (New York: Reinhold Publishing Corporation, 1963), p. 379.

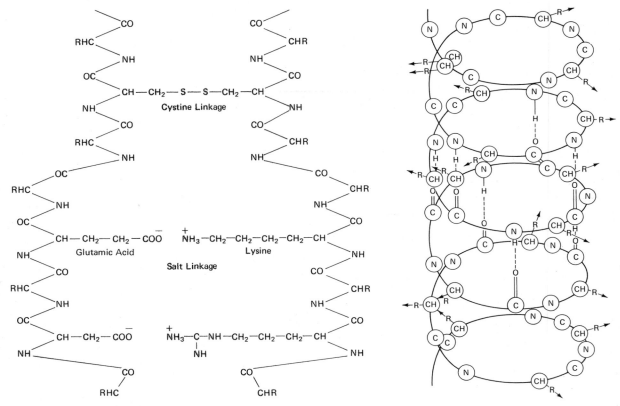

R = Amino Acid or Other Reactive Groups

left: **Figure 8.2** Diagram of the molecular structure of wool, after Astbury.

right: **Figure 8.3** Diagram of the molecular arrangement of wool, after Pauling and Corey.

Figure 8.4 Photomicrograph of Merino wool, longitudinal view. [*E. I. DuPont de Nemours & Company*]

(Fig. 8.3). It is believed that the helical and amorphous molecular arrangement is responsible for the high elongation of wool fiber.

The cystine linkage and the helical form now postulated are exceedingly important in determining the elastic recovery and resilience of wool. Furthermore, recent work suggests that the cystine linkage and intramolecular hydrogen bonding are responsible for the shaping and setting characteristics of wool fibers and fabrics.[2]

Microscopic Properties

Under microscopic observation, the length of the wool fiber clearly shows a scalelike structure (Fig. 8.4). The size of the scale varies from very small to comparatively broad and large. As many as two thousand scales are found in 1 inch of fine wool, while coarse wool may have

[2]J. B. Speakman in *Textile Research Journal,* August 1964.

as few as seven hundred per inch. Fine wool does not have as clear and distinct scales as coarse wool, but they can be identified under high magnification. The scales of coarse wool are easily seen under magnification as low as 100 power.

A cross section of wool shows three distinct parts to the fiber (Fig. 8.5). The outer layer is called the *epidermis* or *cuticle* and is composed of the scales. These scales are somewhat horny and irregular in shape, and they overlap, with the tip pointing toward the tip of the fiber. They are similar to fish scales. The major portion of the fiber is the *cortex* (composed of cortical cells) and extends toward the center from the cuticle layer. Cortical cells are long and spindle-shaped and provide fiber strength and elasticity. The cortex accounts for approximately 90 percent of the fiber mass (Fig. 8.6).

In the center of the fiber is the *medulla*. The size of the medulla varies and in fine fibers may be nearly invisible. This is the area through which food reached the fiber during growth, and it contains pigment that gives color to fibers.

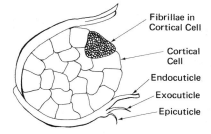

Figure 8.5 Photomicrograph of Merino wool, cross section. [*E. I. DuPont de Nemours & Company*]

Physical Properties

Shape and Appearance   Wool fibers vary in length from a short $1\frac{1}{2}$ inches to about 15 inches. According to one authority, fine wools are usually from $1\frac{1}{2}$ inches to 5 inches long; medium wools from $2\frac{1}{2}$ to 6 inches; and long (coarse) wools from 5 to 15 inches in length.[3]

The width of wool also varies considerably. Fine fibers such as Merino have an average width of about 15 to 17 microns, while medium wool averages 24 to 34 microns and coarse wool about 40 microns. Some wool fibers are exceptionally stiff and coarse. These are called *kemp* and average about 70 microns in diameter.

The wool fiber cross section may be nearly circular, but most wool fibers tend to be slightly elliptical (or oval) in shape. Wool fibers have a natural crimp, a built-in waviness (Fig. 8.7).

[3]J. Gordon Cook, *Handbook of Textile Fibers,* Vol. I (London: Merrow Publishing Co., 1968), p. 102.

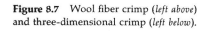

Figure 8.6 Sketch of the cross section of wool fiber, showing fiber morphology. The medulla is not illustrated.

Fibrillae in Cortical Cell

Cortical Cell

Endocuticle

Exocuticle

Epicuticle

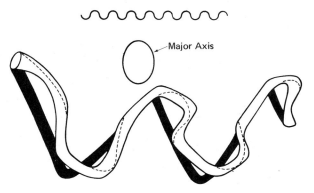

Major Axis

Figure 8.7 Wool fiber crimp (*left above*) and three-dimensional crimp (*left below*).

The crimp increases the elasticity and elongation properties of the fiber and also aids in yarn manufacturing. It is three-dimensional in character and not only moves above and below a central axis but also moves to the right and left of the axis.

There is some luster to wool fibers. Fine and medium wool tends to have more luster than very coarse fibers. Those fibers with a high degree of luster are silky in appearance.

The color of the natural wool fiber depends upon the breed of sheep. Most wool, after scouring, is a yellowish-white or ivory color. Other fibers may be gray, black, tan, or brown.

Strength   The strength of wool is 1.0 to 1.7 grams per denier when dry; when wet, it drops to 0.8 to 1.6 grams per denier. Compared to many other fibers, wool is weak, and this weakness restricts the kinds of yarn and fabric constructions that can be used satisfactorily. However, if yarns and fabrics of optimum weight and type are produced, the end-use product will give commendable wear and retain shape and appearance. Fiber properties such as resiliency, elongation, and elastic recovery compensate for the low strength.

Resiliency   The resiliency of wool is exceptionally good. It will readily spring back into shape after crushing or creasing. However, through the application of heat, moisture, and pressure, durable creases or pleats can be put into wool fabrics. This crease or press retention is the result of molecular adjustment and the formation of new cross-linkages in the polymer. Besides resistance to crushing or wrinkling, the excellent resilience of wool fiber gives the fabric its loft, which produces open, porous fabrics with good covering power, or thick, warm fabrics that are light in weight. Wool is very flexible and pliable, so it combines ease of handling and comfort with good shape retention.

Density   The specific gravity of wool is 1.30 to 1.32. The fiber is comparatively light or low in density and produces fabrics that are warm but comfortable.

Moisture Absorption   The standard moisture regain of wool is 13.6 to 16.0 percent. Under saturation conditions, wool will absorb more than 29 percent of its weight in moisture. This ability to absorb is responsible for the comfort of wool in humid, cold atmospheres. As part of the moisture absorption function, wool produces or liberates heat.[4] However, as wet wool begins to dry, the evaporation causes heat to be absorbed by the fiber, and "chilling" may be experienced, though the chilling factor is slowed down as the evaporation rate is reduced. Research by A. D. B. Cassie indicates that human subjects do not

[4]Cook, p. 105.

become aware of the chilling action.[5] The property of moisture sorption and desorption is referred to as *hygroscropic behavior*.

Wool accepts color easily because of its good ability to absorb. However, despite this property, the fiber has a hydrophobic surface and tends to shed liquid.

Dimensional Stability   When wool yarns and fabrics are subjected to mechanical action, such as agitation or abrasion combined with heat and moisture, they tend to become entangled and matted. This causes yarns to decrease in length and increase in diameter, which, in turn, results in fabrics that are dimensionally smaller in length and width but thicker than the original. This is called *felting shrinkage* and is used to advantage in making more compact, fuller, and more attractive fabrics than those that are just removed from the loom (see pp. 293–294). Wool fibers are made into felt fabrics by utilizing this type of shrinkage action. Obviously, improper control of felting shrinkage would result in highly unsatisfactory products.

Wool fabrics are quite likely to shrink after weaving or knitting as a result of yarn and fabric relaxation. This is called *relaxation shrinkage*. The elasticity and elongation of wool means that it will be extended during yarn manufacture. Sometimes fibers and yarns are held in a partially extended state after yarns and fabrics have been made. Moisture tends to release the tension of yarns, and they will return to their original length, causing the fabric or product to shrink.

In sum, wool fabrics cannot be considered dimensionally stable. They can shrink as a result of mechanical action, felting, or a combination of relaxation and felting.

Thermal Properties

Wool burns slowly in the presence of flame with a slight sputtering. It is self-extinguishing; that is, the fiber stops burning when removed from the source of fire. A crisp, brittle, black, bead-shaped residue is formed as wool burns, and the odor given off may be compared to the smell of burning hair, meat, or feathers.

When wool is heated in boiling water for a long period of time, the fiber becomes weak and somewhat stiff. At dry temperatures above 132°C (270°F), it slowly decomposes and turns yellow; at temperatures greater than 300°C (572°F), wool chars.

Chemical Properties

Wool protein is particularly susceptible to damage by alkaline substances. Solutions of 5 percent sodium hydroxide will dissolve wool.

[5] Alexander and Hudson, Chap. 4.

Wool is considered resistant to action by mild or dilute acids, but strong concentrated mineral acids, such as sulfuric and nitric, bring about the decomposition of wool fibers. Dilute solutions of sulfuric, however, are used in processing wool to "burn out" or *carbonize* vegetable matter adhering to the fibers.

Solvents used in cleaning and stain removal for wool fabrics have no deleterious effects.

The ultraviolet rays of the sun cause breakage of the disulfide bond of cystine, which results in photochemical oxidation.[6] This produces degradation and, if exposure is prolonged, eventual destruction. However, wool has better resistance to sunlight than cotton.

If properly protected and stored, age has no destructive effect on wool.

Biological Properties

Microorganisms   Wool has good resistance to bacteria and mildew. However, both organisms may attack stains left on wool. If wool is stored in an atmosphere where moisture is present, mildew will form, and eventually it will destroy the fiber. Rot-producing bacteria will bring about the destruction of wool that has been subjected to moisture and soil for long periods of time.

Insects   Because wool is a protein and may be considered a modified food product, it becomes an appetizing meal for several types of insects. The larvae of the clothes moth and of the carpet beetle are the most common predators on wool as a source of food. It has been estimated that these insects damage several million pounds of wool fabric each year. Various treatments suggested to prevent this damage include

1. spraying the fabric with chemicals that will kill the insects. These finishes need frequent renewing, since dry cleaning removes them.
2. applying chemicals that react with the wool molecule and make the fiber unpalatable to the moth. This process is durable to laundering or dry cleaning and is the most successful.
3. using substances in close proximity to wool that give off odors that are noxious to the insects.

The problem of insect damage to wool is serious enough that considerable research is currently in progress to develop techniques that will prevent destruction. Some of the methods under investigation include

1. development of animal (sheep) feeding techniques that will produce insect-resistant fibers.

[6]Alexander and Hudson, p. 265.

2. development of sprays that are resistant to cleaning or laundry.
3. development of additional chemical treatments that will modify the wool protein so it becomes unpalatable and undesirable to the insects.

WOOL IN USE

Woolen and worsted fabrics are widely used throughout the world, and they have special acceptance in the United States for their many desirable properties (Fig. 8.8). They are naturally crease resistant, flexible and elastic, absorbent, warm, and comfortable. Wool fabrics tailor well, press easily, and can be shaped to conform to the body.

A problem with wool fabric is the tendency for shrinkage (see p. 95). The scale structure increases felting, because the scales tend to interlock and hook together, holding the fibers in a closely entangled and matted state.

Wool fiber is considered to be bicomponent: the cortex is composed of two parts, *ortho* and *para*. These two parts react differently to certain stimuli and influence the crimp of the fiber, which, in turn, is related to its crease resistance. Moisture affects the bicomponent property, so the crimp decreases when wet and increases when dry.

Wool is rather demanding in terms of care. It can be dry cleaned and pressed easily, but laundering is difficult, because wool cannot tolerate agitation or temperature change. If handling is kept to a minimum, if the wool is not hand-wrung or agitated, and if warm water is used throughout both washing and rinsing, wool can be hand-laundered without damage. Chlorine bleaches cannot be used on wool, for they break the cystine linkage and cause the fiber to disintegrate. When bleaching is required, hydrogen peroxide or similar bleaching compounds, can be used. Mild soaps and detergents are recommended for laundering wool fabrics.

Wool is easily dyed and has good colorfastness when proper dyestuffs are used. Acid and chrome mordant dyes are successful and are stable to dry cleaning.

Following wear, woven wool garments should be placed on hangers and brushed carefully. This removes surface dust and soil and permits wrinkles to hang out. Knitted sweaters should be aired and then folded and stored in drawers to prevent sagging and misshaping from hangers. Frequent cleaning reduces insect damage, and unclean wools should never be stored, for they attract both insects and fungi, such as mildew.

Finishes to control shrinkage and to help make wool washable have been introduced recently. One process softens the cementing material between the scales long enough to permit the scales to become tightly sealed to the cortex. This prevents the interlocking action associated with untreated wool and thus reduces felting. However, if not properly controlled, this finish may remove the scales completely and result in

Figure 8.8 A machine washable and dryable 100 percent wool jersey by Fablon. [*Wool Bureau, Inc.*]

weakened fibers. A process developed by the Western Regional Utilization Laboratory of the USDA involves a technique called *interfacial polymerization*. Fibers are coated with a microscopically thin layer of a polymeric chemical, frequently a polyamide, which coats the fiber surface and seals the scales so the fibers do not become entangled. This prevents felting action, reduces shrinkage, and creates a washable wool fabric. Many textile research scientists are working on developing additional ways to make wool a washable product.

Several other finishes are applied to wool to enhance its insect-repellant, water repellant, and wash-and-wear qualities. These are discussed in Part V.

Newer man-made fibers have replaced wool in many end-uses. The per-capita consumption of wool in the United States has continued to decline (see Table 1.1, p. 6) and now accounts for only slightly more than 5 percent of the fibers used. However, many people enjoy its pleasant hand, and it is still a desirable choice for warm fabrics and for durable apparel and home furnishings.

Wool Products Labeling Act

The Wool Products Labeling Act was passed by Congress in 1939. Its purpose as stated by the law is

> to protect producers, manufacturers, distributors and consumers from the unrevealed presence of substitutes and mixtures in spun, woven, knit, felted, or otherwise manufactured wool products.

This legislation requires that any product containing wool (except upholstery and floor coverings) have a label affixed that specifies the wool as new, reprocessed, or reused. The term *wool* as defined by the act applies to a fiber of animal origin grown naturally as the coat of a living animal. It does not specify the sheep as the source, and other animal fibers may be called wool. However, the vast majority of products labeled as wool are composed of fibers from sheep. Furthermore, the term *wool* indicates fiber that is being used for the first time in the complete manufacture of a wool product. Fibers processed as far as the yarn state, recovered (pulled apart into fiber form), and reconstructed in new yarns may be labeled as wool. *Virgin wool* is a designation used by well-known manufacturers to indicate new wool that has been made into yarns and fabrics for the first time; the term *lamb's wool* connotes wool clipped from sheep less than eight months old. Other terms used by the Wool Products Labeling Act are defined as follows:

> *Reprocessed wool* refers to fibers reclaimed from scraps of fabrics that were never used. These come from cutting rooms, fabric samples, and similar

sources. The fabrics are converted or *garnetted* (shredded) back into a fibrous state and then used in making new yarns and fabrics.

Reused wool refers to fibers reclaimed from fabrics that have been worn or used. Old rags and clothing collected by rag dealers comprise the major source for this wool. Reused wool is sometimes called *shoddy*.

The quality of reused wool and reprocessed wool is usually lower than that of new fibers. During the garnetting process the fibers are frequently damaged by tearing and breaking. However, the warmth property is maintained, and fabrics of reused wool are satisfactory for interlining fabrics, inexpensive blankets, and similar products. Reprocessed wool has warmth, some resiliency, and good durability. In fact, new wool of poor quality may give less wear than reprocessed or reused wool of high original quality.

All products containing wool must be labeled with the fiber content and the percentages of the various types of wool. The Wool Products Labeling Act is a protection to the consumer against the purchase of fabrics designed to imitate wool in appearance but that would not possess the desirable qualities of a true wool product.

SPECIALTY AND FUR FIBERS

Fibers from such animals as the goat, camel, alpaca, and llama are referred to as *specialty fibers*. These fibers are available in limited quantities and are desired for particular characteristics.

MOHAIR

Mohair is the fiber of the Angora goat (Fig. 8.9). Turkey was the major source of this fiber until the nineteenth century, at which time South Africa gained comparable importance. In the late nineteenth and early

Figure 8.9 Angora goat. [*United States Department of Agriculture*]

Figure 8.10 Casement in 56 percent mohair with wool by Westgate. [*The Mohair Council*]

twentieth century, ranchers in Texas and California began to raise the Angora goat, and since the mid-twentieth century the United States has produced more Angora goats and consumed more mohair than any other country.

Angora goats are sheared twice a year. The fibers are fine and silky in appearance and measure 4 to 6 inches in length. Occasionally, they are sheared only once a year, and the fibers are 9 to 12 inches long. Mohair is similar to wool in both physical and chemical properties. Its major advantages include remarkable resistance to wear and abrasion, a high degree of luster, excellent resiliency, and adaptability to complex yarns and textured fabrics. Suiting and sportswear fabrics, upholstery, rugs, and draperies (Fig. 8.10) may be of mohair or of a blend including mohair fibers.

CASHMERE

Cashmere is the fiber of the cashmere (Kashmir) goat, which is raised in Asia. Tibet, India, Persia, and parts of China are major sources. The fiber is combed from the animal, and the yield per goat averages about 4 ounces of good fiber. The short fibers—1 to $3\frac{1}{2}$ inches—are very soft and fine; longer fibers—2 to 5 inches—are somewhat coarse and stiff.

Although similar to wool in most properties, cashmere does differ in that it is more easily damaged by alkalies than sheep's wool.

The yearly production of true cashmere is very small, so the fiber is expensive. However, consumers who wish a very soft and attractive fiber will pay high prices. Fabrics made of cashmere are warm and comfortable. Furthermore, the fiber is highly adaptable, in that it can produce either fine or thick yarns, which, in turn, can be constructed into thick, medium, or rather lightweight fabrics.

Cashmere fabrics are considered luxury items. However, if properly maintained, they give good service, and they are appropriate to both warm and cool climates.

CAMEL HAIR

The Bactrian, or two-humped camel, is the source of camel-hair fiber. This breed serves as a means of transportation in Asia in the desert regions of China, Tibet, and Mongolia. The animal sheds about 5 pounds of fiber each year, which is used in textile products. The outer camel fibers are coarse and utilized only in low-quality merchandise, but the fine, short underhairs are as soft and fine as top-quality wool. In addition, camel hair possesses thermal properties similar to those of wool that keep the wearer warm in cold weather.

In certain countries camel hair is made into fabrics and worn by individuals who traverse hot desert areas. When the desert temperatures are greater than 100°F, as frequently happens, the camel-hair fabric will partially protect the wearer from the intense heat.

Fine camel-hair products require the same careful handling and care as cashmere and mohair. Camel hair is used in coating fabrics, sportswear, and knitted products. The natural tan to reddish brown color is very attractive and is retained for many choice fabrics.

Coarse camel-hair fibers are used in industry for special belting, ropes, and artist's brushes. Natives use the coarse fibers to make blankets.

ALPACA

The alpaca is a member of the camel family and is native to South America. It thrives in the Andes Mountain regions of Peru, Bolivia, Ecuador, and Argentina. The fiber is sheared from the animal once every two years. The fine fibers, which are separated from the coarse guard hairs, are used in fabric manufacturing.

Alpaca is similar to camel hair and offers excellent warmth and insulation. The fibers are strong and glossy and make fabrics similar in appearance to mohair. Alpaca fabrics are used for suits, dresses, plush upholstery, and linings. The natural fiber color ranges from white to brown and black, and a variety of attractive fabrics can be created without additional dyestuffs.

LLAMA

Also a member of the camel family, the llama produces fibers similar to those of the alpaca and is found in the same geographical area. The fibers are sheared once a year. They are soft, strong, and relatively uniform in length and diameter but somewhat weaker than alpaca or camel hair.

South American Indians weave most of the pure llama fabrics. Some fiber is sold to wool manufacturers for blending with sheep's wool, other specialty fibers, or man-made fibers.

HUARIZO AND MISTI

The huarizo and the misti are cross breeds between llama and alpaca. The huarizo is the product of a llama father and an alpaca mother, while the misti has an alpaca father and a llama mother. Fibers from both animals are fine, lustrous, and silky. They make fabrics of good durability and quality.

VICUÑA

The most valuable and most prized hair fiber is that taken from the vicuña. This small animal (about the size of a large dog) is found in the Andes Mountains at elevations of approximately 16,000 feet. It is a member of the llama family and, thus, of the South American camel family. The vicuña is extremely wild, and attempts to domesticate it have been relatively unsuccessful so far, but ranchers hope soon to be able to raise the animal in captivity.

Vicuña fiber is one of the softest known to man. It is fine and lustrous, has a lovely cinnamon brown or light tan color, and is strong enough to make very desirable fabrics. Moreover, it is very light in weight and very warm. Choice uses of the fiber are for coats, suit fabrics, and soft shawls or capes.

As yet, the fiber can be obtained only by killing the animal, and the Peruvian government limits the yearly kill. Each animal yields about 4 ounces of fine fiber plus 10 to 12 ounces of shorter, less choice fiber. The total yearly production is just a few thousand pounds, so garments and fabrics of vicuña compare in price with good fur coats.

Currently, attempts are being made to crossbreed the vicuña and the alpaca. It is hoped that the resulting animals (paco-vicuña or vicuña-paco) can be domesticated as easily as the alpaca and that fibers comparable to true vicuña will be obtainable.

GUANACO

An animal similar to the llama, the guanaco also is quite wild and is found in the Andes Mountains. The fibers are rather like those of alpaca. Attempts to domesticate the guanaco have proved successful, and the fiber is available in quantities comparable to the alpaca and llama.

FUR FIBERS

Fibers used in small amounts in blends with wool or man-made fibers are obtained from several animals more frequently used for fur pelts. These include mink, beaver, fox, chinchilla, muskrat, nutria, raccoon, and rabbit. They are used primarily to add softness, color interest, and prestige value to fabrics.

The angora fiber from the Angora rabbit is frequently used in knitting yarns and in knitted fabrics, because it gives a fluffy, white, silky appearance to products. The Angora rabbit is raised in France, Italy, Japan, and the United States. The fur is combed and clipped every three months. Fibers are smooth, lustrous, fine, and resilient.

Most specialty and fur fibers require careful handling. They are luxury items and should be treated with considerable respect.

SILK

HISTORICAL REVIEW

The history of silk is based on both fact and myth, and it is difficult to isolate truth from fiction. Of course, the fact that silk was discovered and eventually used in making fabrics is the important issue. However, the legends concerning silk are romantic and interesting. The story told most frequently involves a Chinese Empress.

Emperor Huang-Ti, who ruled China sometime between 2700 and 2600 B.C., assigned his Empress, Hsi-Ling-Shi, the task of studying a blight that was damaging the Imperial mulberry grove. Tiny white worms were devouring the leaves, then crawling from leaf to stem to spin shining, pale (almost white) cocoons. Hsi-Ling-Shi gathered a handful of cocoons and carried them into her apartment, where she accidentally dropped one into a basin of hot water.

The empress noticed that the cocoon separated into a delicate cobweblike tangle from which she could draw a slender, tiny filament into the air. She further observed the filament was continuous, and the more she unwound the smaller the cocoon became. Thus, one legend (probably a combination of fact and fiction) records the discovery of silk fibers.

For approximately three thousand years China successfully held the secret of silk and maintained a virtual monopoly of the silk industry. About A.D. 300 Japan obtained the formula. The story is told that the Japanese sent several Koreans on a mission to China to try to learn the secret. The delegation brought back four Chinese girls, and these girls instructed the Japanese in the art of *sericulture*—the raising and production of silk.

Gradually silk production spread westward across Asia into India and eventually into Persia. Once again, according to legend, Emperor Justinian of Constantinople was so enamored of silk that about A.D. 550 he persuaded two monks who had lived in China and knew the techniques of sericulture to bring silk cocoons and eggs back to his country and begin a silk industry there. Before the development of the silk industry in Byzantium, silk trade between East and West was of great importance, and silk fabric was a luxury available only to royalty.

During the late seventh century sericulture moved to Italy, Spain, and France. Cities such as Florence, Milan, Genoa, and Venice were famous for their silks throughout the Middle Ages. About 1480, silk weaving began in Tours, France, and in the sixteenth century the silk industry reached England.

Shortly after the colonization of America, James I tried to have a silk industry established in what is now the United States, but his attempts and those that followed were never successful. During the nineteenth century it became apparent that sericulture could be profitable only in countries where labor was both plentiful and cheap. Thus, the silk industry, especially sericulture, was confined chiefly to China, Japan, Italy, and India. The weaving of fine silk fabrics is a skilled craft in France, primarily in Lyons.

Throughout the development of the industry, silk has maintained a position of great prestige and is still considered a luxury fiber. Silk is often called the "queen of fibers," a title well deserved by virtue of its association with royalty, the care required in its culture, and the properties and characteristics with which it has been endowed.

SERICULTURE (GROWTH AND PRODUCTION)

Silk is produced by the larvae of several moths, but the *Bombyx mori* is the only one that is raised under controlled conditions. These larvae live on mulberry leaves only, and each tiny larva consumes an extremely large number of leaves. Raising these insects is a skilled occupation and requires countless hours of labor. The present-day industry is carefully controlled in relation to disease prevention, and modern silk "factories" are as clean and frequently as sterile as the better hospitals (Fig. 8.11).

The female *Bombyx mori* lays eggs once a month. She may lay as many as seven hundred. Each egg is about the size of a pinhead, and each has a small dot on one end that is soft and permits the larva to hatch easily. The eggs are carefully screened and tested to ensure freedom from disease. They can be kept for long periods of time in cold storage without damage. When a supply of fresh young mulberry leaves is ready, the eggs are warmed slightly, and three to seven days later the worm hatches and begins to feed on the tender leaves. The new larvae are about $\frac{1}{4}$ inch long. Although the larvae of the *Bombyx mori* are called silkworms, the are technically caterpillars.

During the growth cycle of the larva, it moults (sheds its outer skin) four times. Each time it grows a new skin to fit its larger size. After the fourth moulting, it eats for about ten days, making a total eating period of about thirty-five days. At the end of the growing period the caterpillar has increased about 10,000 times over the weight of the newly hatched larva. It is more than 3 inches long and from $\frac{1}{4}$ to $\frac{1}{2}$ inch in diameter. In this state it is ready to spin its chrysalis case or cocoon.

The larva attaches itself to a specially constructed straw frame, rears its head, and begins to spew the silk liquid, which hardens on contact with air. The larva spins by moving its head in a figure-eight motion and constructs the cocoon from the outside in (Fig. 8.12). As it spins, the larva decreases in size, and upon completion of the cocoon it changes into the dormant *chrysalis.* Except for those to be used for breeding, the cocoons are subjected to heat, which kills the chrysalis. These cocoons can then be stored until they are unreeled in preparation for yarn manufacturing. The cocoons not stifled are permitted to hatch, and the moth emerges and breeds. The female then lays eggs for a new batch of larvae, and the cycle begins again.

The silkworm extrudes the liquid fiber from two tiny orifices or *spinnerettes* in its head. As the liquid emerges into the air, it solidifies into silk filaments (Fig. 8.13). The fibers, coated with a gummy

Figure 8.11   Silk factory. [*Eastfoto*]

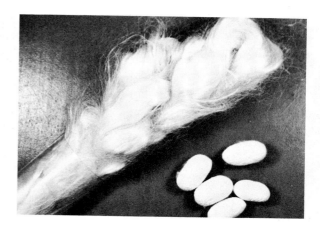

substance, *sericin*, are extruded by two glands in close proximity to the cocoon.

Raising of silkworms is a highly scientific art and, in addition, is carefully controlled in order to maintain healthy larvae and moths to produce a supply of high-quality fiber. Frequent checking of silk eggs, larvae, and moths is done by trained scientists to prevent the spread of disease.

In addition to the sericulture of *Bombyx mori*, several other moths produce larvae that spin cocoons composed of usable silk fiber. Probably the most important in this group are the moths that produce tussah silk. The larvae of the species *Antheraea*, primarily *A. myllita* and *A. pernyi*, feed on oak leaves. They grow wild, and the cocoons must be collected from trees and shrubs. The larvae of this species usually grow much larger than the *Bombyx mori*, sometimes up to 6 inches long; they are greener in color and are covered with the short fuzzy hair characteristic of many caterpillars. Tussah fiber is tan or light brown in color and cannot be bleached white. It is obtained in China and India.

Attacus ricini, a species found in Asia and parts of Central and South America, feeds on leaves of the castor bean plant. Fibers from these larvae are of high quality and nearly pure white, but they are limited in number, so the amount of this fiber is small.

left: **Figure 8.12** Silkworms and cocoon formation. [*International Silk Association*]

right: **Figure 8.13** Silk filaments and cocoons.

PROCESSING

Reeling

Silk filaments are unwound from the cocoons in a manufacturing plant called a *filature*. Several cocoons are placed in hot water to soften the gum, and the surfaces are brushed lightly to find the ends of the filaments. These ends are collected, threaded through a guide, and wound onto a wheel called a reel, hence the name *reeling*.

Figure 8.14   Hand reeling of silk.

Reeling requires extremely skillful operators (Figs. 8.14, 8.15). The fibers are narrower at the beginnings and ends, and experienced workers can join cocoons so the diameter of reeled yarn is of constant size. Only uniform reeled silk will sell for premium prices.

Throwing

As the fibers are combined and pulled onto the reel, twist can be inserted to hold the filaments together. This is called *throwing,* and the resulting yarn is *thrown yarn.* Fibers may be thrown in a separate operation following reeling.

Several types of thrown yarns are available. They differ in the amount of twist and in methods of combining single yarns.

Tram silk is a low-twist ply yarn formed by combining two or three single strands. It is moderately strong and is frequently used for filling yarns.

Organzine is a two or more ply with a medium twist. It is very strong and is used for warp yarns. *Crepe* organzine is a very highly twisted yarn used in crepe fabrics and chiffons.

Singles is a strand of several filaments collected together and held by low, medium, or high twist. A *singles* yarn can be used for either warp or filling or in knitting.

Grenadine is a tightly twisted ply yarn composed of two or three singles. The ply twist is in the opposite direction from the twist of each single strand.

For further discussion of single and ply yarns and direction of twist, see Part III.

Spinning

Short ends of silk fibers from the outer and inner edges of cocoons and from broken cocoons are spun into yarns in a manner similar to that used for cotton.

Degumming

Sericin remains on the fibers during reeling and throwing. Frequently it is left on through the fabric construction processes. Before finishing the gum is removed by boiling the fabric in soap and water. If stiffness is desired in the completed fabric, some of the sericin may be replaced, but this is not usually desirable, since the presence of gum or sericin increases the tendency for silk to water spot.

Silk is constructed into fabrics, finished, and colored by the same general methods used for other fibers (see Parts IV and V).

Figure 8.15   Mechanical reeling of silk.
[*Eastfoto*]

FIBER PROPERTIES

Molecular Structure

Silk is a natural protein fiber. The actual fiber protein is called *fibroin*, while the protein *sericin* is the gummy substance that holds the filaments together. Like wool, silk proteins are composed of amino acids, which, by condensation, form a polypeptide chain. Fibroin differs from wool keratin in that no cystine is involved, and hence there are no sulfur or cystine linkages.

Fibroin is composed of about fifteen amino acids.[7] There is considerable difference of opinion among authorities about the arrangement of the acids; however, most agree that the simple amino acids such as glycine, alanine, serine, and tyrosine make up the largest part of the fiber.

An important aspect of silk fiber is its high degree of molecular orientation, which accounts for the excellent strength of silk products.

Figure 8.16 Photomicrograph of silk, longitudinal view.

Microscopic Properties

Cultivated degummed silk, viewed longitudinally under a microscope, resembles a smooth, transparent rod (Fig. 8.16). Silk in the gum has rough, irregular surfaces. Wild silk tends to be quite uneven and is somewhat dark. It may have longitudinal striations.

Cross-section views of silk show triangular fibers with no markings (Fig. 8.17). Two filaments usually lie with their flat sides together. This can be explained by the fact that two filaments are extruded by each silkworm, and they come together as indicated in Figure 8.18. The two are called *brins*.

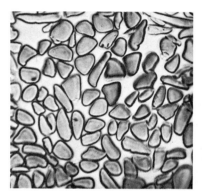

Figure 8.17 Photomicrograph of silk, cross section.

Physical Properties

Shape and Appearance   Silk filaments are very fine and long. They frequently measure about 1000 to 1300 yards and can be as long as 3000 yards. The width of silk is from 9 to 11 microns. The fibers are quite smooth. They have a high natural luster or sheen and are off-white to cream in color. Wild silk is uneven, has slightly less luster, and is tan to light brown in color.

Strength   Silk is one of the stronger fibers used in making fabrics. It has a tenacity of 2.4 to 5.1 grams per denier when dry, and its wet strength is about 80 to 85 percent of the dry.

Elastic Recovery and Elongation   Silk has good elasticity and moderate elongation. When it is dry, the elongation varies from 10 to 25

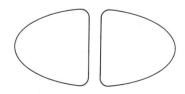

Figure 8.18 Diagram of silk filaments or brins.

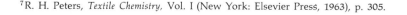

[7]R. H. Peters, *Textile Chemistry*, Vol. I (New York: Elsevier Press, 1963), p. 305.

percent; when wet, silk will elongate as much as 33 to 35 percent. At 2-percent elongation the fiber has a 92-percent elastic recovery.

Resiliency   Silk has medium resiliency. Creases will hang out relatively well but not as quickly or as completely as in wool.

Density   The specific gravity of silk is given by Kaswell as 1.34,[8] by Cook as 1.25,[9] and by Wolfgang as 1.33.[10] This variation may be accounted for by different sources of fibers used for analysis. In any case, silk is less dense than cellulosic fibers. The creation of lightweight fabrics is possible because of the fine, strong filament fibers.

Moisture Absorption   Silk has a relatively high standard moisture regain of 11.0 percent. At saturation the regain is 25 to 35 percent. The absorption property of silk is helpful in the application of dyes and finishes, but, unlike many fibers, it will absorb impurities in liquids such as metal salts. These contaminants tend to damage silk by weakening the fiber or even causing ruptures to occur.

Dimensional Stability   Silk fabrics have good resistance to stretch or shrinkage when laundered or dry cleaned. Crepe fabrics will shrink when wet but can be steamed back to size easily.

Thermal Properties

Silk will burn when directly in the path of flame. After removal from the flame it will sputter and eventually extinguish itself. It leaves a crisp, brittle ash and gives off an odor like that of burning hair or feathers.

Heated to about 135°C (275°F), silk will remain unaffected for prolonged periods; however, if the temperature is raised to 177°C (350°F), rapid decomposition occurs. Silk scorches easily if ironed with temperatures above 300°F, and white silk will turn yellow if pressed with a hot iron.

Chemical Properties

Effect of Alkalies   Silk is damaged by strong alkalies and will dissolve in heated caustic soda (NaOH); however, silk reacts more slowly than wool, and frequently the identity of the two fibers can be determined by the speed of solubility in NaOH. Weak alkalies such as soap, borax,

[8] E. R. Kaswell, *Handbook of Industrial Textiles* (New York: Wellington Sears, 1964), p. 321.

[9] Cook, p. 159.

[10] William Wolfgang, *Man-made Textile Encyclopedia*, J. J. Press, ed. (New York: Textile Book Publishers, 1959), p. 145.

and ammonia cause little or no damage to silk unless they remain in contact with the fabric for a long time.

Effect of Acids   Silk protein, like wool, can be decomposed by strong mineral acids. Medium concentrations of hydrochloric acid (HCl) will dissolve silk, and moderate concentrations of other mineral acids cause fiber contraction and shrinkage. The molecular arrangement in silk permits rapid absorption of acids but tends to hold the acid molecules so they are difficult to remove. This accounts for some of the acid damage to fibroin that does not occur to keratin. Organic acids do not damage silk and are used in some finishing processes. Some authorities maintain that the *scroop* of silk—a rustling or crunching sound—which used to be considered a natural characteristic, is developed by exposure to organic acids.[11]

Effect of Organic Solvents   Cleaning solvents and spot-removing agents do not damage silk.

Effect of Sunlight, Age, and Miscellaneous Factors   Sunlight tends to accelerate the decomposition of silk. It increases oxidation and results in fiber degradation and destruction.

Silk requires careful handling and adequate protection in storage to withstand the ravages of age. Oxygen in the atmosphere causes a gradual decomposition of silk, and unless it is stored in carefully sealed containers, the fiber will lose strength and eventually be destroyed.

Silk is a poor conductor of electricity, which results in the buildup of static charges. Like other protein fibers, it has a lower thermal or heat conductivity than cellulosic fibers.[12] This factor, coupled with certain methods of construction, creates fabrics that tend to be warmer than comparable fabrics of cellulosic fibers.

Biological Properties

Silk is resistant to attack by mildew and is relatively resistant to other bacteria and fungi. It is decomposed by rot-producing conditions.

Silk has good resistance to the clothes moth, but carpet beetles will eat it. Destruction attributed to moths has usually been caused by carpet beetles.

SILK IN USE

Silk has been the "queen of fibers" for centuries. As in the past, it is still used for luxury fabrics and for high-fashion items. However, its durability makes it practical for modern-day usage.

[11]Cook, p. 161.
[12]Hearle and Peters, p. 556.

Figure 8.19 Handwoven tussah silk. [*Jack Lenor Larsen, Inc.*]

Dry cleaning is the preferred method of care for silk fabrics and products. However, if handled carefully, some silk fabrics can be hand laundered. A mild soap or synthetic detergent in warm, not hot, water should be used for silk, and it should receive minimal handling. Thorough rinsing is required, and the best method for extracting the water is to roll the garment in a towel and then hang it in a cool place, out of the sun, to dry. Silk should be ironed or pressed with medium to low ironing temperatures, and steam is acceptable.

If silk requires bleaching, hydrogen peroxide or perborate bleach, not chlorine types, must be used for chlorine is destructive to silk. One problem with silk fabrics is that body perspiration tends to weaken the fibers and frequently will alter the color. Many deodorants and antiperspirants contain aluminum chloride, which is damaging to silk. It is often advisable to wear protective dress shields.

Several factors are involved in the demand for silk. It offers an incredible variety in fabric and yarn structure. In dyeing, many beautiful fabrics can be produced (Fig. 8.19). Probably no other fiber is so widely accepted and suitable for various occasions. It is versatile and can be used in sportswear; men's and women's suits; lingerie; dress, blouse, and shirt fabrics; and decorator fabrics for homes or offices.

Because of its strength, silk is durable, and with proper care it will withstand years of wear. Except for extremely hot days, silk fabrics are comfortable and maintain a neat and attractive appearance.

Many silk fabrics cost more than similar fabrics of man-made fibers. However, the consumer who has formed an attachment to silk is willing to pay high prices. Silk combines strength, flexibility, good moisture absorption, softness, warmth, luxurious appearance, and durability, so it creates choice products for the discerning consumer.

Man-made
Protein Fibers:
Azlons

The generic name assigned to man-made protein fibers by the Textile Fiber Products Identification Act is azlon. The act defines *azlon* as

> a manufactured fiber in which the fiber-forming substance is composed of any regenerated natural protein.

There are no azlon fibers currently in production in the United States, but a few are still being manufactured in other countries. Despite the present inactivity, the fibers have sufficient importance to merit brief discussion.

HISTORICAL REVIEW

The first man-made protein fiber, called *Vandura silk,* was introduced in 1894. It was of scientific interest only, since it was partially soluble in water, but it did lead the way to further discoveries of comparatively stable protein fibers.

In 1904 Todtenhaupt disclosed a process for making fibers from casein, but the product lacked adequate pliability. Between 1924 and 1935 Antonio Ferretti carried out research on man-made protein fibers that resulted in the production of usable textile fibers from casein. Snia

Viscosa of Italy made commercial amounts of the fiber in 1935 and named it Lanital. Other casein fibers—such as Tiolan in Germany, Lactofil in Holland, Fibrolane in England, Merinova in Italy, and Aralac in the United States—soon followed.

Concurrently with the development of casein protein fibers, scientists found ways to convert the protein from soybeans, peanuts or groundnuts, and corn (maize) into fibrous form. The most successful of the man-made proteins include Fibrolane and Merinova from casein, Vicara from the protein zein in corn, and Ardil from the protein in peanuts. At the present time, only Merinova, Enkasa, and Wipolan are in commercial production, and the amount of fiber sold yearly is small.

Production of azlon fibers has been discontinued in the United States for several reasons. The raw materials for azlons came from food sources, and the supply was variable; early azlon fibers, especially Aralac, gave off a disagreeable odor when wet; and some antipathy existed against these fibers because potential food sources were used for textiles. But the most powerful reason for discontinuing production was the fact that other man-made fibers with equal softness but better strength qualities than azlons had been introduced.

Despite the declining popularity of azlons, the technology of protein fiber production is still being utilized. Fibrous protein from soybeans is one ingredient of imitation bacon-flavored products.

PRODUCTION

The steps in manufacturing protein fibers are similar for all varieties. Protein is extracted from its original source by various techniques and then processed into a spinning solution that can be extruded through the spinnerette. Following extrusion, the fiber is coagulated in a chemical bath. To produce sufficient strength for use in fabrics, protein fibers are stretched slightly to improve molecular orientation. Nonetheless, molecular arrangement remains highly amorphous compared to that of most other fibers. Finally, protein fibers are treated by a hardening bath to impart dimensional stability and chemical resistance.

FIBER PROPERTIES

Since man-made protein fibers are of little general interest, the following discussion will cite only important contributions, major fiber properties, and the possible future of a selected few products.

Fibrolane

A casein fiber still produced in England, Fibrolane is characterized by exceptional softness and warmth, high moisture regain, and good resistance to age, sunlight, and insect damage.

Merinova

Merinova, a product of Italy, has the same general properties as Fibrolane. Its specific gravity is 1.3, its tenacity 0.9 to 1.1 grams per denier. Merinova has high elongation and good elastic recovery.

The fiber swells and disintegrates in strong acids and alkalies, but it has good resistance to weak acids and alkalies. Merinova tolerates organic solvents and can be dry cleaned satisfactorily.

Ardil

Ardil was made from the protein in peanuts or groundnuts. Production was suspended in 1957. While weaker than casein fibers, it has approximately the same elongation and elastic recovery. Ardil possesses good resistance to chemicals, both alkalies and acids, except for strong concentrated alkalies. Neither insects nor microorganisms damage the fiber.

Vicara

A product of the Virginia-Carolina Chemical Corporation, Vicara was produced between 1948 and 1958. The company bought the plant in which Aralac had previously been manufactured and converted the equipment to meet the processing requirements of the protein zein from corn. Zein is, in part, a by-product of chemical processes involved in the manufacture of starch and corn syrup.

The specific gravity of Vicara is 1.25. The moisture regain is 10 percent, with a saturation regain of 16 to 18 percent. It is estimated that the fiber can absorb up to 40 percent moisture when subjected to rain or laundering.

The dry tenacity of Vicara is 1.2 to 1.5 grams per denier, with a wet strength of .65 to .8 gram per denier. Elongation is 30 to 35 percent when dry and about 40 percent when wet. It has good elastic recovery and good resiliency.

An important property of Vicara is its resistance to both acid and alkali. In addition, it is not damaged by moths or by mildew. Prolonged exposure of Vicara to sunlight results in loss of fiber strength.

MAN-MADE PROTEIN FIBERS IN USE

Protein fibers on the current market, as well as those that have been discontinued, are very similar to wool in many respects. Azlon fibers do not felt as wool fibers do, so fabrics that are composed of blends of azlon and a normally washable fiber can be considered launderable. Blends of azlon and wool fibers should receive care procedures recommended for wool. Because azlon fibers are weak, any fabric that contains them should be handled gently during care.

The major use for the azlons is in blends with stronger fibers. They will add softness and warmth to fibers that are stiff and cool. Azlons increase the speed and ease with which wool fiber felts, and yet they reduce the shrinkage of wool in finished fabric. Loftiness and resilience are added qualities in blends of azlons and cellulosic fibers. Of considerable importance to many countries is the fact that man-made protein fibers contribute properties similar to wool but at a much lower cost.

Despite the apparent trend away from these fibers, they were and are of value, both historically and economically.

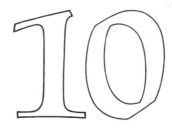

Nylon
(Polyamide)
Fibers

HISTORICAL REVIEW

The history of nylon is a story of scientific research and development. In 1927 the DuPont Company decided to give a small group of scientists unrestricted funds to do fundamental research. DuPont hoped this work would yield completely new scientific information that could ultimately lead to chemical advancements. However, they were willing to gamble even though the project might prove valueless.

Wallace Carothers, a brilliant young organic chemist from an eastern university, was selected to head this unusual research team. In 1928 the men began work, and they selected for study long-chain molecules such as those found in natural fibers, rubber, and plastic products. The team of chemists created certain types of giant molecules called *macromolecules, polymolecules,* or *polymers.* From these developments valuable scientific data were acquired, including the fact that linear polymers could be man-made. These linear polymers are composed of relatively small molecules linked end-to-end, much like a chain of paper clips.

In 1930 the chemists discovered an unusual characteristic in one of the substances under investigation. It was found that when a glass

"Nylon!"

Figure 10.1 The transformation of *Nylon* into a household word. [Drawing by Richard Decker; Copyright © 1940, 1968. *The New Yorker Magazine, Inc.*]

rod in contact with some viscous material in a beaker was pulled away slowly, the substance adhered to the rod and formed a fine filament, that hardened as soon as it was exposed to cool air. Furthermore, it was observed that the cold filaments could be stretched several times their extruded length to produce a flexible, strong, and attractive fiber. After this discovery the chemists turned to developmental research to find ways to produce such a fiber in a practical and economic manner.

The next few years were devoted to improving the polymer, finding efficient methods for manufacturing it, developing necessary mechanical equipment for its production, and, perhaps of most importance, finding possible uses for the new fiber. In 1938 a pilot plant commenced operation, and in 1939 a large-scale plant was put "on stream" by DuPont at Seaford, Delaware. It has been estimated that the first pound of nylon fiber produced at Seaford cost $27 million.

Nylon, in the form of knitted hosiery, was test-marketed in various parts of the United States in late 1939 and early 1940. Introduction of the fiber to the general public was well planned and coordinated; it was a classic example of successful mass marketing (Fig. 10.1). Throughout the nation, heralded by uniform advertising campaigns, nylon stockings were launched on May 15, 1940. They were a tremendous and immediate success. Despite the temporary absence from the consumer market during World War II, nylon hosiery has retained its early success and popularity.

The first nylon was and is referred to as type 6,6. The numbers derive from the fact that in each of the two chemicals used in making this nylon, there are six carbon atoms. Nylon type 6,10—composed of one chemical with six carbon atoms and one with 10—was developed simultaneously, and DuPont used it in making bristles for tooth-brushes, paint brushes, and similar products.

Since the war, other types of nylon, such as type 6, type 11, type 7, type 4, type 8, and several based on aromatic compounds, have been or are in the process of development for consumer use. In each case, the numbers refer to the number of carbon atoms in the basic chemical compound used in forming the polymer.

DuPont has stated that the real importance of nylon lay in the fact that, for the first time, man had returned to basic elements to create a molecule tailored specifically for use as a textile fiber.[1] He had stopped trying to imitate the silkworm and had struck out on his own to create a fiber with his own intelligence. To a large degree nylon freed man from the vagaries of animal and vegetable life and the capriciousness of nature.

The Textile Fiber Products Identification Act defines *nylon* as

a manufactured fiber in which the fiber-forming substance is any long chain synthetic polyamide having recurring amide groups

$$\left(\begin{matrix} -\underset{\underset{O}{\|}}{C}-NH- \end{matrix}\right)$$

as an integral part of the polymer chain.

MANUFACTURING AND MOLECULAR STRUCTURE

Nylon 6,6

Nylon 6,6 is made by the linear condensation polymerization of hexamethylene diamine and adipic acid (Fig. 10.2).

Early advertising of nylon proclaimed that the fiber was made from coal, air, and water. This is, of course, an oversimplification, but it is true that the elements found in nylon are also found in coal, air,

[1] DuPont Corporation, *Fibers by DuPont*, p. 7.

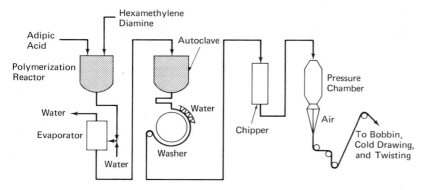

Figure 10.2 Flow diagram showing the steps in the manufacture of nylon 6,6.

$$HOOC(CH_2)_4COOH + NH_2(CH_2)_6NH_2 \longrightarrow H_2O + HOOC(CH_2)_4CONH(CH_2)_6NH_2$$

Adipic acid   + Hexamethylene $\longrightarrow$ Water +   Hexamethylenediamine
diamine   adipamide

Polymer Repeat
$$[-OC(CH_2)_4CONH(CH_2)_6NH-]n$$
n = times repeated in final molecule = 50–80+

Figure 10.3 The chemical reactions in the production of nylon 6,6.

and water, that is, carbon, hydrogen, oxygen, and nitrogen. Modern methods of manufacture involve the production of adipic acid from phenol and of hexamethylene diamine from adipic acid. The phenol is usually obtained from benzene, which, in turn, is derived from distillation of coal tar or petroleum.

Specific amounts of the acid and the diamine are combined in solution to form a salt called 1-aminohexamethylene adipamide or *nylon salt*. This nylon salt is purified and then polymerized in an autoclave under an atmosphere of nitrogen. It is extruded in a ribbon form and chipped into small flakes or pellets. The polymer then is melted and extruded through a spinnerette into cool air, where the nylon filaments are formed. After cooling, the filaments are stretched or cold-drawn to orient the molecules in the fibers and develop fiber strength and fineness.[2] According to data from some manufacturers, nylon fibers are heated slightly to increase ease and speed of drawing.

The 6,6 polymer is composed of hundreds of molecules of adipic acid and hexamethylene diamine hooked together end-to-end. Nylon 6,6 has a molecular weight of 12,000 to 20,000. If below 6000, the polymer does not form fibers, and between 6000 and 10,000 any fibers formed are weak and brittle. The degree of polymerization for nylon 6,6 varies from about 50 to over 80. The chemical reactions involved in producing nylon 6,6 are given in Figure 10.3.

Well-known trademarks for nylon 6,6 include DuPont nylon, "501," Antron, Astroturf, Blue "C" by Monsanto, Cadon, Cumuloft, and Actionwear. Several trademarks are controlled by certification standards and are authorized for use only when the fabric satisfies the fiber producer's specifications. Among these are "501," Blue "C," and Cumuloft.

Nylon 6

Nylon 6 is manufactured by polymerization of caprolactam, a cyclic form of omega caproic acid, which has the following formula:

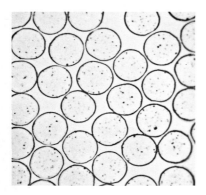

Figure 10.4 Photomicrograph of nylon 6,6, longitudinal view. [*E. I. DuPont de Nemours & Company*]

Figure 10.5 Photomicrograph of nylon 6,6, cross section. [*E. I. DuPont de Nemours & Company*]

Figure 10.6 Photomicrograph of nylon 6, longitudinal view. [*E. I. DuPont de Nemours & Company*]

[2]R. H. Peters, *Textile Chemistry*, Vol. I (New York: Elsevier Press, 1963), p. 440.

$NH_2(CH_2)_5COOH$. The cyclic form of caprolactam is

$$CH_2(CH_2)_4CONH,$$

or it can be structured as follows:

$$
\begin{array}{c}
CH_2 \\
H_2C \quad\quad CH_2 \\
H_2C \quad\quad CH_2 \\
OC\!\!-\!\!NH
\end{array}
$$

The unit forming the nylon polymer repeat is indicated thus:

$$[-(CH_2)_5CONH-]n \qquad n = 200\pm$$

Like nylon 6,6, the nylon 6 polymer is formed under pressure in an autoclave. It is extruded and chipped into pellet or flake form, washed, dried, and melt-spun through a spinnerette. The filaments are cold drawn.

The most common trade names for nylon 6 are Ayrlyn, Caprolan, Crepeset, Enka, Nytelle, Touch, Celon, Grilon, and Perlon.

FIBER PROPERTIES OF NYLON 6,6 AND NYLON 6

Microscopic Properties

Nylon filaments are smooth and shiny. When viewed in cross section, nylon is usually perfectly round. Exceptions to this are trilobal nylons such as Antron, "501," Cadon, and Cumuloft. Longitudinal magnification shows relatively transparent fibers of uniform diameter with a slight speckled appearance (Figs. 10.4–10.9).

Physical Properties

Shape and Appearance   Nylon is a man-made fiber, and therefore the diameter and length of the filaments or staple fibers are determined by the manufacturer and by the ultimate end-use. It is transparent when produced and can be made bright or dull.

Strength   One of the major advantages of nylon fibers is their strength. While high-strength types are available, even regular nylon is stronger than most natural fibers. The tenacity of regular nylon is 4.6 to 5.8 grams per denier, and high-strength nylon frequently has a tenacity of 8.8 grams per denier. Wet nylon retains much of its strength, but the wet tenacity may drop to between 4.0 and 5.1 grams per denier and to 7.6 for high-tenacity fiber.

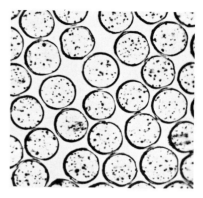

Figure 10.7 Photomicrograph of nylon 6, cross section. [*E. I. DuPont de Nemours & Company*]

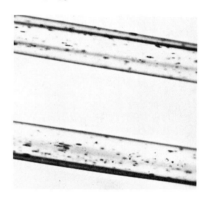

Figure 10.8 Photomicrograph of trilobal nylon—Antron®, longitudinal view. [*E. I. DuPont de Nemours & Company*]

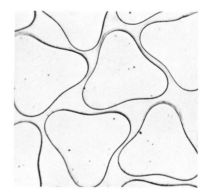

Figure 10.9 Photomicrograph of trilobal nylon—Antron®, cross section. [*E. I. DuPont de Nemours & Company*]

Elastic Recovery and Elongation   Nylon is a highly elastic fiber with relatively good elongation (Table 10.1). The elastic recovery properties assure excellent shape retention of nylon fabrics.

Resiliency   Nylon has good to very good recovery from creasing or wrinkling.

Density   The specific gravity of nylon 6,6 and nylon 6 is 1.14. If nylon and cotton were to be made into two identical fabric constructions, the nylon fabric would be considerably lighter because of the fiber's lower density.

Moisture Absorption   Compared with natural fibers, nylon has rather low moisture absorbency. Nylon 6,6 has a percentage moisture regain at standard conditions of 4.2 to 4.5; for nylon 6 the range is 3.5 to 5.0. At conditions of 95-percent relative humidity, nylon fibers will absorb approximately 8 percent moisture. Nylon 6 has a slightly higher absorbency than nylon 6,6 so it will accept dyes more easily.

The low moisture regain of nylon results in fabrics that dry quickly after laundering. However, the low moisture absorption, plus poor electrical conductivity, causes an accumulation of static electric charges on nylon.

Dimensional Stability   Because nylon is heat sensitive or thermoplastic, it can be heat-set during processing so that it will retain its shape during use and maintenance. The fiber will stretch under stress but will return to its original size following release of the stress. (For details concerning heat setting see pp. 267–268.) Application of temperatures higher than those used for heat setting may cause fiber deformation and shrinkage. Therefore, in order to maintain dimensional stability, it is important to avoid high temperatures for most nylon products.

Table 10.1   Elastic Properties of Nylon Fibers[a]

| | Breaking Elongation | | %Recovery at |
Fiber	dry	wet	4% Extension
Regular 6,6—filament	26–32	30–37	100
Regular 6,6—staple	37–40	42–46	100
High-tenacity 6,6	19–24	21–28	100
Regular 6—filament	23–42	27–34	100
Regular 6—staple	23–50	31–55	100
High tenacity 6	16–19	19–22	100

[a] 1970 Properties of Man-made Fibers, *Textile Industries* (chart); and Cook, *Handbook of Textile Fibers*, Vol. 2, pp. 257–8, 297–8.

Thermal Properties

Nylon 6,6 melts at approximately 250°C (480°F) and nylon 6 at 210°C (400°F). All nylon can withstand temperatures to 149°C (300°F) for long periods of time without damage, but when temperatures exceed 300°F for a few hours, the fibers will discolor. If temperatures approach 177° to 205°C (350°–400°F), the fiber softens, and discoloration and loss of strength occur quickly. Low temperatures have no adverse effect on nylon. It can withstand temperatures below 0°F without damage.

Safe pressing temperatures are considered to be between 300° and 350°F. Temperatures above 350°F should be used carefully to avoid glazing of fibers and yarns. Nylon 6 should not be ironed at temperatures above 300°F.

Nylon melts away from a flame and forms a gummy gray or tan ash that hardens as it cools. The fiber will burn if held in an open flame, but it does not support combustion. The smoke is white or grayish in color, and the odor is likened to that of cooking celery or green beans.

Chemical Properties

Effect of Alkalies   Nylon is substantially inert to alkalies.

Effect of Acids   Mineral acids, such as hydrochloric, nitric, and sulfuric, will cause nylon to disintegrate or dissolve almost immediately. Even dilute solutions of hydrochloric acid will destroy the fiber. Organic acids, such as formic, will dissolve nylon in concentrations of 88 to 90 percent. Acid fumes in the air in industrial regions have been known to bring about fiber disintegration.

Effect of Organic Solvents   Most organic solvents have little or no effect on nylon. Phenol, metacresol, and formic acid dissolve the fiber, but solvents used in stain removal and dry cleaning do not damage it.

Effect of Sunlight, Age, and Miscellaneous Factors   Sunlight has a destructive effect upon nylon, and there is a marked loss of strength after extended exposure. If nylon is left in direct sun for several weeks, it may actually decompose. Bright nylon has better resistance to sunlight than delustered fiber. Special dyes have been developed for nylon that inhibit sunlight damage. These are important when the fiber must be subjected to sun for long periods of time.

Age appears to have no effect on the fiber. If stored away from light and other deleterious influences, nylon will last for many years.

Nylon is a very tough and very flexible fiber. In addition, it has excellent resistance to abrasion. However, the product itself is abrasive and will wear other fibers.

Soaps, synthetic detergents, and bleaches do not damage nylon fibers.

Biological Properties

Nylon is highly resistant to attack by most insects and microorganisms. However, some insects normally found outdoors, including ants, crickets, and roaches, will eat nylon if they are trapped in folds or creases. Microorganisms producing mildew may attack finishes used on nylon but do not damage the fiber itself. Bacteria have no effect on nylon.

OTHER NYLONS

Many other nylons, or polyamide fibers, are either being manufactured in various countries or are in the developmental stages. In addition, the older nylon fibers have undergone various modifications, and second- and third-generation fibers are on the market.

The code system based on the number of carbon atoms in the monomers used in making the fiber polymers is still in effect, but in addition there are letters combined with numbers to identify several of the new nylon fiber polymers. A few of the newer nylons will be cited briefly.[3]

Nylon 11   Rilsan, or nylon 11, is produced in Europe, Asia, and South America. It is a polymer of aminoundecanoic acid, which has the formula $NH_2(CH_2)_{10}COOH$, and the polymer repeat can be represented as

$$[-HN(CH_2)_{10}CO-]_n$$

The fiber properties are similar to those of nylon 6 and 6,6, except that nylon 11 has a lower moisture regain—1.18 percent; lower specific gravity—1.04; and a lower melting point—189°C (374°F). It has been stated that nylon 11 does not discolor—turn yellow or gray—as quickly as do types 6,6 and 6. The low density makes it adaptable to bulky yarns.

Nylon 4   Nylon 4 is made by polymerizing 2-pyrrolidone. This fiber is being developed by Radiation Research in the United States. The

---

[3] For detailed discussions of nylon fibers and polyamides, the reader is referred to J. Gordon Cook, *Handbook of Textile Fibers*, Vol. I (London: Merrow Publishing Co., 1968), pp. 212–333; Adeline Dembeck, *Guidebook to Man-made Textile Fibers and Textured Yarns of the World*, 3d ed. (New York: United Piece Dye Works, 1969), pp. 136–158; and H. F. Mark, N. G. Gaylord, and N. M. Bikales, *Encyclopedia of Polymer Science and Technology*, Vol. 10 (New York: Wiley-Interscience, 1969), pp. 347–460, 483–597.

polymer repeat is

$$[-(CH_2)_3CONH-]_n$$

The fiber has a high moisture absorbency and a higher melting point than nylon 6.

Nylon 5   Polyvalerolactum, or nylon 5, has been investigated by several American and foreign manufacturers as a potential fiber. Its properties resemble those of nylon 6,6. The fiber repeat is

$$[-NH(CH_2)_4CO-]_n$$

Nylon 7   Polyheptanoamide, nylon 7, is made in the Soviet Union, under the trade name Enant. The polymer repeat is

$$[-(CH_2)_6CONH-]_n$$

This fiber is similar to nylon 6,6 and nylon 6, but it has a higher melting point and a lower moisture regain than either of the common fibers.

Nylon 6T   Polyamide fibers made from hexamethylene diamine and terephthalic acid have a high melting point—370°C (698°F); a slightly higher density—1.21; and a higher modulus with lower elongation than nylon 6,6 or nylon 6.

Nomex, or nylon 6T, was developed to withstand extremely high temperatures. It shows a slight loss of strength after long-term exposure to temperatures above 200°C (392°F). The fiber ignites with difficulty at temperatures near 370°C (698°F) and is self extinguishing. It is resistant to nuclear radiation and is finding considerable use in space suits, space vehicles, military applications, and protective clothing in industrial plants. Nomex has excellent resistance to acids and to abrasion, and these properties should enhance its use in work clothing. Nomex is not generally adaptable to consumer goods, because it tends to yellow in a short time, and it is difficult to dye, so discoloration cannot be masked, and fashion needs cannot be satisfied.

Para ACM 9 Nylon and Para ACM 12 Nylon   Para ACM 9 and para ACM 12 are made from bis (p-aminocyclohexyl) methane and non-anedioic acid (9) or dodecanedioic acid (12). These nylons are reported to have high wet recovery from folding or stress and thus have excellent wash-and-wear properties. Reports also claim resistance to oily stains, high comfort factors, excellent hand, and shape retention. It has been suggested that para ACM 12 is the polymer used to make DuPont's Qiana.

Qiana nylon was introduced by DuPont into high-cost apparel items during the late 1960s. In 1970 the price was reduced slightly, and the fiber is now available in medium- to upper-price merchandise. It resembles silk in hand and appearance, and yet it has excellent wash-and-wear properties. The fibers take a variety of dyes in clear, brilliant colors—solids or prints—and they do not discolor. Qiana is used optimally in weaves and knits. The fiber was given a couture introduction and has been considered a luxury item, so it is marketed to appeal to the affluent.

Nylon 6,10   Nylon 6,10 is made from hexamethylenediamine and sebacic acid. It is widely used for bristles.

Nylon 22N   A new anti-static nylon by Monsanto, called 22N, has just been introduced to the public. It is a modified fiber with built-in chemical and physical properties that reduce the static buildup to a point near that of cotton. The fiber has a silklike luster, good covering power, resistance to soiling, and superior whiteness retention. It is recommended for apparel.

Other Nylons   Other nylons on the market or in development include nylon 8 from caprylamide, nylon 9 from aminononanic acid, nylon 12 from lauryl lactam, nylon 6,8 from hexamethylenediamine and suberic acid, and nylon MXD-6 from metaxylylene adipamide.

NYLON MODIFICATIONS

There are a number of special types of nylon 6,6 and nylon 6 currently in production. Among them, nylon 420 has been developed for use in blends with cotton. The 420 is similar to cotton in its initial modulus of elasticity (resistance to stretching or load necessary to cause a small amount of elongation). It adds strength and abrasion resistance to the nylon-cotton blend.

In addition to developing new polymers, fiber scientists have been active in modifying existing forms of nylon. These modifications are frequently referred to as second- and third-generation fibers, and they take several forms: change of cross-section shape; texturizing yarns by special processes; combining two or more types of nylon to form bicomponent fibers; and chemically modifying the polymers by cross-linking molecules and by graft polymerization.

Cross Section   Original nylon fibers were round in cross section; modified cross sections include triangular, irregular, trilobal, and other multilobal shapes. These changes are made possible by altering the shape, arrangement, and number of orifices in the spinning jet.

Modified cross sections can produce many desirable qualities, such as increased cover; a crisp, silklike, firm hand; reduced pilling; increased bulk; sparkle effects; and heightened resistance to soil. Trilobal nylon is available in various fiber deniers or diameters and is used in apparel or home-furnishing fabrics.

Textured Nylon   Stretch yarns of nylon, manufactured by various texturizing processes, are widely used alone or in blends to produce many of the stretch fabrics on the current market. Other texturizing processes produce bulky yarns. These products will be discussed more fully in Chapter 28.

Bicomponent Fibers   Extruding two filaments of different composition results in bicomponent fibers (Fig. 10.10). The spinnerette openings can be either side by side or one inside the other. The filaments must be compatible in order for them to form a strong bond at the interface. Filaments produced by this method usually differ in such characteristics as shrinkage and thermal behavior, which enables the processor to introduce fiber crimp by utilizing the shrinkage differential. Such fibers have greater bulk, improved fit, retained sheerness, and attractive appearance.

Cantrece is a bicomponent nylon and is considered a third-generation product. This fiber is composed of two nylon components with different shrinkage potentials. Upon subjection to heat a latent crimp is developed, and one component shrinks more than the other. The fiber has good resilience and fit retention and is used primarily for women's hosiery. Crepeset nylon is designed to develop a special crimp during the fabric finishing processes. It is frequently used in knit structures.

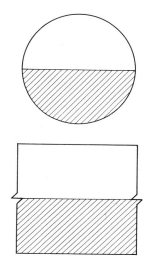

Figure 10.10   Diagram of a simple bicomponent fiber. [*Allied Chemical*]

Cross-linking   Nylon molecules can be cross-linked by adding a chemical characterized by reactive groups at both ends of the chemical molecule. This chemical reacts with either the terminal amine group or the terminal carboxyl group on the nylon molecule to hook adjacent fiber molecules together. Such modifications increase the fiber modulus and are especially useful in fibers intended for vehicle tires.

Graft polymerization   Nylon fibers can be modified by grafting other chemicals onto the fiber molecules. Various chemicals have been or will be employed in future developments. One important property altered by graft polymerization is moisture absorbency, which is heightened, and, in turn, increases the wet crease recovery and diminishes static buildup.

NYLON IN USE

Nylon is widely used in fabrics for apparel, home furnishings, and industry (Fig. 10.11). It has proved to be the leading fiber in the

Figure 10.11   Qiana nylon in a jersey print. [*E. I. DuPont de Nemours & Company*]

Figure 10.12 Carpet and wall covering in 100 percent nylon. [*Bigelow-Sanford, Inc.*]

manufacture of hosiery and is of considerable importance in the lingerie market. For outerwear, it is used in blends with several different fibers to contribute dimensional stability, elastic recovery, shape retention, and abrasion resistance.

Much carpeting and upholstery are made of nylon (Fig. 10.12), for it wears well, is easy to clean, and does not require special protection against moths and carpet beetles. Trilobal nylons such as Cumuloft and "501" are popular in carpeting because they resist soiling (or do not show soil quickly) and crushing, and retain their attractive appearance.

Nylon is easy to launder. It can be washed safely at all laundry temperatures, drip dried, spun dry, or dryer dried. However, at low or medium low temperatures (below 110°F), the fabric will be less likely to wrinkle and, thus, will be easy to iron. In some cases it will probably not require any ironing to regain a neat appearance. Drying in dryers at low temperatures, followed by prompt removal, frequently will produce smooth and neat products.

Bleaches can be used on nylon, and soaps and synthetic detergents will not damage the fiber.

Problems encountered in the laundering of nylon include the fact that nylon fibers tend to scavenge color and soil during the washing process. This produces gray, yellow, or discolored items that may be difficult to restore, by cleaning, to their original appearance. White nylon fabrics are particularly vulnerable to discoloration by improper care techniques. There is some evidence that colored detergents may discolor white nylon if not thoroughly rinsed away.

Several special 6,6 nylons, such as DuPont's type 91, have had a fluorescent dye added to the spinning melt to retard discoloration of white nylon during its period of use.

Modern dyestuffs, if wisely selected and properly applied, produce fast colors on nylon fabrics. However, it is extremely important that great care and proper techniques be used in applying the dyes to avoid streaky or uneven dyeing.

Pilling, the formation of tiny balls of fiber on the surface of the cloth, is a severe problem with fabrics made of spun nylon yarns and to a lesser degree with filament fiber fabrics. Pills form much more slowly on filament fiber fabrics. Since the nylon fibers are extremely resistant to abrasion, the pills are not rubbed off after they form as they would be on the surface of fabrics with low abrasion resistance.

Because of the low moisture absorbency of nylon, fabrics must have adequate spaces or interstices between yarns to permit water vapor passage and thus ensure the wearer's comfort. Various finishing techniques that improve surface absorbency and wicking have been developed. These finishes increase comfort by drawing moisture away from the body. One technique involves coating yarns with nylon 8, a polyamide with good absorbency.

The capacity of nylon to be heat-set makes it possible to build in various surface designs in fabrics, that is, embossed effects. Puckered or crinkled nylon can be made by using a weak solution of phenol to shrink certain areas of the fabric. Metallic substances such as copper and aluminum can be employed to print interesting designs on nylon. These treatments are permanent to laundering and drying if temperatures do not exceed those used in heat setting.

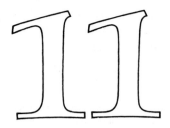

Polyester
Fibers

HISTORICAL REVIEW

The first work on polyester fibers was done by the Carothers team of chemists during the early stages of their fundamental research for DuPont. However, when polyamides appeared to show more promise, they were selected for development, and the polyesters were set aside. In fact, the early polyesters were poor in quality and threatened to be too expensive to produce.

While Carothers and his assistants directed their emphasis to the polyamides, chemists in Britain began experimenting with long-chain linear polyester polymers. In 1941 J. R. Whinfield and J. T. Dickson of Calico Printers' Association introduced a successful polyester fiber. Development of this fiber was delayed by the war, and public announcement of the discovery was withheld until 1946. The Imperial Chemical Industries, Ltd. (ICI) purchased the rights to manufacture the fiber for all countries except the United States, where DuPont obtained the manufacturing privilege.

DuPont's polyester fiber was made available to the American consumer in small quantities in 1951 and was known successively as Fiber V, Amilar, and finally Dacron (pronounced day'-kron). A large manufacturing plant was completed at Kinston, North Carolina, in 1953, and since that time Dacron polyester has become one of the most widely used and desirable of all synthesized fibers.

While the fiber known as Dacron was becoming popular in the United States, the same fiber, called *Terylene,* was gaining status in England and several European countries. In recent years other companies have entered the polyester market. Outside the United States, through licensing arrangements with ICI, polyesters with such names as Terlenka, Diolen, Terital, Tergal, Tetoron, and Teriber were placed on the market. Simultaneously, DuPont was granting licenses to other American firms. Celanese introduced Fortrel polyester—then called Teron—in 1958, and Beaunit Mills produced Vycron polyester in 1959. Tennessee Eastman developed a polyester, Kodel, which was released in 1958, but Kodel was different from other polyesters, so no licensing arrangement was required. DuPont's patent controls have now expired, and in the past few years polyesters have been marketed by several fiber producers.

Polyester fibers found immediate consumer acceptance because of their ease of maintenance and excellent crease resistance. When the fiber was first previewed for the press, several products were exhibited, and the most impressive item was a man's business suit. To dramatize the superior wrinkle resistance and easy-care properties of the new fiber, the suit was worn for 67 days without ever being pressed. During this time it was submerged twice in a swimming pool and washed once in an automatic washing machine. When shown at the press preview it was still presentable.

The TFPIA defines a *polyester* fiber as

a manufactured fiber in which the fiber-forming substance is any long chain synthetic polymer composed of at least 85% by weight of an ester of a dihydric alcohol and terphthalic acid (p—HOOC—C_6H_4—COOH).

PRODUCTION AND MOLECULAR STRUCTURE

Polyesters are the product of the reaction between a dihydric alcohol and dicarboxylic acid. The generic definition specifies the presence of terephthalic acid; however, some manufacturers are using dimethyl terephthalate, because they believe it is more easily purified than the acid. Most polyesters use ethylene glycol as the dihydric alcohol. (In the manufacture of Kodel II, 1, 4 cyclohexanedimethanol is used.) The reaction of the ethylene glycol and terephthalic acid is shown in Figure 11.1.

Dimethyl terephthalate is more frequently used than terephthalic acid, because it is more easily obtained in pure form. The resulting polymer is of superior quality. The reaction that occurs when dimethyl terephthalate is used is similar to that shown in Figure 11.1, except that methanol—methyl alcohol (CH_3OH)—rather than water is a by-product of the reaction.

Ethylene glycol + Terephthalic acid $\longrightarrow$

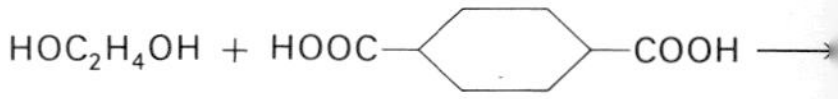

HOC_2H_4OH + HOOC—⬡—COOH $\longrightarrow$

(HOOC—C_6H_4—COOH)

Polyester, ethylene terephthalate + Water
HOOC—C_6H_4—COOCH$_2$CH$_2$OH + H_2O

Figure 11.1 Chemical reactions in polyester formation.

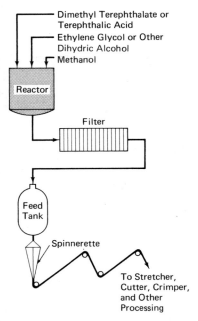

- Dimethyl Terephthalate or Terephthalic Acid
- Ethylene Glycol or Other Dihydric Alcohol
- Methanol

Reactor

Filter

Feed Tank

Spinnerette

To Stretcher, Cutter, Crimper, and Other Processing

Figure 11.2 Flow diagram showing the steps in the manufacture of polyester.

Terephthalic acid or dimethyl terephthalate and ethylene glycol polymerize by condensation reaction to form the polyester polymer. The repeat unit may be represented thus:

$$[—OOC—C_6H_4—COOCH_2CH_2—]_n \qquad n = \pm 80$$

Each fiber molecule or polymer contains about eighty repeat units.

The preceding reactions are used, with some modifications for certain fiber variants, to produce most polyester fibers (Fig. 11.2). One major exception is Kodel II, which employs a different dihydric alcohol and forms the following repeat unit:

As the substances are polymerized, they are extruded from the polymerizing vessel in the form of a ribbon and are cut into chips. The chips are diced and conveyed to a hopper, from which they are fed to the melt spinning tank. The hot solution is forced through the spinnerette and solidifies into fiber form upon contact with air. It is stretched while hot; the stretching contributes strength to the fiber and controls elongation. The higher the amount of stretch, the stronger the fiber, and the lower the elongation.

The fiber molecules are long and thin and lie relatively parallel to each other within the fiber. Polyester fibers are processed into yarns and fabrics in the same manner as other fibers.

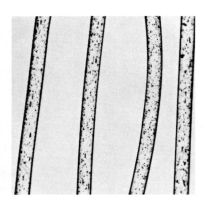

Figure 11.3 Photomicrograph of regular polyester, longitudinal view. [*E. I. DuPont de Nemours & Co.*]

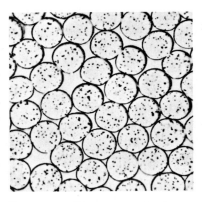

Figure 11.4 Photomicrograph of regular polyester, cross section. [*E. I. DuPont de Nemours & Co.*]

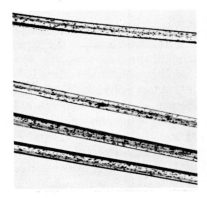

Figure 11.5 Photomicrograph of trilobal polyester (Dacron T-62), longitudinal view. [*E. I. DuPont de Nemours & Co.*]

Table 11.1   Properties of Polyester Fibers

Fiber	Tenacity in Grams per Denier Wet and Dry	Elongation % Wet and Dry	Elastic Recovery at 2% Elongation	Specific Gravity	Moisture Regain at 20°C (70°F); 65% Relative Humidity
Dacron					
regular	2.8–5.2	19–30	97	1.38	0.4–0.8
high	6.0–9.5	10–14	100	1.38	0.4–0.8
Kodel					
Type II	2.5–3.0	24–34	85–95	1.22	0.4
Type IV	4.5–5.5	35–45	75–85	1.38	0.4
Fortrel					
regular	4.5	25–30	100	1.38	0.4
high	7.0–8.5	8–10	100	1.38	0.4
Vycron					
Type 2	5.3–8.0	30–40	93	1.38	0.6
Type 5	5.0–5.5	35–45	98	1.38	0.6
Avlin	4.0–5.0	18–22	90–97	1.38	0.4
Encron					
regular	4.4–5.0	20–30	95–97	1.38	0.4
high	8.5–8.9	11.5–13.0	100	1.38	0.4

FIBER PROPERTIES

Microscopic Appearance

A longitudinal view of polyester fiber exhibits uniform diameter, smooth surface, and a rodlike appearance (Fig. 11.3). The cross section is usually round (Fig. 11.4). Variations are encountered in trilobal polyester (Dacron T-62, for example) and in Vycron, which is somewhat oval in cross section (Figs. 11.5, 11.6). Trevira, a fiber marketed by Hystron Fibers, is pentalobal in cross section.

The inclusion of pigment in the melt solution produces fibers that are dull or semidull in appearance and have a speckled effect, since the pigment causes changes in light reflection.

Physical Properties

Shape and Appearance   Except for the multilobal varieties, polyesters are generally round and uniform. They can be of any length or diameter required by the fiber producer and the yarn and fabric manufacturer. The fiber is partially transparent and white or slightly off-white in color. Optical brighteners are frequently added to produce clear, bright white polyester fibers.

Strength   The tenacity of the polyesters is given in Table 11.1. Regular and high-tenacity fibers are produced by several manufac-

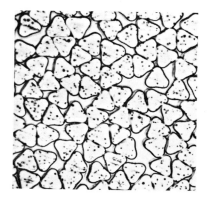

Figure 11.6 Photomicrograph of trilobal polyester (Dacron T-62), cross section. [*E. I. DuPont de Nemours & Co.*]

turers, and strength varies from a low of 2.5 grams per denier for some Kodel II to a high of 9.5 grams per denier for some high-tenacity Dacron. There is no loss of strength when polyester fibers are wet.

Elastic Recovery and Elongation   Table 11.1 shows comparable figures of elastic recovery and elongation. The amount of extension or elongation possible for Dacron is inverse to tenacity. Stronger fibers have lower degrees of elongation.

Elastic recovery of high-tenacity fibers is good. Regular-tenacity polyesters exhibit some variation in elastic recovery, but they are superior to cellulosic fibers and only slightly inferior to nylon.

Resiliency   The property of resilience and recovery from creasing and wrinkling is excellent for polyester fibers. Furthermore, heat setting can stabilize fibers and yarns so they need little or no pressing to retain a smooth appearance. Even if a wearer is caught in a rainstorm, fabrics of polyester will dry, usually without noticeable wrinkles. Pleats or creases heat-set into the fabric during construction will remain neat and distinct.

Density   The specific gravity of most polyester fibers is 1.38. Kodel type II is an exception; it is less dense, with a specific gravity of 1.22. Most polyesters are similar to wool or acetate in their density.

Moisture Absorption   Polyester fibers have low moisture regain of only 0.4 to 0.6 at standard conditions. Even at 95- to 100-percent relative humidity the moisture content is only 0.6 to 0.8 percent. Because of this low regain, moisture has little effect on the strength or elongation of the fiber, and static electric charges are accentuated. Furthermore, the low moisture absorption demands special techniques in dyeing and finishing. Polyesters, like cotton and linen, have a high degree of wickability. This wicking property, coupled with certain yarn and fabric constructions, can produce end-products that carry exterior moisture through to the inside, or body perspiration through to the outside. However, when this is undesirable, other yarns and fabric constructions can be substituted to inhibit the movement of moisture through the fabric.

Dimensional Stability   If polyester is properly heat set, it will not shrink or stretch when subjected to boiling water, boiling cleaning solvents, or ironing temperatures that are lower than the heat-setting temperature, which is usually above 195°C (385°F) and may be as high as 220°C (425°F) for some products. Polyester that has not been heat-set may shrink at elevated temperatures, and if subjected to boiling water for one hour, the fiber will shrink 11 to 12 percent. For satisfactory end-use, heat setting is essential.

The producers of Kodel have stated that their fiber has outstanding dimensional stability without heat setting or other special processing. Nonetheless, heat setting is recommended.

Thermal Properties

Polyester fibers melt at temperatures from 238° to 290°C (460°–554°F) depending upon type. As the fiber melts, it forms a gray or tawny-colored bead that is hard and noncrushable. Polyesters will burn and produce a dark smoke and an aromatic odor. However, they do not burn as rapidly as many other fabrics. In light fabric constructions the fibers melt and drip away from the source of ignition, preventing the propagation of flame.

Heat does not discolor polyester. Filament Dacron can be ironed safely at temperatures between 275°F and 300°F. However, staple fibers tend to glaze at these temperatures, and, thus, it is suggested that yarns or fabrics of staple Dacron be ironed at 250°F. There is minimal loss of strength following exposure to temperatures below 300°F. After thirty days at 300°F, Dacron fibers exhibit a 24-percent loss of strength; a 78-percent loss results when fibers are held at 347°F for only eleven days.

Other polyesters respond satisfactorily to the same ironing temperatures as Dacron. Kodel is in this category, but the fiber producer states that ironing temperatures as high as 400°F are safe, if this does not exceed the heat setting temperature.

Heat setting of yarns and fabrics is essential if they are to possess the easy-care, wrinkle-free properties associated with the fiber. Permanent pleats in polyester fabrics, once heat set in the desired location, will hold as long as processing or maintenance temperatures do not exceed the heat-set temperature.

Chemical Properties

Effect of Alkalies   Polyester has good resistance to weak alkalies at hot and at room temperatures. It exhibits only moderate resistance to strong alkalies at room temperatures and is degraded at elevated temperatures.

Effect of Acids   Weak acids, even at the boiling point, have no effect on polyesters unless the fiber is exposed for several days. They have good resistance to strong acids at room temperature. Prolonged exposure to boiling hydrochloric acid destroys the fiber, and 96 percent sulfuric acid causes disintegration.

Effect of Organic Solvents   Polyester is generally resistant to organic solvents. Chemicals used in cleaning and stain removal will not damage it, but hot meta-cresol will destroy the fiber, and certain mixtures of phenol with trichlorophenol or tetrachlorethane will dissolve polyesters. Oxidizing agents and bleaches do not damage polyester fibers.

Effect of Sunlight, Age, and Miscellaneous Factors   Polyester exhibits good resistance to sunlight when behind glass, so it is satisfactory for window coverings, but prolonged exposure to sunlight weakens the fiber. Polyester fibers are not affected by aging.

The fiber resists abrasion very well. Soaps, synthetic detergents, and other laundry aids do not damage it. It can be safely laundered in automatic washers and dried in controlled-temperature dryers. Control of both laundering and drying temperatures is essential to prevent the formation of undesirable wrinkles. (See the discussion regarding nylon, p. 126.)

Polyester fabrics should be laundered in warm water. Best results are obtained when driers with durable-press controls are used, for these cycles cool the fabric during the last few minutes of the drying period. The garments generally can be worn with little or no ironing.

One of the most serious faults with polyester is its *oleophilic* quality. It adsorbs oily materials easily and holds the oil tenaciously.

Biological Properties

Insects will not destroy polyesters if there is other food available. However, if trapped, beetles and other insects will cut their way through the fabric as a means of escape.

While microorganisms will not harm the fiber, they may attack finishes that have been applied. Usually, any discoloration is easily removed, since it does not penetrate the fiber.

MODIFIED AND TRADEMARKED FIBERS

The trademarks for polyester fibers made in the United States in 1970 included Dacron, Blue "C," Avlin, Encron, Fortrel, Fybrite, Kodel, Quintess, Vycron Tough-Stuff, and Trevira. These products are primarily polyethylene terephthalate (PET) and, therefore, properties of the various fibers are similar. Polyester fiber trademarks protected by quality specifications include Avlin, Blue "C," Encron, Fortrel, Kodel, Trevira, and Vycron Tough-Stuff.

Among the polyester fibers produced abroad that may be encountered by American consumers are Crimplene, Tergal, Terlenka, Terylene, Tetoron, and Toray.

Polyester fibers are available in a variety of types, some of them second- and third-generation fibers. In early 1971 there were 8 types of Avlin, 41 types of Dacron, 24 types of Fortrel, 15 types of Kodel,

15 types of Trevira, and 11 types of Vycron. Each type has at least one special characteristic that alters its behavior in some way.

These new fibers are created by altering the cross section shape from round to multilobal, by changing fiber formulas, or by varying the physical processing. The shape changes produce fibers with different hand and appearance than regular polyester, while chemical modifications result in fibers that are dye selective; have altered strength characteristics and, as a result, altered pilling behavior; and are more crush resistant.

Cross sectional shape change is usually affected by altering the spinning jet. Chemical modification involves the addition of small amounts of selected chemicals. For example, fiber dyeability is altered by the inclusion of chemicals that add sulphonic groups and by the substitution of isophthalic acid for a portion of the terephthalic acid. Other fiber changes can be obtained by modifying the spinning speed and degree of molecular orientation, which affects physical properties such as strength and pilling.

Trilobal and pentalobal cross sections are responsible for several desirable properties of polyester fibers. Yarns and fabrics constructed from such fibers are characterized by a silklike hand and appearance. In addition, the fabric appears to soil less quickly, because dirt lodged in the valleys between the lobes is inconspicuous. Luster and surface sheen please the eye, and the fibers have improved covering power.

Special fibers processed to impart increased loft and bulk are available for use as fiberfill in pillows, sleeping bags, garments, and home furnishings. These fibers have outstanding resiliency, they maintain their whiteness, they do not absorb odors, and they are nonallergenic.

Polyester is widely used in blends with cellulosic fibers. To provide polyester that blends particularly well, the fiber may be altered to give it initial modulus similar to cotton, so processing is easier, and yarns are more uniform.

Crimp can be set in the fiber during manufacture to increase bulking and resilience.

Research scientists are currently seeking ways of modifying fiber formulas to reduce the oleophilic quality of polyester and produce fibers with inherent soil release properties. Constant research by polyester manufacturers is evident in the new fiber variants that are introduced each year.

POLYESTER IN USE

The most important characteristics of polyester fibers are wrinkle-free appearance and ease of care. The fabrics require little or no ironing; they are easy to launder and quick to dry. Fabrics of 100 percent filament polyester are used for apparel for men, women, and children (Fig. 11.7). There are limitless varieties available.

Figure 11.7 Denim-look knit of 100 percent Dacron polyester. [*E. I. DuPont de Nemours & Company*]

Figure 11.8 Sail cloth. The mainsails are of Dacron polyester, the spinnakers of nylon. [*E. I. DuPont de Nemours & Company*]

Blends of wool, cotton, rayon, or linen with polyester fibers are popular with both men and women. In blended fabrics polyester fibers contribute easy maintenance, strength and durability, abrasion resistance, relatively wrinkle-free appearance, shape and size retention; protein or cellulosic fibers enhance dyeability, comfort, absorbency, and reduce static charges.

Polyester fabrics are subject to pilling, especially when staple fibers are used in yarns and fabrics. However, some of the newer polyesters exhibit less pilling than older varieties, because their lower tenacity (with little or no increase in elongation) diminishes abrasion resistance and therefore reduces pilling.

Polyesters are used in industry for such items as laundry bags, calender sheeting, press covers, conveyor belts, fire hose, sail cloth (Fig. 11.8), fish netting, ropes, and protective clothing.

The care of 100-percent polyester fabrics is minimal. It is advisable to pretreat heavily soiled areas with a lubricating detergent and then wash by machine or by hand, using warm water with a good soap or synthetic detergent. Garments can be drip dried or dried in a dryer if removed before any wrinkles are set. White polyester fabrics should be washed separately.

It is safe to launder blends if the article has been labeled as washable. For best results with blends label information should be kept and followed. The item may be washable by hand, machine washable, or dry cleanable. Bleaches can be used unless one of the other fibers in a blend will be damaged.

Frequent complaints concerning polyester fabrics arise from the oleophilic property. One solution to this problem involves applying a liquid hair shampoo to the stain before laundering. This usually will ensure lift off of oil and grease. Spray spot removers and liquid synthetic detergents are often effective for tenacious stains.

In polyester blends various fiber manufacturers have established desirable minimum amounts of polyester for use with different fibers. No less than 50 percent polyester, and preferably 65 percent, is recommended with all cellulosic fibers. At least 50 percent polyester should be used with wool or with synthesized fibers such as acrylics or modacrylics. When blended with cotton and nylon, there should be 65 to 75 percent polyester and nylon combined.

Polyester fibers seem to be the most desirable and satisfactory choice for blended fabrics with durable-press finishes and for the easy-care, wrinkle-free, textured woven and knit fabrics so compatible with modern living.

Acrylic and Modacrylic Fibers

ACRYLIC FIBERS

HISTORICAL REVIEW

The success of nylon was so extraordinary that many manufacturers began experimenting with other chemicals in an attempt to find new fiber-forming polymers.

Early in World War II, DuPont, partly as a result of the fundamental research of Carothers, developed an acrylic fiber, Fiber A, that showed potential as a substitute for either wool or silk. In the staple form the fiber resembled wool; in filament form it resembled silk in texture and appearance. For the duration of the war the fiber was used in limited amounts for government purposes, and after cessation of hostilities the DuPont organization completed development of Fiber A. They gave it the trademark *Orlon,* introduced pilot-plant quantities to a test segment of the population, and evaluated the results. Interest was keen, and DuPont decided to build a full-scale plant in Camden, South Carolina, where production of Orlon began in 1950.

Many questions had to be answered during the development of the fiber, such as what solvents to use in manufacture, end-use potential, and recommended care techniques. While DuPont was developing Orlon, other companies were also working on acrylic fibers.

In Germany, postwar scientific efforts included considerable investigation of textile fibers, and Farbenfabriken developed two acrylic fibers, Pan and Dralon.

In the United States the increased public acceptance of man-made fibers encouraged other companies to enter the industry. Chemstrand Corporation of Decatur, Alabama, was formed in 1949 as a joint operation of Monsanto Chemical and American Viscose Corporation. In 1950, Chemstrand began production of an acrylic fiber, Acrilan, in pilot-plant quantities. A broad advertising campaign introduced the new acrylic to the public in 1952, and the full-scale plant went "on stream" in 1953. However, the first quantities of the fiber were somewhat inferior, so the company made minor changes in formulations and manufacturing techniques. An improved product appeared in 1954. This latter fiber has been extremely successful.

American Cyanamid Corporation had begun working with acrylic and other polymers several years before World War II, but they did not direct their efforts toward the production of textile fibers until the early 1950s. They, too, saw the potential of acrylic fibers and in 1958 introduced commercial quantities of their fiber, Creslan. Also in 1958 the Dow Chemical Company released a new textile fiber called Zefran. While this is basically acrylic, the Dow chemists prefer to call it an acrylic-alloy fiber.

Since 1958 there have been no new acrylic fibers introduced by American firms. Producers in other countries have continued work on developing new acrylic fibers, but the U.S. chemical industry has directed its talents toward different types of fibers. At the present time the acrylic fibers produced in the United States include Orlon, Acrilan, Creslan, and Zefran. Besides these four, such fibers as Courtelle, Crylor, and Toraylon are sometimes found on the U.S. market in imported products.

The Textile Fiber Products Identification Act defines an *acrylic* fiber as

a manufactured fiber in which the fiber-forming substance is any long chain synthetic polymer composed of at least 85% by weight of acrylonitrile units

$$\left(\begin{array}{c} -CH_2-CH- \\ | \\ CN \end{array} \right)$$

PRODUCTION AND MOLECULAR STRUCTURE

Acrylic fibers are polymers formed by additional polymerization of at least 85 percent by weight of a chemical called *acrylonitrile* or *vinyl cyanide*, $H_2C=CHCN$. To polymerize vinyl cyanide the double bond ($=$) between the first two carbon atoms is broken, and the molecules attach themselves to each other in a linear chain. The structural formula is shown in Figure 12.1.

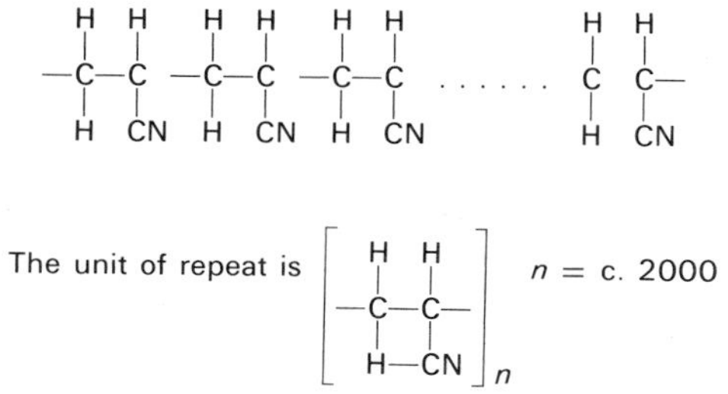

$$-\underset{\underset{\text{H}}{|}}{\overset{\overset{\text{H}}{|}}{\text{C}}}-\underset{\underset{\text{CN}}{|}}{\overset{\overset{\text{H}}{|}}{\text{C}}}-\underset{\underset{\text{H}}{|}}{\overset{\overset{\text{H}}{|}}{\text{C}}}-\underset{\underset{\text{CN}}{|}}{\overset{\overset{\text{H}}{|}}{\text{C}}}-\underset{\underset{\text{H}}{|}}{\overset{\overset{\text{H}}{|}}{\text{C}}}-\underset{\underset{\text{CN}}{|}}{\overset{\overset{\text{H}}{|}}{\text{C}}}\ldots\ldots\underset{\underset{\text{H}}{|}}{\overset{\overset{\text{H}}{|}}{\text{C}}}\;\underset{\underset{\text{CN}}{|}}{\overset{\overset{\text{H}}{|}}{\text{C}}}-$$

The unit of repeat is $\left[-\underset{\underset{\text{H}-\text{CN}}{|}}{\overset{\overset{\text{H}\quad\text{H}}{|\quad|}}{\text{C}-\text{C}}}-\right]_n$ $n = $ c. 2000

Figure 12.1 The structural formula of acrylics.

It had been known for many years that acrylonitrile would polymerize to form high polymer compounds, but the resulting fiber was characterized by insolubility and degradation at relatively low melting temperatures. Chemists finally discovered solvents that would dissolve the polymer, the most satisfactory being dimethyl formamide and dimethyl acetamide.

The dissolved polymer is extruded through spinnerettes into a heated spinning container where the filaments solidify, the solvent is recovered, and the solid filaments, while still hot, are stretched to introduce molecular orientation and fiber fineness (Fig. 12.2).

Orlon is manufactured by DuPont. Originally it was a pure polymer of acrylonitrile, but today most types of Orlon are modified by the addition of up to 14 percent of a second component. It is known that this second compound produces the property differences of the various Orlon fibers. While patent literature indicates several possible compounds, it does not specify which are used in which types of Orlon.

Acrilan acrylic, a product of Monsanto's Textiles Division (formerly Chemstrand), is composed of 85 percent acrylonitrile, plus other compounds depending upon the type of Acrilan. A study of patent literature indicates that one type is made from acrylonitrile and vinyl acetate

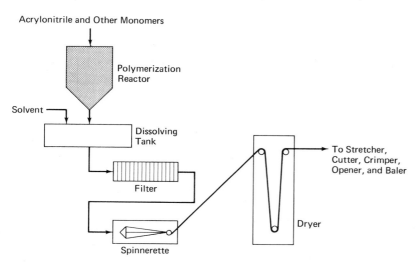

Figure 12.2 Flow diagram showing the steps in the manufacture of acrylic fibers.

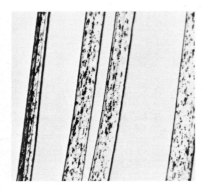

Figure 12.3 Photomicrograph of Orlon®, longitudinal view. [*E. I. DuPont de Nemours & Company*]

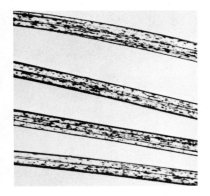

Figure 12.4 Photomicrograph of Orlon® Sayelle (bicomponent fiber), longitudinal view. [*E. I. DuPont de Nemours & Company*]

Figure 12.5 Photomicrograph of Acrilan, longitudinal view. *E. I. DuPont de Nemours & Company*]

$(CH_3—COOCH=CH_2)$, and the evidence suggests that one other variety derives from acrylonitrile and vinyl pyridine. All compounds used possess a double bond that permits polymerization by cleavage of that bond.

The manufacture of Acrilan is similar to that of Orlon, except for the probable use of dimethyl acetamide as the solvent, and it appears that the filaments are spun into a bath rather than dry spun. After coagulation, the fibers are stretched and permanently crimped.

Creslan acrylic is manufactured by the American Cyanamid Company. While actual information concerning its production is not available, it is believed that Creslan is a copolymer of acrylonitrile and acrylamide. It has been suggested that the fiber is wet spun using a frigid coagulating bath.

Although Dow Chemical Company maintains that their Zefran fiber is an acrylic or nitrile alloy and not a true acrylic fiber, the Federal Trade Commission contends that the polymer conforms to the generic definition of acrylic fibers. Zefran is composed of acrylonitrile and a second compound, which is said to be vinylpyridine. This latter fact is neither supported nor denied by the manufacturers.

FIBER PROPERTIES

Microscopic Properties

Acrylic fibers, viewed longitudinally, show uniform diameters, a rod-like appearance, and some irregularly spaced striations or parallel lines (Fig. 12.3–12.7). Cross-section microscopic views exhibit minor differences among the various kinds. Orlon possesses a dumbbell shape for most types (Fig. 12.8), while the bicomponent varieties 21 and 24 have cross sections that are mushroom- or acorn-shaped (Fig. 12.9). Fine Acrilan fibers have nearly round cross sections (Fig. 12.10), while fibers of high denier tend to be somewhat bean-shaped. The cross sections of Creslan and Zefran are nearly round (Figs. 12.11, 12.12).

In general, acrylic fibers have a grainy or pitted appearance. Those that have been delustered show the characteristic spotted effect caused by the pigment breaking up or reflecting light rays.

Physical Properties

Shape and Appearance   Like all man-made fibers, acrylics can be controlled in terms of length and diameter. In general, the fiber is marketed in staple or tow form and is used as staple fiber. Acrylics are available in bright, semidull, or dull lusters.

The first Orlons were creamy or ivory-colored, but production improvements have enabled the manufacturers to market pure white

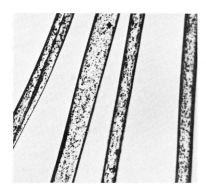

Figure 12.6 Photomicrograph of Creslan, longitudinal view. [*E. I. DuPont de Nemours & Company*]

Figure 12.7 Photomicrograph of Zefran, longitudinal view. [*E. I. DuPont de Nemours & Company*]

Figure 12.8 Photomicrograph of Orlon®, cross section. [*E. I. Dupont de Nemours & Company*]

fibers. Zefran is and always has been pure white. The natural color of the other acrylics varies from off-white to cream.

The introduction of Zefkrome marked the first appearance of a producer-dyed acrylic fiber. It is available in standard colors. Development of fiber-dyed acrylics by other manufacturers may follow in the future.

Strength   The tenacities of acrylic fiber when dry and when wet are given in Table 12.1. It will be noted that the strength of Acrilan, Orlon, and staple Creslan is similar wet or dry. Zefran is slightly stronger, as is filament Creslan.

The tenacity of acrylic fibers is not high, but it is adequate for a variety of end-uses. Filament Creslan is employed primarily in home furnishings, seldom in apparel. The higher strength makes it practical for such applications.

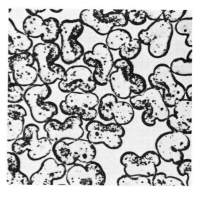

Figure 12.9 Photomicrograph of Orlon® Sayelle (bicomponent fiber), cross section. [*E. I. DuPont de Nemours & Company*]

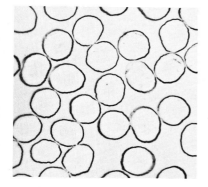

Figure 12.10 Photomicrograph of Acrilan, cross section. [*E. I. DuPont de Nemours & Company*]

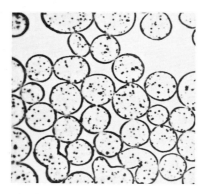

Figure 12.11   Photomicrograph of Creslan, cross section. [*E. I. DuPont de Nemours & Company*]

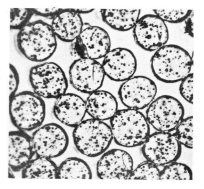

Figure 12.12 Photomicrograph of Zefran, cross section. [*E. I. DuPont de Nemours & Company*]

Table 12.1   Acrylic Fiber Properties

Fiber	Tenacity Grams per Denier		Elongation %		Elastic Recovery at 2% Elongation	Specific Gravity	% Moisture Regain*
	dry	wet	dry	wet			
Orlon	2.2–2.6	1.8–2.1	20–28	26–34	97	1.16	1.5
Acrilan	2.0–2.7	1.6–2.2	34–50	34–60	99	1.17	1.5
Zefran	3.3–4.2	2.9–3.6	30–36	30–36	99	1.18	1.5–2.5
Creslan							
staple	2.0–3.0	1.6–2.7	40–55	48–72	75–80	1.18	1.0–1.5
filament	3.8–4.2	3.0–3.8	22–25	25–30	80–85	1.18	1.0–1.5

* Standard conditions, 20°C (70°F), 65 percent relative humidity.

Elongation and Elastic Recovery   The elongation of acrylic fibers varies from 20 to 55 percent (see Table 12.1). When fibers are wet, the elongation increases for all acrylics except Zefran, which remains the same both wet and dry.

The elastic recovery of acrylic fibers is medium to medium high. At 2-percent elongation, the immediate elastic recovery of Orlon is 97 percent; Acrilan and Zefran recover 99 percent, while Creslan has a much lower recovery of 75 to 85 percent. Recovery drops rather sharply as elongation is increased. For example, at 3-percent elongation Creslan has an elastic recovery of 67 percent. Other acrylics may not be so low, but, in general, acrylics have poorer elastic recovery than many other fibers.

Bicomponent acrylic—one variety is made by DuPont and marketed under the Orlon Wintuk trademark—has good elastic recovery. Furthermore, after laundering, bicomponent acrylic can be tumbled dry and will recover its original size.

Resiliency   Acrylic fibers have good resiliency. They resist wrinkling, and undesired creases hang out rather quickly. In bulky fabrics acrylic fibers are especially resilient and lofty. These fibers retain their shape well.   *as good as wool or better*

Density   Acrylic fibers have low density: the specific gravity varies from 1.16 to 1.18 (see Table 12.1). The low density and irregular cross-section shape of some acrylics result in a fiber that has excellent covering power and exceptional bulk.

Moisture Absorption   The standard moisture regain of acrylic fibers varies from 1.0 to 2.5 percent. The saturation regain is also *low*; it increases only 1 to 2 percent over the standard regain. This contributes to difficulty in dyeing.

Dimensional Stability   With proper pretreatment (such as heat setting) and with appropriate care, acrylic fibers show little dimensional change. However, the application of excess heat and steam will cause shrinkage and a loss of loft or bulk in textured yarns. This can be extremely evident when knitted sweaters receive improper care.

Certain texturizing processes used in Orlon acrylic develop a crimp in the fiber, and in time the crimp tends to disappear. This results in the product stretching out of shape. Bicomponent acrylic resists this loss of crimp to a high degree; if stretch occurs, proper laundering and dryer drying will return the fabric to its original size and shape.

Thermal Properties

Acrylic fibers have good resistance to heat. The fibers are thermoplastic and respond to heat-setting procedures. Degradation or decomposition of the fiber occurs before true melting, although when held near a flame the fibers will shrink or appear to melt and pull away. Sticking temperatures are given as between 232° and 255°C (450–490°F). It is recommended that iron settings be kept below 325°F, for higher temperatures may cause yellowing and discoloration.

Upon exposure to fire, acrylic fibers burn with a yellow flame, and form a gummy, hot residue that drips away from the burning fiber. This residue is hot enough to ignite combustible substances upon which it may fall. When cold, the residue is hard, black, and irregular in shape.

Acrylics can be dried in dryers if the drying temperatures do not exceed 300°F. It is recommended that these fibers not be subjected to boiling water, because the combination of heat and moisture results in excessive shrinkage.

Chemical Properties

Effect of Alkalies   Acrylic fibers have good resistance to weak alkalies, but concentrated alkalies, in general, cause rapid degradation. Zefran is moderately resistant to alkalies, but it tends to yellow at high temperatures when in an alkaline environment. Though less susceptible than other acrylics, Acrilan will exhibit evidence of damage from concentrated alkaline substances.

Effect of Acids   Acrylics have good resistance to most mineral and organic acids. Weak or dilute acids appear to have no destructive effect, but strong or concentrated acids will cause a loss in fiber strength, and cold concentrated nitric acid dissolves acrylic fibers.

Effect of Organic Solvents   Solvents used in cleaning and stain removal have no effect on acrylic fibers, and bleaches can be used if directions are followed. Oxidizing agents other than bleaches used in processing fibers do not destroy acrylics.

The resistance of acrylic fibers to organic solvents actually delayed their commercial production, for it was difficult to find a practical solvent for manufacturing purposes. The chemicals now in use, such as dimethyl formamide, are not commonly encountered.

Effect of Sunlight, Age, and Miscellaneous Factors.   Acrylic fibers have excellent resistance to sunlight and all other climatic elements. Age does not appear to affect acrylics; after prolonged storage there is no noticeable change in fiber properties. Detergents and soaps have no deleterious effect.

Static electricity will build up in acrylics, and this factor is increased when the humidity is low.

Biological Properties

Mildew and bacteria do not attack acrylic fibers, and none of the common household pests, such as moths and carpet beetles, will eat or damage them.

VARIETIES AND TRADEMARKS

While the various types of Orlon have many properties in common, they may differ in physical shape, appearance, and dyeability. These differences indicate minor modifications in chemical structure.

In 1970 DuPont offered sixteen types of Orlon to yarn and fabric manufacturers; Monsanto produced eighteen kinds of Acrilan, Cyanamid made eleven varieties of Creslan, and Dow manufactured four types of acrylic fibers. These fibers varied in such factors as dye receptivity, bulkiness, texturizing characteristics, properties related to spinning, and crimping behavior. The standard Orlon fiber is Type 42, which is dyeable with basic dyes. This is frequently blended with Type 44, dyeable with acid dyestuff, to produce interesting cross-dyed effects (see Chap. 29).

Among the most important additions to the acrylic family were bicomponent fibers. Bicomponent fibers are made of two different chemical formulations extruded simultaneously, so they form a fiber with a mushroom- or acorn-shape cross section. When processed into fabrics or yarns that meet fiber-producer quality standards, Orlon bicomponents may be certified and labeled *Wintuk* or *Sayelle*.

Each component differs in properties. When dry, one component curls and gives a spiral crimp to the fiber. The other may have a higher moisture regain and may accept a deeper shade of color because of easier dye penetration. It is essential that the fibers be dried without any tension so that the spiral crimp develops properly and the yarn returns to its original size. Products bearing the trademarks Sayelle or Wintuk should be dried in an automatic dryer, where the tumbling action can restore the fabric to proper size.

A new type of Orlon, Type 43, has been designed to have a soft, luxurious hand similar to that of cashmere. This fiber is used in fine-gauge or semibulky knits and is pill resistant. In certified products it is labeled *Nomelle*.

Acrylic fibers manufactured by Monsanto are trademarked Acrilan when the end product meets quality standards. Monsanto acrylic fibers, available in fifteen types, include bright fibers, semidull fibers, basic-dye dyeable and acid-dye dyeable fibers, bicomponent fibers, fire-retardant fibers for carpets, and special fibers for indoor-outdoor carpeting. In addition to the Acrilan trademark, Monsanto also uses *Anywear, Bi-Loft,* and *Glace* as certified trade names.

American Cyanamid, with nine varieties of acrylic fibers on the market, permits these fibers to be trademarked Creslan if the end-use conforms to company performance standards. In general, the types of Creslan available are similar to fibers made by other companies. Several kinds of Creslan will accept disperse dyes as well as acid or basic dyestuffs.

Zefran acrylic fibers are manufactured by Dow-Badische. Type 200 is basic dyeable, can be bright or semidull, and is usually recommended for floor coverings; Type 100 has special dyeing characteristics, is usually bright, and is intended for apparel; Type 400 is producer-dyed (in a manner similar to solution dyeing) and may be trademarked *Zefkrome*. Most of Dow's acrylic fibers accept a variety of dyestuffs. They are stronger than other acrylics and resist higher temperatures.

In addition to the trade names used by the four fiber producers, numerous yarn and fabric manufacturers purchase acrylic fibers, process them through texturizing machines (especially *Turbo* equipment), and market the product under their own trade names. Some of these trade names include Glenspun, Pharr-mist, Shag-paca, Signette, and Templon.

Figure 12.13 A chenille knit of 85 percent Orlon acrylic and 15 percent nylon. [*E. I. DuPont de Nemours & Company*]

ACRYLIC FIBERS IN USE

Acrylic fibers respond well to handling and are considered easy-care fibers. They have good wash-and-wear properties and will accept permanent pleats and creases if they are heat-set at right angles to the yarns.

All types of acrylic fibers are used in knitted and woven fabrics. Blends of acrylic fibers with wool, cotton, other cellulosic fibers such as rayon, and fibers such as nylon (Fig. 12.13), are common. Acrylic fibers have low density, and they are soft. These properties contribute to producing fabrics that are bulky, soft, and light in weight compared to fabrics of similar construction made of natural fibers.

End-uses such as blankets, carpeting, and upholstery are excellent for acrylics because of the fiber's rapid recovery from deformation, its light weight, and the ease of maintenance (Fig. 12.14).

Acrylic fibers are found in items of apparel where shape retention and easy care are important considerations. They are popular in sportswear. Their light, bulky, soft properties make them prized in ski clothes, children's snow suits, and sport shirts.

Deep-pile fabrics frequently have acrylic fibers in their construction. Acrylics are used, also, in a number of industries.

Many acrylic fabrics can be safely washed in home laundry equipment and dried in home dryers with variable temperature controls, but fragile fabrics of acrylic should be laundered by hand as for any fine fabric. The consumer should heed labels attached to the product that give care information.

MODACRYLIC FIBERS

HISTORICAL REVIEW

Dynel, the first modacrylic fiber, was introduced to the public in 1950 by Union Carbide Corporation. This was followed in 1956 by Verel, a modacrylic fiber produced by Tennessee Eastman. Dynel resulted from attempts to develop a fiber made with vinyl chloride and a second monomer that could withstand higher temperatures than previous vinyl chloride fibers. Both of the modacrylic fibers have had comparative success in selected end-uses. Union Carbide introduced a filament modacrylic, Aeress, in the latter part of 1963 but removed it from the market soon afterward. Several other filament modacrylics are imported from Japan.

The Federal Trade Commission defines a <u>modacrylic fiber</u> as

Figure 12.14 Sportface. A nonwoven tennis surface of acrylic fiber. [*J. P. Stevens & Co., Inc.*]

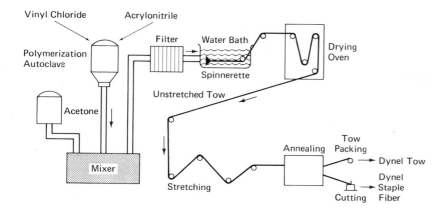

Figure 12.15   Flow diagram showing the steps in the manufacture of Dynel modacrylic fiber. [*Union Carbide Corporation*]

a manufactured fiber in which the fiber-forming substance is any long chain synthetic polymer composed of less than 85% but at least 35% by weight of acrylonitrile units

$$\left[\begin{array}{c} -CH_2-CH- \\ | \\ CN \end{array} \right]$$

except fibers qualifying under category (2) of Paragraph j (rubber).

PRODUCTION AND MOLECULAR STRUCTURE

Dynel is an addition polymer composed of two compounds: 60 percent vinyl chloride (CH_2CHCl) and 40 percent acrylonitrile (vinyl cyanide) (Fig. 12.15). Both of these compounds have double bonds in their structure; as these bonds are broken by various methods the molecular units hook together into a long chain. The two substances are polymerized in an autoclave, and a white powder is formed. The polymer powder is dissolved in acetone, filtered to remove impurities and undissolved polymer, then extruded into a water bath, where the fibers coagulate. The fibers are carried to drying ovens, and stretched while hot to impart desired characteristics and crimp. They are marketed in staple or tow form.

There is very little information available concerning the production of Verel. It has been suggested that vinylidene chloride is polymerized with acrylonitrile and a small amount of a modifier, perhaps methyl acrylamide.

FIBER PROPERTIES

Microscopic Properties

The longitudinal view of Dynel is characterized by striations (Fig. 12.16). It is clear and transparent. The cross section is irregular and

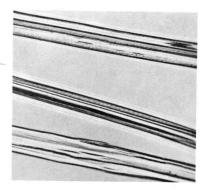

Figure 12.16   Photomicrograph of Dynel, longitudinal view. [*E. I. DuPont de Nemours & Company*]

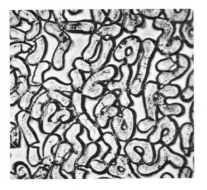

Figure 12.17 Photomicrograph of Dynel, cross section. [E. I. DuPont de Nemours & Company]

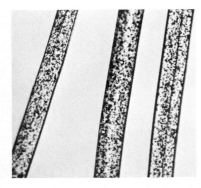

Figure 12.18 Photomicrograph of Verel, longitudinal view. [E. I. DuPont de Nemours & Company]

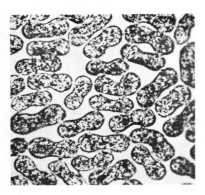

Figure 12.19 Photomicrograph of Verel, cross section. [E. I. DuPont de Nemours & Company]

somewhat flat (Fig. 12.17). The fiber may be curved into a U shape. DuPont describes the cross section as an irregular ribbon shape.

The longitudinal view of Verel exhibits faint striations and a grainy effect (Fig. 12.18). The cross section is peanut-shaped (Fig. 12.19).

Physical Properties

Shape and Appearance   Dynel modacrylic is naturally a cream color rather than a pure white, but it can be bleached white if desired. At the present time, some Dynel is being solution dyed in standard colors. The fiber is crimped during processing and is used in fabrics in staple form only.

Verel modacrylic is a white fiber. It is available smooth or with built-in crimp. Both Dynel and Verel are offered with controlled shrinkage properties for interesting fabric surfaces and textures.

Fabric and yarn manufacturers use modacrylic fibers in staple or tow form. Filament and tow modacrylic is widely used in the manufacture of wigs.

Strength   The tenacities of the two modacrylic fibers are given in Table 12.2. Verel is slightly weaker than Dynel, but both fibers have adequate strength for the end-uses to which they are normally applied.

Table 12.2   Modacrylic Fiber Properties

Property	Dynel		Verel	
	dry	wet	dry	wet
Tenacity	3.0	3.0	2.5–2.8	2.4–2.7
Elongation percent	30–42	30–42	35	35
Elastic recovery immediate recovery at 2-percent elongation	100%		79%	
Specific gravity	1.30		1.37	
Moisture regain percent standard	0.4		3.0–4.0	
saturation	1.0		3.0–4.0	

Elongation and Elastic Recovery   Both modacrylic fibers have an elongation of approximately 35 percent, and this is not altered when the fiber is wet. Dynel has an immediate elastic recovery, at 2-percent extension, of 100 percent. At 10-percent extension, it has a 95-percent recovery. Thus, it is considered to have excellent elasticity.

The elasticity of Verel is quite poor compared with that of Dynel. However, over a period of time it recovers between 90 and 95 percent.

Thermal Properties

Modacrylic fibers do not support combustion. They will burn when placed directly in a flame, but they self-extinguish as soon as the flame source is removed. The fibers do not drip while burning, and the residue that remains is hard and black.

Dynel is more sensitive to heat than Verel. It softens and shrinks at relatively low temperatures, while Verel will resist temperatures to 149°C (300°F) or higher. Recommended ironing temperatures for Verel are 300°F or less with steam. Some fabric producers recommend that Dynel not be ironed, and it usually requires very little ironing. When pressing is necessary, temperatures should not exceed 250°F, and a press cloth should be used.

Both modacrylics can be laundered in home equipment with warm water. Verel can be safely dried in home dryers if they have adjustable temperature controls, but dryer drying of Dynel is not recommended.

Chemical Properties

Effect of Alkalies   Modacrylic fibers possess good resistance to most alkalies. Prolonged exposure in some concentrated alkaline solutions results in fiber discoloration but little or no reduction of strength.

Effect of Acids   Modacrylics have good to excellent resistance to acids. Verel exhibits little or no damage from acids at all concentrations. However, hot strong mineral acids cause discoloration and some strength loss on Dynel.

Effect of Organic Solvents   Most organic solvents have no deleterious effect on either Dynel or Verel, but acetone dissolves both fibers. Elevated temperatures are required for acetone to dissolve Verel, while Dynel is affected by the solvent at both room temperatures and high temperatures.

Aniline causes Dynel to disintegrate but does not noticeably affect Verel. Some solvents used in removing paint stains stiffen Dynel and may cause shrinkage. Verel is not damaged by paint solvents except for acetone. Other organic solvents for cleaning and stain removal can be used on the fibers.

Effect of Sunlight, Age, and Miscellaneous Factors   Verel has excellent resistance to sunlight, but Dynel may discolor after prolonged exposure. In addition, the sun may cause Dynel to suffer a loss in strength.

Age appears to have no effect on either modacrylic fiber. These fibers are not damaged by bleaches, if properly used, and all types of detergents are safe.

Figure 12.20 "Fake" fur of modacrylic fibers in giraffe pattern. [*Collins & Aikman*]

Figure 12.21 Braided chignon of Dynel. [*Union Carbide Corporation*]

Both Dynel and Verel build up static electric charge. The fibers are considered nonallergenic.

Biological Properties

Modacrylic fibers are highly resistant to microorganisms and insects. Tests indicate that moth larvae will starve to death rather than eat their way through a Dynel netting to reach desirable food.

TYPES AND TRADEMARKS

Union Carbide manufactures several types of Dynel modacrylic: crimped or uncrimped, white or solution dyed, stabilized or with regular or high-shrinkage potential, and different diameters.

Verel modacrylic is available in staple form with the same basic varieties as cited for Dynel. Fibers with additional crimp are produced for such end-uses as pile fabrics, paint rollers, and bulky knitting yarns.

Modacrylic fibers for wigs and other hairpieces are marketed under such trade names as Kanekalon, Excelon, and Teklan.

MODACRYLIC FIBERS IN USE

The major end-uses for modacrylic fibers include "fake" furs (Fig. 12.20), blankets, knitted goods, wigs and hairpieces (Fig. 12.21), draperies, and industrial materials. The fibers produce fabrics that are soft, resilient, stable in size, easily laundered or dry cleaned, low in pilling, and resistant to fire damage. This last factor (as well as the self-extinguishing property) is of value in home furnishings and in deep-pile apparel fabrics, where safety is of major importance. Modacrylic fiber is widely used in blends with acrylics for carpeting. It is gaining significance in this area, particularly in contract carpeting, because of its fire resistance.

The modacrylics are frequently blended with other fibers for apparel. Pleats and creases can be heat-set in modacrylic fabrics or blends if there is sufficient modacrylic in the blend or if it is combined with other heat-sensitive fibers.

In the care of modacrylic fibers, extreme caution should be used to avoid temperatures that might damage the fiber or fabric. Fabrics can be laundered in home equipment only when there are variable temperature controls.

The solubility of modacrylics in acetone must be stressed. Since some fingernail polish removers contain this chemical, they could cause irreparable damage if spilled or mistakenly used on these fibers.

Label information regarding care should be carefully followed in order to avoid disappointment with modacrylic fiber products.

Olefin Fibers

One of the newest and fastest growing areas in fiber development is the category of olefin fibers. The TFPIA definition for _olefin_ is

> a manufactured fiber in which the fiber-forming substance is any long chain synthetic polymer composed of at least 85% by weight of ethylene, propylene, or other olefin units, except amorphous (non-crystalline) polyolefins qualifying under category (1) of Paragraph j (rubber) of Rule 7.

Such a definition gives the chemist considerable leeway, for he can manufacture fibers from any of the hydrocarbons of the _alkene group_— those hydrocarbons that have a double bond in the basic chemical component. At the present time, the fiber industry is concentrating its efforts on polymers from ethylene ($CH_2{=}CH_2$) and propylene ($CH_3CH{=}CH_2$).

Polyethylene fibers were first developed by Imperial Chemical Industries, Ltd., in England. Polypropylene fibers were the product of the research staff of Montecatini of Italy. In 1969 there were approximately fifty American companies producing olefin fibers of some type.

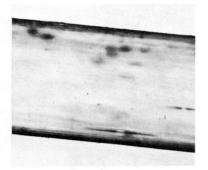

Figure 13.1 Photomicrograph of poly-ethylene, longitudinal view. [*E. I. DuPont de Nemours & Company*]

PRODUCTION

There are two general processes used in manufacturing polyethylene: the high-pressure system and the low-pressure method. The high-pressure system polymerizes the ethylene gas in autoclaves at 200°C (392°F) under pressure equal to 10 tons per square inch. The low-pressure method polymerizes the gas at temperatures between 55 and 65°C (130°–150°F) by using pressure of only 450 pounds per square inch, with a catalyst and a hydrocarbon solvent. This process is less expensive and results in polymers more suitable for textile fibers than those created by high-pressure process.

To produce fiber filaments the polymer is melt spun at temperatures of 300°C (570°F) into a current of cooling gas. The filaments are cooled and then "cold drawn" or stretched to six times the spun length. This drawing process introduces molecular orientation and makes the fibers fine and pliable.

The molecular structure of polyethylene is a linear polymer of ethylene units:

$$-CH_2CH_2CH_2CH_2CH_2 \ldots \ldots CH_2-$$

The repeat unit is $(-CH_2CH_2-)_n$

Propylene, to be a good fiber, must polymerize in a regular form. This is called an *isotactic* form and retains the methyl side chain on the same side of the backbone chain as follows:

$$\begin{array}{cccccc} -CHCH_2CHCH_2CHCH_2CHCH_2 \ldots \ldots CHCH_2- \\ | \quad\quad | \quad\quad | \quad\quad | \quad\quad\quad\quad | \\ CH_3 \quad CH_3 \quad CH_3 \quad CH_3 \quad\quad CH_3 \end{array}$$

The repeat unit is
$$\left[\begin{array}{c} -CHCH_2- \\ | \\ CH_3 \end{array} \right]_n$$

FIBER PROPERTIES

Microscopic Properties

Both polyethylene and polypropylene fibers resemble glass rods in longitudinal and cross-section views (Figs. 13.1–13.4). They are even, clear, and usually round or elliptical. Some polypropylene fibers are given various forms for special end-uses, and these are irregular in cross section.

Physical Properties

Shape and Appearance   The ethylene polymers are smooth, white, and waxy in both appearance and feel. The low-pressure type is somewhat less waxy than the high-pressure fiber.

Figure 13.2 Photomicrograph of poly-ethylene, cross section. [*E. I. DuPont de Nemours & Company*]

The propylene polymer is less waxy than polyethylene, but it, too, is smooth and white. The multifilament yarns are soft and have a pleasant hand.

Strength   The high-pressure polyethylene fibers have comparatively low tenacity of 1.5 to 3 grams per denier, wet and dry. Low-pressure fibers are stronger; their tenacity varies with the degree of polymerization from about 4 to 7 grams per denier, wet or dry.

The tenacity of polypropylene fibers also varies with the degree of polymerization from about 3.5 to 8.0 grams per denier, wet and dry. In both instances the stronger fibers have a higher degree of polymerization and molecular orientation.

Elastic Recovery and Elongation   The elongation of the low-density ethylene has a wide variation from 20 to 80 percent. This type has excellent elastic recovery: 100 percent at 2-percent elongation, 95 percent at 5-percent elongation, and 88 percent at 10-percent extension. High-density polyethylene has an elongation from 10 to 45 percent, and elastic recovery is also excellent, with a 100-percent recovery at from 1- to 10-percent elongation. Both types develop permanent deformation if stretched more than 10 percent.

The elongation of polypropylene fibers is influenced also by the degree of polymerization and molecular orientation; it varies from 15 to 50 percent. The denier of the fiber further influences the elongation, with fine fibers tending to have higher degrees of stretch. The elastic recovery of polypropylene is excellent. At up to 5-percent elongation the fibers have an elastic recovery of 100 percent; at 10-percent elongation the recovery is from 95 to 100 percent, while at 15-percent elongation the fibers recover more than 90 percent. This is one of the major reasons why this fiber was suggested for women's hosiery.

Resiliency   Both polyethylene and polypropylene have good resistance to crushing. Evidence indicates that polypropylene fibers give good service in floor coverings. However, to obtain optimum performance from polypropylene carpeting, the structure should be characterized by low, uncut pile and tightly twisted yarns.

Density   Olefin fibers are lighter than water and will float. The specific gravity of polyethylene fibers varies, depending on the type, from 0.92 to 0.96. Polypropylene fibers have a specific gravity of 0.90 to 0.91. With such a low density the fibers can be used to make lightweight fabrics.

Moisture Absorption   Both types of olefin fibers have practically no moisture absorption or standard regain. The producers of Meraklon polypropylene state that their fiber will absorb up to 0.1 percent

Figure 13.3 Photomicrograph of polypropylene, longitudinal view. [E. I. DuPont de Nemours & Company]

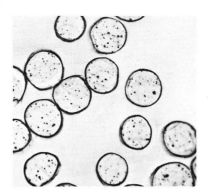

Figure 13.4 Photomicrograph of polypropylene, cross section. [E. I. DuPont de Nemours & Company]

moisture when thoroughly wet. Other manufacturers give from 0.0 to 0.1 percent as standard regain and total absorbency.

Dimensional Stability   Unless the olefin fibers are properly heat treated, they will shrink when exposed to warm temperatures. This fact has been used to advantage by some manufacturers as a means of producing interesting designs and fabrics. The fabric called *Trilok* is a good example. It is highly textured because of the shrinkage of olefin yarns and is an excellent fabric for insulation, for sound baffles, and for special protective clothing.

If olefin yarns are preshrunk and properly treated, the resulting fabrics will retain their size and shape as long as they are not subjected to temperatures greater than about 120°C (250°F).

Thermal Properties

The olefin fibers burn slowly and give off a sooty, waxy smoke. A hard tan or fawn-colored residue is formed. Probably the most serious disadvantage of olefin fibers is their heat sensitivity. They shrink at temperatures as low as 75°C (165°F). They can be preshrunk and stabilized, as stated above, but even then the fibers soften or melt at low temperatures.

Polyethylene fibers melt at 105° to 125°C (225°–255°F); polypropylene fibers will withstand temperatures up to 138°C (280°F) without deterioration and will not melt until they reach a temperature of about 170°C (335°F). The lowest temperatures possible on a group of irons tested by a leading fabric manufacturer ranged from 185° to 225°F;[1] thus the polyethylene fibers are used, generally, in products that do not require ironing. Polypropylene fabrics should be ironed at the lowest iron setting, and it is advisable to use a press cloth between the fabric and the iron.

Chemical Properties

Effect of Alkalies   Olefin fibers are highly resistant to alkaline substances.

Effect of Acids   The polyethylene fibers have excellent resistance to acids, except for strong oxidizing acids that cause a loss in strength and, after prolonged exposure, complete deterioration. Polypropylene has the same general resistance to acids as polyethylene, although some of the fibers appear to be weakened by chlorosulfonic acid.

---

[1] Burlington Industries, *Textile Fibers and Their Properties* (Greensboro, N.C.: Burlington Industries, 1970), p. 50.

Effect of Organic Solvents Chlorinated hydrocarbons cause swelling and may result in degradation of all olefin fibers. The fibers should not be dry cleaned if perchlorethylene is the cleaning solvent used; however, Stoddard solvent has no deleterious effect on the fiber. Since the olefins can be laundered easily and require little or no ironing, problems with cleaning solvents should be minor.

Effect of Sunlight, Age, and Miscellaneous Factors Both polyethylene and polypropylene lose strength and slowly degrade after prolonged exposure to sunlight. In this respect they are similar to nylon. However, age has no effect on olefins. Soaps and synthetic detergents do not appear to be harmful, and bleaches, if properly diluted, can be used.

The olefins are subject to staining and spotting by oils and greases, and polyethylenes stain more quickly than do polypropylene fibers. However, the stains usually can be removed if prompt action is taken following the staining.

These fibers are usually nonallergenic. Manufacturers claim that olefins do not pill badly. They have a high degree of cohesiveness, which results in their holding together exceptionally well in yarns. Olefins do have static electric buildup, but it is considered to be less than on nylon or polyester fibers.

Biological Properties

Olefin fibers have good resistance to microorganisms such as mildew and bacteria and to insects such as moths, beetles, and other household pests.

OLEFIN FIBERS IN USE

Polyethylene fibers are widely used in industrial fabrics, but they have not been well accepted for apparel or home furnishings. One reason for this is the fact that polyethylene fibers cannot be dyed by normal dyeing techniques, although pigments can be added to the molten polymer before extruding. However, the primary uses of the fiber do not require coloration.

Polypropylene fibers have a variety of industrial applications and, in addition, have attained widespread acceptance in home-furnishing and apparel fabrics. Several manufacturers have made fabrics for suits and dresses, and blends of polypropylene with wool, cotton, and rayon are being used in knitted fabrics for sportswear. Polypropylene has a woollike hand and feel. This property, coupled with the low density, has resulted in polypropylene coating and blanket fabrics.

Figure 13.5 Grasslike synthetic turf of poly-propylene olefin fiber. [*Ozite Corporation*]

Figure 13.6   Sno-Mat.   [*Dillon-Beck Manufacturing Corporation*]

One of the best known uses for polypropylene fibers is in the production of carpeting and carpet tiles (Fig. 13.5). The fiber has been employed in needle-punched indoor-outdoor carpeting with great success. Indoor tufted carpeting of polypropylene is also popular. The fiber is easily cleaned and resists crushing, two factors that add to its desirability for floor coverings. The introduction of the polypropylene fibers in indoor-outdoor carpeting and carpet tiles has led to a tremendous increase in the amount of carpet to be found in kitchens, work areas, patios, and even in and around swimming pools.

Pigment can be added to the polypropylene solution to color the fiber. Furthermore, manufacturers have developed techniques for modifying the polypropylene fiber so that a variety of dyestuffs can be employed to color end-use products. These techniques include the addition of small quantities of dye-receptive compounds to the fiber solution and the grafting of dye-receptive molecular units onto the filament during manufacture. The success of these methods is evident by the wide choice of colors in polypropylene yarns and fabrics.

Olefin fibers are relatively low in cost. The increase in production since 1965 has been considerably greater than was expected. Many polypropylene and polyethylene manufacturers do not trademark their fibers, partly because of the proportion of sales to other industries where trade names are not important. For the general consumer, however, the following names are most frequently seen: Herculon, Marvess, Poly-Bac, Polycrest, Polyloom, Typar, Patlon, and Vectra.

One unusual application of olefin fibers might be of interest to some readers. Specially shaped filaments locked into a durable but flexible base are used to form artificial ski slopes. Sno-Mat (Fig. 13.6) is one of the better-known names.

Olefin fibers and fabrics launder well, dry quickly, and require little or no ironing. Floor coverings are easily cleaned, and stains wipe off with a sponge or cloth and water. Detergents can be used if stains are stubborn. As mentioned previously, the olefins respond to laundering better than to dry cleaning.

Other Organic Synthesized Fibers

The fibers discussed in this chapter are not so well known or so widely used by the general public as most other textile fibers. However, occasionally they are encountered in fabrics for home furnishings and apparel, so it is important for the textile student to know something about them.

SARAN

The first saran fibers appeared on the market in 1940. The word *saran*, like nylon, subsequently became a generic term. The Federal Trade Commission defines *saran* as

> a manufactured fiber in which the fiber forming substance is any long chain synthetic polymer composed of at least 80% by weight of vinylidene chloride units ($—CH_2—CCl_2—$).

Dow Chemical introduced saran fiber, and it was spun, under license, by several other companies, including Firestone and National Plastic Products. According to recent data the only saran fiber produced in the United States in 1970 was made by Enjay Fibers.[1] Small

[1] *Textile Organon,* June 1970.

quantities are still manufactured in other countries, but the total world production of saran is relatively small. Many factories that formerly made saran have converted to olefin, probably because of economic factors. Saran is a relatively expensive fiber.

PRODUCTION

Saran is manufactured by polymerizing 80 percent or more vinylidene chloride and 20 percent or less vinyl chloride or vinyl cyanide, in the presence of heat and a catalyst. The repeat is

$$CH_2{=}CCl_2 + CH_2{=}CHCl + CH_2{=}CCl_2 \longrightarrow$$

vinylidene   vinyl   vinylidene
chloride   chloride   chloride

$$(-CH_2CCl_2CH_2CHClCH_2CCl_2-)_n \qquad n = 80+$$

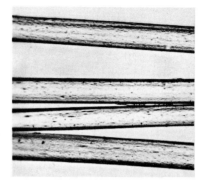

Figure 14.1 Photomicrograph of saran fiber, longitudinal view. [*E. I. DuPont de Nemours & Company*]

The polymer is melt spun into cool air and quenched in water to solidify the fibers rapidly. The filaments are cold drawn or stretched. While there are dyestuffs on the market that can be used with adequate results on saran fibers, yarns, or fabrics, the preferred method for coloring is to add pigments to the melted polymer before extrusion.

FIBER PROPERTIES

Microscopic Properties

Regular saran fibers are transparent, even, smooth, and almost perfectly round in cross section (Figs. 14.1, 14.2). For a time a saran fiber called *Rovana* was marketed. This fiber had an unusual cross section in that it resembled a flat piece of rubber that had been folded twice. A strawlike effect was obtained.

Physical Properties

Shape and Appearance   Regular saran fibers are even and somewhat silky in appearance. They are off-white with a faint yellow tint. Filament fibers are smooth and highly lustrous; staple fibers are somewhat lustrous and have a built-in crimp that becomes an inherent part of the fiber and closely simulates the natural crimp of wool in both appearance and performance. This crimp enables the fibers to cling together in yarn manufacture.

The flat cross section of Rovana saran filaments results in flat yarns that have extremely good covering power.

Strength   The tenacity of saran, wet or dry, varies from 1.4 to 2.4 grams per denier. Rovana has excellent tear resistance because of its unusual shape and form.

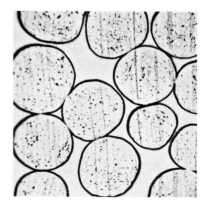

Figure 14.2 Photomicrograph of saran fiber, cross section. [*E. I. DuPont de Nemours & Company*]

Elastic Recovery and Elongation   The elongation of saran ranges from 15 to 30 percent, wet or dry, and the fiber has excellent recovery. At 1-percent elongation recovery is 100 percent; at 3 percent, recovery is 98.5 percent; and at 10 percent, recovery is 95 percent.

Resiliency   Saran fibers have good resiliency.

Density   The specific gravity of saran is 1.7. Because of this comparatively high density the fibers produce fabrics that may be uncomfortable in wearing apparel but are effective in home-furnishing fabrics.

Moisture Absorption   Saran fibers have less than 0.1-percent moisture absorption even after twenty-four hours' immersion in water. This makes dyeing difficult and explains why solution pigmenting is preferred.

Dimensional Stability   If properly processed and if high temperatures are avoided, saran fibers have excellent size and shape retention.

Thermal Properties

Saran is practically nonflammable. The fibers will melt and burn slowly if held in a flame, but they do not support combustion, and as soon as the fibers or fabrics are removed from a direct source of flame, they self-extinguish. This property makes saran fibers desirable in drapery fabrics for buildings where nonflammable materials are required by law.

A major disadvantage of saran fibers is their heat sensitivity: they soften at temperatures of 115°C (240°F) and melt at about 177°C (350°F). The low softening point indicates that ironing is not advisable. Warm water is recommended for laundering.

Chemical Properties

Effect of Alkalies   Saran fibers are resistant to most alkaline substances. Sodium hydroxide causes rapid deterioration, and ammonium compounds may cause discoloration. Other alkalies do not damage the fiber.

Effects of Acids   Acids at practically all strengths and temperatures have no deleterious effect on saran fibers.

Effect of Organic Solvents   Saran has good resistance to organic solvents at room temperatures. Substances such as acetone, carbon tetrachloride, and alcohol may cause a loss in fiber strength at temperatures of 65°C (150°F) or above, but most cleaning solvents and stain-removal agents can be used safely at room temperatures.

Effect of Sunlight, Age, and Miscellaneous Factors   Sunlight causes white or light-colored saran products to darken, but there is no effect on strength. The fibers are also impervious to age. Saran upholstery on outdoor furniture will last for several years, even in tropical climates. In one test patio chairs upholstered in saran were exposed to the weather conditions of Pennsylvania for six years and of southern California for five years. The white yarns on the chairs turned tan, but otherwise the fibers retained adequate strength. After eleven years the upholstery finally became worn.

Saran fibers are usually tough and durable. Thanks to the low moisture regain, stains wipe off easily and quickly. Soaps and synthetic detergents do not damage the fibers.

Saran does develop static electric charges.

Biological Properties

Saran fibers are immune to attack by moths, other household insects, mildew, and bacteria.

SARAN IN USE

While saran fibers are rather dense and therefore undesirable for apparel, they are prized for furnishing fabrics such as upholstery, draperies, and outdoor furniture (Fig. 14.3). Saran is often employed for automobile upholstery and for similar purposes in commercial vehicles. The staple fiber is preferred for draperies, while filament fiber is more satisfactory for furniture, since it is cleaned easily and does not provide interstices between fibers for dirt and stains to settle. However, the smooth surface of the fiber enables soil to be removed easily even from staple yarns. The fibers can be cleaned with detergents and lukewarm water.

Other uses for saran fibers include wigs for manikins and dolls, luggage coverings, window and patio screening, dust mops, and a wide variety of industrial fabrics.

In recent years olefin fibers have replaced saran in many end-uses, because the cost of producing saran makes it somewhat impractical.

VINYON

The first vinyon fibers were made experimentally in 1933 by the Carbide and Carbon Corporation, but commercial quantities were not available until 1939. At that time the American Viscose Corporation began to convert the polymer made by Carbide and Carbon into

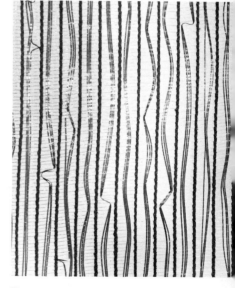

Figure 14.3 Knit casement with extruded saran yarn warp. [*Jack Lenor Larsen, Inc.*]

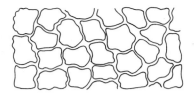

Figure 14.4   Sketch of the microscopic cross section of vinyon.

filament fibers. This fiber, a true synthetic, was introduced the same year as nylon.

Since that time, improvements have been made in the original fiber, and manufacturers in several countries are producing vinyon. The trade names usually encountered are Avisco Vinyon HH, Clevyl, Fibravyl, PCU, Rhovyl, and Voplex.

Vinyon is defined by the Federal Trade Commission as

a manufactured fiber in which the fiber forming substance is any long chain synthetic polymer composed of at least 85% by weight of vinyl chloride units ($-CH_2-CHCl-$).

Conferences of fiber manufacturers from both hemispheres have considered an international labeling program. The suggested name for vinyon fibers is Chlorofibre.

PRODUCTION

Vinyon fibers are either polymers of vinyl chloride or copolymers of vinyl chloride and a second vinyl compound, usually vinyl acetate. These chemicals are polymerized under pressure or by means of catalysts; the process is addition polymerization. The polymer is dissolved in an appropriate solvent and spun into a coagulating medium— water, warm air, or other acceptable environment.

FIBER PROPERTIES

Vinyon is a long chain of vinyl chloride units ($-CH_2CHCl-$). If the polymer is composed of two compounds, the second one occurs periodically in the linear chain.

The fiber has round, dog-bone, or dumbbell-shape cross sections (Fig. 14.4) and smooth, even, relatively clear longitudinal views. Some types of vinyon show faint striations.

Vinyon is white and somewhat translucent. The strength varies from a low of 0.7 gram per denier to 3.8 grams per denier, wet and dry. This range of tenacity is caused by differences in molecular arrangement and polymerization processes.

There is a tremendous spectrum of elongation, from a low of 12 percent for strong fibers to a high of 125 percent for weak fibers. The fibers have specific gravities of 1.34 to 1.43. They absorb very little, if any, moisture.

An undesirable property of vinyon fibers is their extreme heat sensitivity. They cannot be ironed and will soften at temperatures above 65°C (150°F). Vinyon does not support combustion but will burn in a direct flame and melt easily. A newer type of vinyon fiber that is spun from syndiotactic polyvinyl chloride has recently been introduced. This fiber has increased heat resistance.

Vinyon fibers have excellent chemical resistance. Acids and alkalies have almost no effect, and solvents used in cleaning do no damage (except for acetone and other ketones, aromatic hydrocarbons, and ether). Soaps and synthetic detergents have no undesirable effect on the fiber.

VINYON FIBERS IN USE

Vinyon fibers are widely used in industry. Because of their low softening and melting temperatures, they seldom occur in fabrics for apparel or furnishings that require pressing.

Recently the fiber has been adapted to women's handbags and hats. Manufacturers of nonwoven fabrics occasionally use vinyon as the bonding agent.

VINAL

The Federal Trade Commission defines *vinal* as

> a manufactured fiber in which the fiber forming substance is any long chain synthetic polymer composed of at least 50 percent by weight of vinyl alcohol units ($-CH_2-CHOH-$), and in which the total of the vinyl alcohol units and any one or more of the various acetal units is at least 85 percent by weight of the fiber.

Vinal fibers are manufactured in Japan. Keowee Mills in Easley, South Carolina, is licensed to produce and market the fiber in the United States, where it is sold either under the trade name Kuralon or under the generic term *vinal.*

The polymer from which the fiber is spun, called *Poval,* is made from vinyl alcohol. Vinyl alcohol is an unstable material, so a polyvinyl alcohol is made indirectly by the hydrolysis of polyvinyl acetate. This polyvinyl alcohol is solution spun to form the fibers.

As extruded, the fibers are water soluble and must be treated with formaldehyde to make them insoluble. The treatment develops cross-links in the fibers, which increases stability.

Under magnification vinal fibers are smooth, somewhat grainy, and characterized by faint striations. The cross section may be bean-shaped, U-shaped, or nearly round. A somewhat flat U is the most common (Fig. 14.5). The fibers are white, but can be pigment dyed in solution.

Vinal fibers vary in tenacity from 3.5 to 6.5 grams per denier; in elongation they range from 15 to 30 percent. The stronger fibers have low elongation and a high degree of molecular orientation. Vinal fibers are about 25 percent weaker when wet than when dry. The specific gravity of vinal is about 1.26. The fiber has a standard moisture regain of 5 percent and when saturated will absorb up to 12 percent moisture.

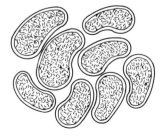

Figure 14.5 Sketch of the microscopic cross section of vinal.

Vinal does not support combustion but melts at about 220°C (425°F). It softens at 200°C (390°F). While this is somewhat lower than the softening point of some fibers, it is higher than that of many other vinyls and adequate for a number of end-use requirements.

Vinal has good chemical resistance. Alkalies have little or no effect, and commonly encountered solvents do very little damage. Weak acids do not harm the fiber, but concentrated acids cause disintegration. Formic acid, phenol, cresol, and hydrogen peroxide dissolve vinal fibers. Chlorine bleaches, if required, can be used safely. Vinal has a high tolerance of sea water.

The fiber has excellent resistance to microorganisms and insects, especially rot-producing bacteria.

Except for limited industrial applications, the fiber has not been used in the United States in any sizable quantity. In Japan and some other countries it is employed in protective apparel—raincoats, jackets, hats, umbrellas, suiting fabrics, lining fabrics, socks, and gloves. It is combined with cotton or rayon in blends that are said to be very attractive and silky. Industrial uses include fishing nets, filter fabrics, tire cord, tarpaulins, and bristles.

Although the fiber is usually produced so that it is insoluble in water, limited amounts of the water-soluble product are manufactured for special fabrics, in which the soluble vinal is dissolved, leaving unusual designs. In addition, the soluble fiber has certain surgical applications.

NYTRIL

Nytril is defined by the Federal Trade Commission as

> a manufactured fiber containing at least 85 percent of a long chain polymer of vinylidene dinitrile ($-CH_2-C(CN)_2-$) where the vinylidene dinitrile content is no less than every other unit in the polymer chain.

The B. F. Goodrich Chemical Company developed nytril fibers and introduced the first commercial product in 1955. They gave the fiber the name Darlan, which was soon changed to *Darvan*. In 1960 Celanese Fibers Company purchased the rights to manufacture Darvan, and in 1961 Celanese joined with a German firm, Farbewerke Hoechst, in an agreement to manufacture the fiber in Europe. The same fiber marketed abroad is called *Travis*. For unexplained reasons Celanese discontinued production of Darvan in the United States in late 1961. Despite the fact that the fiber is not currently available, it has many interesting characteristics that should be noted. Furthermore, it may return to the American market as long as the company continues producing the fiber in Europe. Darvan (or Travis) is the only nytril fiber that has been developed.

This nytril fiber is a copolymer of vinylidene dinitrile and vinyl acetate. The unit of repeat in the polymer is

$$\left[\begin{array}{cccc} H & CN & H & H \\ | & | & | & | \\ -C & -C & -C & -C- \\ | & | & | & | \\ H & CN & H & O \\ & & & | \\ & & & CO \\ & & & | \\ & & & CH_3 \end{array} \right]_n$$

Figure 14.6 Photomicrograph of nytril fiber, longitudinal view. [*E. I. DuPont de Nemours & Company*]

The two compounds are copolymerized, and the polymer precipitates as a powder or small flake. This powder or flake is dissolved in dimethyl formamide and extruded through spinnerettes into a water bath. Then the filaments are hot stretched, crimp is inserted, and the fibers are cut into staple lengths.

Darvan is even in width and has distinct striations in longitudinal microscopic view (Fig. 14.6). The cross section is somewhat ribbon-shaped with a shallow curve (Fig. 14.7).

The producer states that nytril fiber has all the easy-care virtues of other fine man-made fibers, the beauty of the best natural fibers, plus no pilling. An actual examination of fiber properties may raise questions about such a statement, but the fiber does have some strong qualities.

Fiber tenacity is 1.75 grams per denier when dry, and 1.5 when wet; elongation is 30 percent, wet and dry; elastic recovery is only 70 percent at 3-percent extension, moisture regain is 2.2 percent; and specific gravity is 1.18. Darvan is resilient and, if properly treated, retains its size and shape. It has good wrinkle recovery and press retention. The fiber presents problems with static buildup and usually requires a static-reducing finish.

Nytril fibers soften at about 175°C (340–350°F), and recommended safe ironing temperatures are 300° to 325°F. They burn in a manner very similar to that of acrylic fibers.

Darvan has good resistance to dilute acids, but strong acids cause a loss of strength. Dilute alkalies at low or room temperatures have no damaging effects on the fiber. However, warm dilute or strong alkalies result in a severe loss of strength, and prolonged exposure causes degradation. The fiber has excellent resistance to organic solvents and agents used in stain removal.

Nytril fiber is not affected by sunlight, weather, or age. It is easily laundered with various detergents. Microorganisms and insects do no damage.

Darvan is one of the softest fibers made, and it is excellent in knitting yarns and pile fabrics. It produces choice fabrics with good wash-and-wear properties when blended with wool, rayon, cotton, or

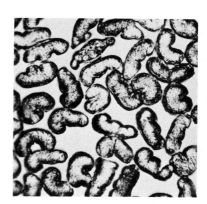

Figure 14.7 Photomicrograph of nytril fiber, cross section. [*E. I. DuPont de Nemours & Company*]

synthesized fibers such as nylon. Its excellent resistance to weather, coupled with its softness, make it highly desirable for coatings and suit fabrics. The return of Darvan to the American market would be highly desirable.

TEFLON

The consumer is acquainted with Teflon as a coating for cooking utensils or in plastic forms, but the product also has been made into fibers and presents several attractive properties for selected end-uses. Teflon was not available to the general public in fiber form when the labeling legislation was enacted, so no provision was made for a generic name or inclusion in the TFPIA.

Teflon is made of tetrafluoroethylene ($CF_2{=}CF_2$), which polymerizes under pressure and heat in the presence of a catalyst. Because of the inertness and high melting point, the fiber is formed by carefully controlling the polymerization process so that the polymer itself is ribbonlike in shape, with a high length-to-width ratio. The particles can thus be used as fiber without further extrusion.

Teflon is of medium strength (1.6 grams per denier), with low elongation and good pliability. It is more dense than fibers used in apparel: the specific gravity is 2.3. It has no moisture regain. Teflon fibers will not burn; they withstand temperatures to 260°C (500°F). They are chemically inert and unaffected by acids, alkalies, or organic solvents. Microorganisms and insects have no destructive effect.

Teflon fibers are more appropriate to home-furnishing materials and industrial fabrics, rather than apparel, because of their high density. As yet, the fiber is difficult to dye and this further limits its end-use potential. Continued research will undoubtedly result in a successful fiber for certain products. Teflon's extreme resistance to chemicals, sunlight, weather, and age will create many desirable uses if the color problem can be solved.

STYRENE

Polystyrene is primarily used for industrial fabrics, but manufacturers have predicted eventual adaptation of the fiber to home-furnishing fabrics. At the present time, it is used for bristles in brushes sold on the consumer market. As with Teflon, the fiber is not included in the labeling legislation. It is a linear polymer of styrene ($CH_2{=}CH{-}C_6H_5$).

Polystyrene has low density (specific gravity 1.08) and relatively poor elongation and elastic recovery properties. It is softened by temperatures above 90°C (200°F) and shrinks when exposed to temperatures greater than 88°C (190°F). It burns and gives off a heavy black sooty smoke. The fiber has excellent resiliency. Its chemical resistance is variable: excellent resistance to alkalies and weak acids;

medium resistance to most organic solvents; poor resistance to strong acids; no resistance to ketones such as acetone, aromatic hydrocarbons such as benzene and toluene, or chlorinated hydrocarbons such as carbon tetrachloride.

Dyes are added to the solution before the fibers are extruded. When compared with other products on the market, the future of polystyrene fibers for general use would appear questionable because of their low softening temperatures.

Elastomeric
Fibers

Elastomers are elastic, rubberlike substances. They can be prepared in various forms, but discussion here is limited to fibrous forms used in textile products. All elastomers are characterized by extremely high elongation (usually between 500 and 800 percent) and excellent elastic recovery.

RUBBER

The Federal Trade Commission defines *rubber* as

a manufactured fiber in which the fiber-forming substance is comprised of natural or synthetic rubber, including the following categories:

1. a manufactured fiber in which the fiber-forming substance is a hydrocarbon such as natural rubber, polyisoprene, polybutadiene, copolymers of dienes and hydrocarbons, or amorphous (non-crystalline) polyolefins.

2. a manufactured fiber in which the fiber-forming substance is a co-polymer of acrylonitrile and a diene (such as butadiene) composed of not more than 50 percent but at least 10 percent by weight of acrylonitrile units

$$\left[-CH_2-\overset{\displaystyle |}{\underset{\displaystyle CN}{CH}}- \right]$$

The term *lastrile* may be used as a generic description for fibers falling within this category.

3. a manufactured fiber in which the fiber-forming substance is a polychloroprene or a copolymer of chloroprene in which at least 35 percent by weight of the fiber-forming substance is composed of chloroprene units

$$\left[-CH_2-\underset{\underset{Cl}{|}}{C}=CH-CH_2- \right]$$

The thick gummy liquid obtained from trees of the Hevea species has been used for many hundreds of years. However, not until the nineteenth century did scientists become aware of the unusual characteristics of this substance. In 1839 Charles Goodyear discovered that the properties of rubber were greatly changed when it was heated with sulfur. Strength and elasticity were increased, and cold temperatures no longer hardened it or made it brittle. Goodyear's process is now known as *vulcanizing;* it set the scene for the development of rubber in many forms. Consumption of rubber remained small until the beginning of the twentieth century, when the growing automobile industry began to require large quantities for tires.

Rubber in fiber form originated in the 1920s as a result of research by U. S. Rubber Company. Scientists discovered that liquid rubber (latex) could be extruded in round forms of minute fineness, which had high elongation and elastic recovery. These early fibers were not used alone but served as a central core for other fibers, such as cotton, which were wrapped around them. To some extent this is still true today.

Synthetic rubbers were first developed in the early 1930s, but it was not until the natural rubber supply was cut off by World War II that synthetic rubber gained consumer acceptance. Even then it was used chiefly by the Government until after 1946. Today, both natural and synthetic rubber have good markets and are used for many products.

The properties that make rubber desirable in selected end-uses include

- a high degree of elasticity
- flexibility and pliability
- strength
- toughness
- impermeability to water and air
- resistance to cutting and tearing
- resistance to many chemicals
- a specific gravity of about 1.0

The properties that introduce problems in end-use are

- sensitivity to temperatures greater than 93°C (200°F), which cause deterioration and loss of pliability

- deterioration by sunlight
- loss of strength and elasticity through aging
- damage from body oils
- damage caused by petroleum solvents

In general, synthetic rubbers have fewer of these drawbacks than does natural rubber.

To prepare latex for fiber, the manufacturer mixes it with certain ingredients (classified information) to reduce air and light deterioration; it is then vulcanized by a special process. The fine rubber filaments (or latex) are covered with some other textile fiber, such as cotton, rayon, or nylon. The technique used in covering the rubber core controls the degree of elongation and elastic recovery of the final yarn. The fiber used in covering the rubber is important, for it influences the comfort, absorbency, hand, and appearance of the end-product.

Rubber yarns contribute support and improved fit to end-use products. Fabrics with rubber are comparatively crease resistant and require a minimum of ironing.

Rubber can be laundered if the water temperature does not exceed 140°F. Built soaps or synthetic detergents can be used, and they remove oily dirt better than unbuilt detergents. Drying in a dryer at medium temperatures is considered safe by some authorities, but others maintain that air drying is the only acceptable method. Probably the most important fact about cleaning rubber items is that they should be laundered after each wearing to reduce damage from body oils. It is best to avoid dry cleaning.

Rubber yarns are used in foundation garments, swimwear, surgical fabrics (such as elastic bandages and support hosiery), underwear for both sexes, elastic yarns for decorative stitching, shoe fabrics, tops of men's and children's socks, and elastic tape. Various forms of natural and synthetic rubber are used in textile finishing (see Part V).

Trade names frequently encountered for rubber fibers or yarns include Contro, Globe, Lactron, Lastex, and Laton.

Foam latex, while not a textile, requires the same careful handling as fibers. It is used for upholstery filler, mattresses, pillow forms, and similar products. Polyurethane (considered by some authorities to be a synthetic rubber[1]) has replaced latex foam in many end-uses. It is less subject to deterioration by factors that damage other rubbers, such as light, dry cleaning solvents, and age. Polyurethane foams are manufactured from the same chemicals as spandex fiber.

During the 1960s one chemical company sought to establish a new generic term for an elastomeric fiber. As a result, the FTC re-

---

[1]Maurice Morton, *Introduction to Rubber Technology* (New York: Reinhold Publishing Corporation, 1959), p. 31.

defined rubber and specified the name *lastrile* as a substitute generic term for one special group of rubber fibers. To date there has been no commercial development or production of lastrile fibers. Consequently, there is no published literature that describes properties, production, and characteristics of such a fiber.

SPANDEX

The Federal Trade Commission defines *spandex* as

> a manufactured fiber in which the fiber forming substance is a long chain synthetic polymer comprised of at least 85 percent of a segmented polyurethane.

The first spandex was introduced by the DuPont Corporation as Fiber K in 1958. For several months the fiber was evaluated by the trade for commercial end-uses. Plans for volume production and a new trade name, Lycra, were announced to the public in late 1959.

United States Rubber patented a spandex elastomer in 1956, but commercial quantities were not produced until about 1961. The amount of this fiber, trademarked Vyrene, is still limited compared to Lycra.

During the 1960s several new spandex fibers were released in limited quantities. These products can be recognized by such trade names as Duraspan, Estane V. C., Glospan, Interspan, Numa, Unel, and Fulflex.

Various textile authorities predicted, during the mid-1960s, the ultimate inclusion of spandex fibers in a large percentage of all fabrics manufactured. At that time many news items indicated that all fabrics would be endowed with stretch properties, either from spandex yarns or from other processes used to create fabric stretch. To date this has not occurred. The current interest in knit fabrics has, at least temporarily, overshadowed fibers with inherent stretch properties.

PRODUCTION

Spandex fibers are composed of segmented polyurethane. The structure is achieved by a chemical reaction between a diisocyanate and a second monomer with the formation of a block polymer, in which long chains of a flexible structure are joined to short chains of a stiff structure through urethane linkages. These linkages are stable to acids and alkalies, which permit dyeing, finishing, and laundering. The block-type polymer provides both elasticity and stability to fabrics.

Figure 15.1 Photomicrograph of Lycra®, longitudinal view. [*E. I. DuPont de Nemours & Company*]

The chemical reaction can be represented as follows (with R equal to the particular radical in the compound)[2]:

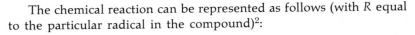

$$(HO-C_2H_4-OH)_n + C{=}C{=}N-R-N{=}C{=}O \longrightarrow$$

ethylene glycol   diisocyanate

$$(-OC_2H_4-O-\overset{O}{\overset{\|}{C}}-\overset{H}{\overset{|}{N}}-R-\overset{H}{\overset{|}{N}}-\overset{O}{\overset{\|}{C}}-)_n$$

polyurethane

Actual composition is variable.

Vyrene is a polymer of a polyester and a diisocyanate. The polyester used is characterized by terminal hydroxyl (OH) groups.[3] It is polymerized with diphenyl-methane diisocyanate. The polymer is further processed to develop cross-links between molecules, which produce stability.

There is little technical information available regarding chemical compounds used in producing other spandex fibers on the market. However, it may be safely conjectured that a diisocyanate is combined with a substance containing terminal hydroxyl groups. The base compounds may vary widely.

FIBER PROPERTIES

Microscopic Appearance

Spandex fibers vary in microscopic appearance, but there are some similarities (Fig. 15.1). As for all man-made fibers, the longitudinal view shows fibers of even diameter. The cross sections exhibit differences in the shape of the filaments, but the way they form into yarns is similar for certain spandex fibers (Figs. 15.1–15.4). Glospan, Lycra, Numa, and Spanzelle (an English product) are formed into multifilament fiber yarns.

Lycra filaments are smooth with slightly dog-bone-shaped contours that fuse together at random intervals. Another spandex fiber has serrated surfaces, which are fused at various points to form scaled or coalesced multifilament yarns. This fusion of filaments serves a twofold purpose: It prevents filaments from slipping in the yarn, and, at the same time, it contributes flexibility and a good hand.

Spanzelle filaments are nearly round. Information is not available about whether these filaments are fused together. Microscopic views

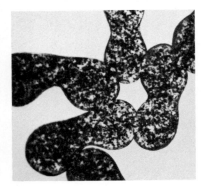

Figure 15.2 Photomicrograph of Lycra®, cross section. [*E. I. DuPont de Nemours & Company*]

[2] William Kirk, Jr., "Lycra Spandex Fiber—Structure and Properties," *American Dyestuff Reporter* (September 16, 1963), p. 59 (P725).

[3] L. M. Boulware, "Vyrene—New Synthetic Elastic Fiber," *Modern Textile Magazine*, Vol. 41 (August 1960), p. 46.

of several other spandex fibers are similar to Lycra or Numa. Vyrene spandex is said to be a monofilament fiber; this may be one of the reasons for the more limited use of this fiber.

Physical Properties

The average tenacity of the spandex fibers is about 0.7 gram per denier. The fibers have an elongation of 500 to 600 percent. Elastic recovery is excellent, despite the fact that spandex is slightly lower in recovery than rubber fibers.

The fibers exhibit minor differences in specific gravity, ranging from 1.21 to 1.25. The moisture regain varies from 0.3 to 1.2 for different fibers. This variation may be the result of the difference in fiber formation; that is, the multifilament coalesced fibers absorb more than the monofilament form.

Spandex fibers are available in fine to heavy denier or diameter. They are normally white in color, but they accept selected dyestuffs easily and evenly.

Figure 15.3 Photomicrograph of Vyrene, longitudinal view. [*E. I. DuPont de Nemours & Company*]

Thermal Properties

Spandex fibers will burn and form a gummy residue. They can be ironed safely at temperatures below 300°F and can be dryer dried without damage. Lycra and Vyrene begin to stick at about 175°C (340°-350°F). Numa spandex does not stick until the temperature approaches 280°C (534°F). Spanzelle is said to be more resistant to heat than Lycra and Vyrene, but data indicate that it sticks at about 180°C (355°F), which is not significantly better than the other products.

Chemical Properties

Spandex fibers have good resistance to most chemicals. Concentrated alkalies at elevated temperatures cause loss of strength and eventual degradation. Strong hypochlorite bleaches may affect spandex by causing a yellowing and a loss of strength. Other chlorine compounds, such as perchlorethylene and hydrochloric acid, may produce some change in color and loss of strength in certain spandex products.

SPANDEX FIBERS COMPARED WITH RUBBER

Major advantages of spandex over rubber include resistance to degradation by ultraviolet light (sunlight); weathering, especially smog; oils, particularly body oils; perspiration; and detergents, natural or synthetic, built or unbuilt. Spandex has a lower density and a higher modulus of elasticity than rubber and results in garments that are lighter in weight but still provide the same degree of figure control.

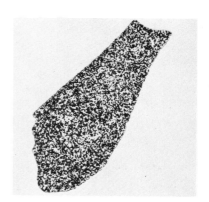

Figure 15.4 Photomicrograph of Vyrene, cross section. [*E. I. DuPont de Nemours & Company*]

Spandex fibers are superior to rubber in their flex-life and dye-ability, as well as in their resistance to oxidation, to abrasion, to damage by dry-cleaning solvents, to bleaching by chlorine and other bleaching compounds, and to damage in machine laundering and drying.

SPANDEX FIBERS IN USE

Spandex fibers can be laundered by hand or machine and dried in the air or in a dryer at medium or low temperatures. Any type of good detergent can be used. Some manufacturers of spandex maintain that chlorine bleaches are safe for the fiber, but DuPont recommends perborate bleaches for their Lycra spandex. All spandex producers state that normal concentrations of chlorine in swimming pools will cause no damage to their fiber. However, this does not guarantee that dye-stuffs will retain their color or that white products will remain white. In addition to discoloration from chlorine compounds, white spandex may yellow as a result of acid fumes, smog, body oils, and perspiration. However, this does not affect the stretch properties to a noticeable degree. Frequent laundering will reduce the discoloration. Colored fibers show very little, if any, color alteration.

Spandex is utilized in the bare filament or uncovered form; in yarn constructions where spandex yarn is wrapped spirally with yarns of other fibers to produce a covered or core yarn; and in core-spun yarns. The latter is of major importance in comfort stretch fabrics. These core-spun yarns are manufactured by feeding staple fibers around the core filament to complete the yarn formation. The end product is a single yarn composed of a filament center and a staple fiber outer layer (Fig. 15.5). The amount of stretch can be predetermined and controlled by yarn spinning machine adjustments.

Fabrics in which stretch and flex are required are ideal for spandex fibers. They provide a high flex-life and can be used in varying amounts to provide comfort or power stretch (see Chap. 28). Spandex combines high holding power with light weight.

End-use items in which spandex fibers can be seen include foundation garments, bras, straps for lingerie, sock tops, hosiery, and medical products requiring elasticity. In addition, stretch fabrics using spandex fibers appear in many items of wearing apparel, some home-furnishing fabrics such as slipcover materials, and some domestic fabrics, particularly fitted sheets. The importance of stretch, including spandex stretch, is discussed in Chap. 28.

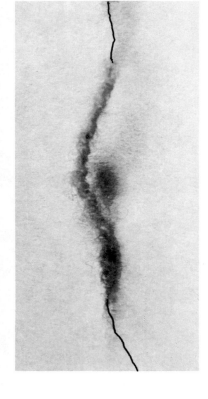

Figure 15.5   Core-spun yarn, showing the staple covering fibers pulled away at each end to reveal the core.

ANIDEX

The FTC added the generic term *anidex* to the TFPIA in 1969. It is defined as

> a manufactured fiber in which the fiber forming substance is any long chain synthetic polymer composed of at least 50% by weight of one or more esters of a monohydric alcohol and acrylic acid.

The one anidex fiber in production in 1970 was ANIM/8. The manufacturer, Rohm and Haas, states that ANIM/8 has the following properties:

1. Its recovery from stretching is superior to any previous elastomeric fiber.
2. It can be bleached successfully with chlorine.
3. It blends well with all natural and man-made fibers.
4. It is equally adaptable to weaving and knitting.
5. It can be used corespun, covered, plied, or bare.
6. It can be dyed, printed, and finished by traditional processes.
7. It takes permanent-press and soil-release finishes.
8. It retains its elasticity after washing and dry cleaning.
9. It has excellent affinity for disperse dyes.
10. It is described as an acrylate.

16

Mineral and Miscellaneous Fibers

NATURAL MINERAL FIBER

ASBESTOS

Asbestos is the only mineral matter used as a textile fiber in the form in which it is obtained from natural sources. The substance is a fibrous vein in serpentine or amphibole rock (Fig. 16.1). It has been known since the days of early Greece and Rome. In fact, the word *asbestos* is of Greek derivation.

The use of asbestos was recorded by Pliny the Elder in the first century A.D. Legends concerning this amazing fiber have been told for centuries. It is said that Emperor Charlemagne delighted in mystifying guests by throwing a tablecloth of asbestos into a roaring fire and then removing it, unharmed and clean, from the flames. A few centuries later, Marco Polo told his friends in Italy about a substance he observed in Siberia that could be woven into attractive textiles that would not burn, even in direct flame. These stories all point out the salient property of asbestos: It is completely resistant to fire.

Early uses for asbestos included such unexpected items as handkerchiefs and cremation fabrics, as well as wicks for oil lamps and the aforementioned table coverings.

Commercial development of the fiber began in the nineteenth century after the discovery of large deposits of asbestos in Canada and South Africa. There are many varieties in the asbestiform group of minerals, but only six are of economic importance. They are chrysotile, crocidolite, amosite, anthphyllite, termolite, and actinolite. Of these, the first is most often used in textile-fiber form (Fig. 16.2). According to a 1959 survey by the United States Department of Interior, chrysotile accounted for 95 percent of all textile asbestos.[1]

Chrysotile is a fibrous form of serpentine rock, which is hydrated magnesium silicate [$Mg_3Si_2O_5(OH)_4$]. The other forms of asbestos, the amphiboles, belong to the hornblende group of rocks. These are complex silicates of iron and other metals such as sodium, aluminum, or magnesium. Since chrysotile is the important asbestos used in textile production, it is the one discussed in detail.

Several properties combine to make this type of asbestos especially adaptable to textiles. The fibers have good strength, flexibility, toughness, low conductivity, and adequate length for spinning into yarns.

Tenacity of chrysotile varies from 2.5 to 3.1 grams per denier. The fiber has a silky texture; it is soft to harsh in hand, and white, amber, gray, or green in color. Asbestos of this type withstands extremely high temperatures. The water bound within the molecular structure of the fiber is lost at 593°C (1100°F), and the fiber will fuse at 1520°C (2770°F) if held at that temperature. Brief exposures to temperatures as high as 3315°C (6000°F) without fiber destruction have been recorded. Chemical resistance of asbestos is good and permits the use of the fiber in various chemical manufacturing processes.

Chrysotile is mined in the following locations, listed in order of importance:

- Canada—provinces of Quebec and Ontario
- Africa—Southern Rhodesia and Swaziland
- United States—Arizona and California
- Italy
- Turkey, Venezuela, and Colombia
- Russia

Crocidolite, a hornblend amphibole, is used in textile products to some extent and in fiber form for various purposes. This asbestos is a complex silicate of iron and sodium. It is more difficult to spin than chrysotile and has a lower heat resistance; however, it is stronger than the serpentine fiber and has better resistance to acids. It is sometimes called *blue asbestos* because of its natural blue color.

Asbestos is mined by either open-pit or underground methods. It is separated from other rock layers by crushing and screening. After

[1]United States Department of Interior, Bureau of Mines, *Asbestos, Materials Survey* (Washington, D.C.: Government Printing Office, 1959).

Figure 16.1   The rock formation from which asbestos fiber is obtained.

Figure 16.2   Asbestos fiber.

Figure 16.3 Fire proximity coverall, hood, and gloves of aluminized asbestos cotton, shown in the "Body Covering" exhibition, Museum of Contemporary Crafts, New York. [*U.S. Navy Clothing and Textile Research Unit, Bob Hanson, and the American Crafts Council*]

final separation from waste rock, the fibers are cleaned, graded by length, and prepared for shipment to textile mills, where they are made into desired textile forms.

The price of asbestos is determined by quality, which, in turn, is based on fiber length. Because of the labor and processing involved, high-quality asbestos is a comparatively expensive fiber.

When asbestos arrives at the mill, the fibers are further cleaned, and those too short for spinning are removed and assigned to nontextile uses. Usable fibers are frequently blended with 5 to 20 percent cotton or rayon to produce adequate yarn and fabric strength. The fibers are carded and processed into yarns in a manner similar to that employed for wool or cotton fibers and woven into fabrics by the same techniques used for other textiles.

Asbestos is employed commercially in different forms obtained at various stages of manufacture and processing. Carded asbestos fibers are used for filtering media, where clarification of liquids is important, and sometimes for insulation. Asbestos roving is important in the electrical industry as the insulative covering around heater cords and heating elements.

Yarns of asbestos are the basis for fabrics of various constructions. The yarns themselves may be plain or have a wire core or a core of some other fiber, such as glass or nylon. Yarns may be singles, plies, or cord structures.

The asbestos cords have many industrial applications, especially in the electrical industry. Other yarns treated to produce exceptionally smooth and uniform surfaces serve as thread in the construction of asbestos curtains (frequently used in theaters and other public buildings), in protective clothing, and in industrial applications.

Fabrics of asbestos are widely used in industry. Protective clothing for firefighters (Fig. 16.3) and workers in industrial plants and fire-smothering blankets used in fire-fighting are often made of asbestos.

In addition to the multitude of industrial and commercial applications, asbestos is the basis for ironing board covers; protective mits for barbecuing or handling hot dishes; and protective pads for tables, stoves, and counter tops.

A new use of asbestos was publicized in the late spring of 1971. The material was incorporated into blends for the manufacture of coating fabrics. The particular fabric involved was composed of 70 percent wool, 22 percent nylon, and 8 percent asbestos. No data was provided concerning the reasons for such a blend, but it attracted widespread attention because the medical profession believed it constituted a serious health hazard. Small pieces of the asbestos fiber broke off easily and separated from the fabric. According to the spokesman for the medical profession, the fiber could be easily ingested into the lungs and could possibly lead to lung cancer. The consumer should be aware that this is an isolated case and not the result of the normal use of asbestos. However, it is important, also, to point out that *every* end-use of a fiber may not be advantageous to consumers.

Asbestos fabrics can be washed if they are handled carefully or simply wiped clean with a sponge. In extreme cases they may be subjected to open flames to burn out the dirt without damaging the fabric. This technique assumes that no flammable fiber is combined with the asbestos, as it frequently is in heat protective mits and pads.

MAN-MADE MINERAL FIBERS

GLASS FIBER

The origin of glass and glass fibers is uncertain. One legend credits Phoenician fishermen with the discovery of glass fiber. It is said that these men noticed pools of a molten substance under fires they built on a sandy beach of the Aegean Sea. (Heat caused the silica in the sand to fuse with the alkali of the wood ash.) Natural curiosity caused the men to poke at the molten material, and as they withdrew the sticks, the glass pulled out in a long, stringlike form. This was, perhaps, the first glass fiber.

During the Middle Ages Venetian artisans developed spun glass, which they used as decoration on blown glass forms, such as goblets and vases. This, too, was a form of glass fiber.

Serious efforts to create glass fibers for textiles began in the late nineteenth century. Edward Drummond Libbey succeeded in attenuating glass into fiber and made sufficient yarn to manufacture fabric for a dress, which was exhibited at the Columbian Exposition of 1893. The garment was attractive and not transparent as the public had anticipated. However, the fibers were coarse and low in strength and flexibility; the future of glass fiber appeared doomed. Fortunately, scientists kept the fiber in mind, and research continued. In 1931 a plan to develop glass fibers was undertaken, and by 1938 research and development had progressed to the point where usable glass textile fibers were produced in commercial quantities. The Owens-Corning Fiberglas Corporation was formed as a joint effort of Owens-Illinois Glass company and Corning Glass Company, both of whom had been working on glass fibers. Owens-Corning was the first producer of glass fiber, marketed as *Fiberglas.*

Today, glass fibers are manufactured by several corporations and sold under such trade names as Uniglass, Pittsburgh PPG, Fiberglas, Beta, Modiglass, Vitron, Ultrastrand, Unistrand, "401," Ferro, Stranglas, and Unifab. The labeling legislation defines *glass* as

a manufactured fiber in which the fiber-forming substance is glass.

Production

The raw materials for glass are primarily silica sand and limestone, with small amounts of other compounds, such as aluminum hydroxide, sodium carbonate (soda ash), and borax. The formulation of the glass depends upon the desired end-use. For example, fibers for electrical applications differ from those used in chemical industries. Most textile glass fiber is made from the electrical end-use formulations (Fig. 16.4).

The selected raw materials are melted in a furnace at temperatures of about 1650°C (3000°F). The compounds combine to form a clear glass marble about ⅝ inch in diameter. The marbles (called *cullet*) are inspected, and any with imperfections are sorted out to prevent flaws in the filaments. The perfect marbles are fed into a small furnace, where they are remelted, and the molten glass falls by gravity through a platinum spinnerette with approximately one hundred orifices. As the melted glass leaves the spinnerette, it solidifies. For filament yarns the fibers are pulled together, lubricated for ease in handling, and wound on tubes in strand form for fabric manufacture. Each marble produces about 100 miles of filament; it weighs less than 1 ounce.

Staple fibers, used more often as reinforcement for plastic or as insulation mats, can be spun into yarns following the same general procedures as for cotton or wool yarns. The staple fibers, however, are produced in a different manner than filament yarns. The fibers are hit by a jet of steam under high pressure as they leave the spinnerette. This breaks the filaments into lengths varying between 5 and 15 inches. The latter are collected on a drum and carded, then

Figure 16.4 Flow diagram showing the steps in the manufacture of glass fiber.

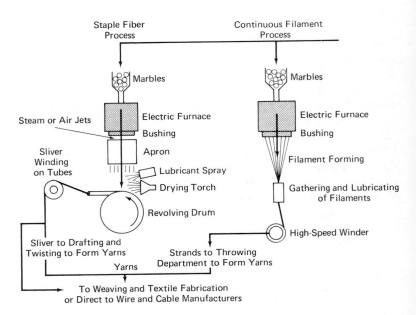

pulled into a sliver and roving. When used as reinforcement for plastic, the fibers are gathered in a thick batt instead of being pulled into a strand form.

During the last few years several manufacturers have installed direct processing equipment. In this method the molten glass is fed directly to the spinnerette, thus avoiding the intermediate cullet step.

Fiber Properties

Glass fiber has high strength—a tenacity of 6.3 to 6.9 grams per denier. Elongation is only 3 percent, but elastic recovery is 100 percent. The fiber has excellent dimensional stability, and it can absorb abnormal stresses without permanent deformation. Specific gravity is 2.54, which is too dense for general apparel use. The fibers do not absorb measurable amounts of moisture and have good resistance to wrinkling. Glass fiber is smooth, even, and transparent. The cross section is circular.

Fibers of glass will not burn. However, they soften at about 815°C (1500°F), and strength begins to decline at temperatures greater than 315°C (600°F). After brief exposure to heat fiber strength returns. This property is utilized in special finishing processes.

Hydrofluoric and hot phosphoric are the only acids that attack glass fibers. However, the fibers are damaged by strong alkalies at cold or warm temperatures and by hot weak alkalies. Organic solvents have no effect. Glass fibers launder easily and dry quickly.

Glass exhibits excellent resistance to age, sunlight, and weather. Microorganisms and insects will not damage it.

Although glass fibers are pliable and flexible, they lack abrasion resistance. When folded, as for hems in draperies, the edge will tend to crack if it is subjected to rubbing against another surface, such as the floor or window sill.

The dyeing and printing of fiber glass was a real challenge to the research chemist. A process called *Coronizing,* by which either solid colors or prints can be applied, proved to be an adequate solution. The fabric is padded (saturated) with a colloidal dispersion of silica, and then heat-set. After heat setting, pigment is applied with a resin solution, dried, treated with a final compound to produce colorfastness, and given a final drying. The Coronizing process softens the yarns so they flatten, and yarn slipping is prevented.

In the early 1960s Owens-Corning introduced a new superfine glass filament, trademarked *Beta.* This fiber has been very successful and is used in draperies, bedspreads, and table coverings. Beta has been suggested for upholstery fabric and apparel, but has not been widely adopted in such applications. Beta glass is one of the finest and softest fibers manufactured. It blends successfully with some other fibers, including modacrylics and olefins.

Glass Fibers in Use

The use of glass fiber has nearly tripled since 1955, but much of this growth has been in fabrics for industry, such as filter cloths, fire blankets, heat- and electrical-resistant tapes and braids, insulating fabrics and filler, and special mail bags. Various plastic materials reinforced through impregnation with glass fibers are frequently called fiber-glass products. This group includes such items as sports car bodies, bodies for sand-dune buggies, poles for pole vaulting, boats, and furniture.

Glass is not used in apparel as yet (though Beta is a potential apparel fiber), because the sharp fiber ends that are found at cut edges frequently cause skin irritations. In household fabrics and home furnishings glass is valued for its drapability, durability, appearance, and fire resistance. Window coverings and tablecloths are often made from glass. Other applications include lampshade fabrics, awnings, screens, ironing-board covers, and recently in some facets of building construction (Fig. 16.5).

Laundering of glass fabrics is preferred, and dry cleaning is not recommended. The fabrics should be handled carefully to prevent excess abrasion. Mild soaps and synthetic detergents can be used, and, if bleaching is necessary, a weak solution of a hypochlorite bleach such as Clorox® is safe for the fiber. Glass fabrics should be laundered *separately*, preferably not in a washing machine, for tiny residues from the fibers may be left in the laundry tub. These can transfer to other items washed in the following load and cause contact dermatitis. The fabrics should be rinsed thoroughly, then rolled or wiped with a towel.

Figure 16.5 U.S. Pavilion at Expo '70 in Osaka, Japan. Roof of translucent Beta fabric (Owens-Corning Fiberglas Corporation) supported by steel cables (David, Brody & Associates, architects).

Window coverings can be rehung immediately, while other fabrics should be laid smooth to dry or hung over padded lines. In sum, the major points to remember are the following: never wring or spin dry glass fibers; do not iron (it is not needed in any case); do not rub; do not use strong alkaline detergents.

METALLIC FIBERS OR THREADS

Metallic threads are truly the oldest form of man-made fiber, dating back to ancient Persia and Assyria. Actually, the first metallic fibers were not true fibers but were the result of slitting very thin sheets of metal into narrow, ribbonlike forms. Even today most metallic yarns are manufactured by variations of this old process.

The Federal Trade Commission defines *metallic* fiber as

> a manufactured fiber composed of metal, plastic coated metal, metal coated plastic, or a core completely covered by metal.

Gold, silver, and aluminum are the metals most often used in textile products. Gold and silver yarns are extremely costly. Occasionally they are pure metal, but since these metals are soft, it is more common for thin strips of the product to be wrapped around a central core of a strong, flexible product, generally silk or very fine copper wire. Silver tarnishes quickly in the air, and gold may discolor, so aluminum has replaced these fibers in Western countries. Some fabrics from the Orient are still made with pure gold or silver threads.

Modern aluminum yarns are made by one of two basic procedures. First, aluminum may be encased in a plastic coating of either a polyester, such as Mylar, or cellulose acetate-butyrate. The second product is cheaper, but the polyester is more desirable. Color is applied either to the plastic coating or directly to the aluminum by an adhesive. The second technique for manufacturing involves mixing finely ground aluminum, color, and polyester together, and then laminating this product to clear Mylar polyester.

Both types of metallic yarns can be obtained in a variety of colors, with gold and silver effects probably the most desirable and most frequently used. The yarns are bright and colorful and do not tarnish. The plastic coating prevents damage from salt water, chlorine, and alkaline detergents.

Metallic yarns are not especially strong, but they are quite adequate for normal decorative purposes. Polyester coated yarns are stronger than the acetate-butyrate coated and have been used in fabrics for evening wear. The yarns are colorfast to light and mild laundering.

The care of metallic yarns is determined to a considerable degree by the type of plastic coating, the core substance in wrapped yarns, and other fibers in the final product. Careful observance of all labels

Figure 16.6 Fabrics of metallic yarns.
[*E. I. DuPont de Nemours & Company*]

is recommended. Temperatures above 70°C (158°F) should be avoided on the acetate-butyrate-coated yarns, because the coating softens, but polyester-coated products safely withstand temperatures to 140°C (285°F). All the metallic yarns currently used in the United States can be laundered, but some do not dry clean satisfactorily.

The chemical and biological resistance of the yarns depends upon the coating material. Cellulose acetate-butyrate coatings react as acetate fibers; Mylar polyester coatings react as polyester fibers.

A recent addition to the metallic fibers available on the consumer market is stainless steel. This fiber may be used as a monofilament fiber (yarn), or it may be combined with other fibers in yarn manufacture. Stainless steel in fabrics contributes strength, tear resistance, abrasion resistance, and thermal conductivity. It also reduces static buildup in floor coverings. In addition, steel fibers may be used as resistance wire to produce draperies, floor coverings, and upholstery fabrics that can radiate heat when properly wired and connected to a power source. One of the better-known trade names is Brunsmet.

The popularity of metallic yarns is affected tremendously by the dictates of current fashion (Fig. 16.6). A wide variety of home-furnishing fabrics—including drapery and curtain materials, upholstery, bedspreads, towels, and tablecloths—contain metallic yarns. Metallics are used to enhance evening gowns, cocktail dresses, hostess gowns, and even sportswear, such as slacks, shorts, and bathing suits. Shoes, handbags, and hats often utilize metallic yarns. The luxury of gold and silver, once available only to kings and queens, may be enjoyed today by almost anyone.

OTHER MINERAL FIBERS

Several other mineral fibers have become important in industry, especially in the aerospace field. Ceramic fibers from compounds such as aluminum silicate and aluminum oxide fall into this category. Among the fibers are avceram, fiberfrax, and sapphire whiskers.

MISCELLANEOUS FIBERS

The textile industry is ever changing. New fibers are constantly introduced and either accepted or rejected. Many fibers are used only in industry for highly technical purposes. The following are important examples of new fibers for both industrial and commercial applications.

Graphite   Continuous-filament yarns of graphite are made by converting fibers such as rayon into pure carbon. The graphite fibers are extremely strong and resist high temperatures. They are widely used in aerospace products. An important trade name is Thornel.

Boron   Fibers from boron and from boron nitride are employed in industries where heat resistance, strength, and flexibility are important. Boron nitride is a flexible white fiber. It has been used in space fabrics, protective apparel, and thermal shields.

Chromel   Alloys of nickel and chromium (chromel) are important to the aerospace industry for space suits and other products. These fibers can be knitted, woven, and braided on standard equipment, and they blend satisfactorily with other fibers. Chromel resists extremely high temperatures, it is nonflammable and static free.

PBI Fiber   Polybenzimidazole, or PBI, resists temperatures greater than 177°C (350°F) for very long periods and temperatures above 538°C (1000°F) for a short time. The fiber has been used in apparel for space and for drogue chutes. It is nonflammable, comfortable, and flexible. One special product is the soft, flexible undersuit worn by astronauts.

Source   Allied Chemical introduced a biconstituent fiber in the late 1960s. This fiber, called *Source,* is composed of a nylon 6 matrix, with polyester fibrils—running parallel to each other and to the longer dimension of the nylon matrix—embedded in each filament (Fig. 16.7). No special generic term has yet been assigned. The fiber is found in carpeting, tires, and some apparel fabrics. It has a silklike quality, wears well, is easy to care for, and resists crushing and creasing. Similar fibers have been introduced in Japan and in Russia.

A-Tell   At the same time that Qiana was introduced in the United States, a fiber described as having similar properties to Qiana and to silk was released in Japan. This fiber, named *A-Tell,* is a polyester-ether made by polycondensation of *p*-hydroxybenzoic acid and ethylene oxide. The announcement of the fiber detailed the following characteristics: no waxiness, resistance to abrasion, resistance to sunlight, no damage by acids or alkalies, dimensional stability, and close resemblance to silk.

Arnel Plus   Filaments of Arnel triacetate and nylon are combined to form Arnel Plus, another third-generation fiber developed from the blending of filaments after the spinnerette stage. The fiber is said to have the strength of nylon, as well as the ease of care, set, texture, bulk, crimp, and hand qualities of triacetate.

Arnel Plus is used in warp knits and woven fabrics for men's and women's apparel. This filament combination is also known as *tricelon* or *tricilon.*

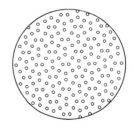

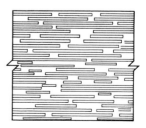

Figure 16.7   Diagram of a biconstituent fiber. [*Allied Chemical*]

Polycarbonate Fibers   Polycarbonates are characterized by the grouping —O—CO—O— as part of the repeating polymeric unit. *Lexan* is the name given to fibers made by General Electric Corporation. The polymer has limited use as a fiber, and data concerning fiber properties are not available.

Polyurea Fibers   Polyureas are spun from linear polymers containing the repeating unit —NH—CO—NH—. These fibers are strong; they have a low specific gravity, low moisture regain, good resistance to chemicals, and average resistance to heat.

In hand polyurea fibers resemble silk, while in mechanical properties they are similar to nylon. Industrial end-uses—such as conveyor belts, tires, nets, and ropes—are most common. The fibers could be used in apparel, but they have not been as yet.

Kynol   Kynol is a flame-resistant fiber made by the Carborundum Company, which describes it as a phenolic fiber. The properties cited include a breaking tenacity of 1.7 grams per denier; a specific gravity of 1.25; 6-percent moisture absorption at 65-percent relative humidity; a yellow-orange color. Kynol has excellent resistance to organic solvents; fair resistance to dilute oxidizing acids and dilute alkalies; and poor resistance to concentrated oxidizing acids and to concentrated alkalies. It resists flame temperatures to 2500°C (4500°F); it is non-melting, and the fiber identity is retained.

Suggested applications for Kynol in woven and knitted fabric structures include protective and flame-resistant clothing, draperies, industrial fire curtains, airplane linings and panels, upholstery, welders' clothing, and uniforms for fire fighters and laboratory workers. Needled felt structures can serve as flame barriers, and lofted felts can be used in linings for flame and thermal garments, military uniforms, and sports clothing. Kynol received international recognition after demonstrations of its safety value in suits for race drivers.

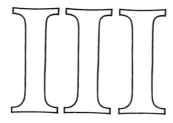

Yarn Structure

To weave or knit a fabric it is necessary to have yarns. Thus, the making of yarns is nearly as old as the manufacture of fabric and definitely predates recorded history. In prehistoric eras man devised simple ways of twisting fibers together to form usable yarns.

It is reasonable to suppose that the first technique for yarnmaking involved rolling fibers together between the palms of the hands, between fingers, or between a hand and another part of the body, such as the thigh. This latter process is still used by some peoples in isolated parts of the world.

The next development was the invention of spindles. Various types of spindles have emerged from archeological digs, but the most common device seems to have employed a distaff, spindle, and a whorl. Loose fibers were tied to a distaff. The spindle was a short stick notched at one end and pointed at the other. The spinning whorl was secured near the pointed end. The spinner, usually the woman of the household, held the distaff under her arm to free both hands for the actual spinning. Fibers were attached to the spindle notch, and the whorl pulled the fibers and spindle downward in a twirling motion. As the spindle dropped, the spinner drew out the fibers and formed them into a thread with her fingers while the whirling spindle twisted them into a tight strand. When the spindle neared the ground, the spinner took it up, wound the finished thread around it, and caught the new yarn in the notch. The process was then repeated.

Figure III.1   Flyer spinning wheel. From *Art and Technique of Hand-spinning* by Allen Fannin. Copyright © by Litton Educational Publishing, Inc. Reprinted by permission of Van Nostrand Reinhold Company.

Yarns were spun by these hand methods until late in the fourteenth century, at which time a crude spinning wheel was developed. The flyer spinning wheel (Fig. III.1), seen in many antique shops and museums today, was introduced in the sixteenth century. It is still used in some parts of the world and is currently enjoying a revival among handcraftsmen. This spinning wheel formed the yarn mechanically, but the power was human. Either a hand-propelled wheel or a foot treadle operated the wheel.

During the eighteenth century multiple spinning frames were developed, and shortly thereafter water power was incorporated. The basic machine-spinning techniques still used today follow the principles of the early multiple frames. A number of men played major roles in the perfection of spinning processes and equipment. Lewis Paul and John Wyatt developed the roller method of spinning in 1737; James Hargreaves invented the spinning Jenny in 1764; Richard Arkwright introduced the water-power spinning frame in about 1770; and Samuel Crompton created the spinning mule in 1779. Theirs are the basic inventions upon which all modern spinning is dependent.

Today yarns are manufactured by several processes. The type of yarn used may produce variations in fabric appearance, durability, maintenance, and comfort.

The following three chapters are concerned with the processes by which yarns are manufactured and the yarn variations available. It is important to note that fiber properties have considerable influence on yarns that have been made from either a specific fiber or from two or more fibers.

Yarn
Construction

BASIC PRINCIPLES

Yarns are composed of textile fibers. The different ways in which these fibers are joined together create the variety of yarn structures available to fabric manufacturers. The term *yarn* has been defined in several ways by various texts and dictionaries, but the definition given by the ASTM is representative:

a generic term for a continuous strand of textile fibers, filaments or material in a form suitable for knitting, weaving, or otherwise intertwining to form a textile fabric. Yarn occurs in the following forms:
a. a number of fibers twisted together
b. a number of filaments laid together without twist
c. a number of filaments laid together with more or less twist
d. a single filament . . . monofilament
e. One or more strips made by the lengthwise divisions of a sheet of material such as a natural or synthetic polymer, a paper, or a metal foil, used with or without twist in a textile construction.

The varieties possible include single yarn, plied yarn, cabled yarn, cord, thread, and fancy yarn. Insertion of the phrase "or strands" following

Figure 17.1   Staple and filament fibers.

"continuous strand" in the foregoing definition will serve to broaden and clarify the use of the word *yarn*, as it is interpreted in this text.

Yarns can be made from short, staple-length fibers or from long filament fibers (Fig. 17.1). If filaments are used, the yarns may be either *multifilament* (composed of several filaments) or *monofilament* (composed of a single filament). The staple fibers may derive from those natural fibers that are available only in short lengths or they may be composed of man-made fibers or silk that have been cut short. In any case, yarns made of short fibers require considerable processing.

YARN PROCESSING

Yarns composed of staple fibers are frequently called *spun yarns*, and this term will be used interchangeably with the term *staple fiber yarns*. Spun yarns usually are manufactured by the cotton system, the woolen system, or the worsted system. Techniques specific to special fibers (such as silk) are mentioned in the chapters dealing with those fibers.

The Cotton System

Sorting and Blending   Cotton or staple fibers of similar length arrive at the yarn processing unit in a large bale. To make cotton yarns, yarns of other fibers of similar length, or yarns that are an intimate blend of two or more fibers with uniform quality, fibers from several bales and sources are combined. These bales are opened, and thin layers of fibers from several bales are fed into the blending machines, which loosen the closely packed fibers, separate them, remove dirt and other heavy impurities either by gravity or centrifugal force, and blend the fibers to form a uniform mixture.

Picking   From the opening and blending equipment, fibers are conveyed to the *pickers*. These machines further clean the fibers and

left: **Figure 17.2**   Card clothing and brushes on a carding machine. [*Pepperell Manufacturing Company, Inc.*]

center: **Figure 17.3**   Card sliver being formed on a carding machine. [*Pepperell Manufacturing Company, Inc.*]

right: **Figure 17.4**   Card sliver coming off the carding rolls.

form them into a *lap* about 45 inches wide of randomly oriented fibers. These laps resemble absorbent cotton in form and shape. The quality of the cotton yarn is dependent to a considerable degree upon the thoroughness of the picking operation and the uniformity of the picker lap.

Carding   The picker lap is next fed into the carding machine. This step continues cleaning of the fibers, removes fibers too short for use in yarns, separates the fibers, and partially straightens them so their longitudinal axes are somewhat parallel. The fibers are then spread into a thin, uniform web. The carding machine requires two layers of *card clothing* (Fig. 17.2), which consists of fine wire pins anchored in a base fabric. These cards are attached to a steel cylinder and flat plates. The sets of pins move in the same direction but at different speeds and tease the fibers into a fine web.

The web moves into a funnel-shaped device, where it is gathered into a soft mass and formed into the *card sliver* (Figs. 17.3, 17.4), a ropelike strand of fibers about ¾ inch to 1 inch in diameter. The card sliver is not completely uniform in diameter, and the fibers are considerably more random in arrangement than a combed sliver. Carded yarns go directly to the drawing machine; combed yarns receive an additional processing before the drawing.

Combing   For high-quality yarns of outstanding evenness, smoothness, fineness, and strength, the fibers are combed as well as carded. In the combing operation several card slivers are combined and then drawn onto the comb machine (Fig. 17.5), where they again are spread into a web form and subjected to further cleaning and straightening. Short fibers are removed. Some of the short fibers are returned to a carding machine for use in lower grades of fabric, while others go into the manufacture of cellulose base plastics, rayon, or acetate fibers. After combing the fibers are pulled from the combing wires and formed into a *comber sliver*. This sliver is much superior to the card sliver and produces yarns of high quality.

Drawing   Depending on the quality of yarn desired, drawing follows either carding or combing. Several slivers are combined and conveyed to the drawing machine (Fig. 17.6), where they are pulled together and drawn out into a new sliver no larger than one of the original single slivers used in the first stage of the operation.

If the yarn is to be an intimate blend of two or more fibers, the slivers will be of different fibers. For example, one sliver of cotton fibers for each sliver of polyester fibers will produce a blend of approximately 50 percent polyester and 50 percent cotton. Therefore, the sliver from the drawing machine represents a blending of the original slivers. As yet, no twist has been introduced into the yarn.

Figure 17.5 Combining card slivers to feed into the combing machine. [*Pepperell Manufacturing Company, Inc.*]

Figure 17.6 Feeding slivers into the drawing frame to form roving. [*Pepperell Manufacturing Company, Inc.*]

Figure 17.7   The roving moving onto spindles. [*Pepperell Manufacturing Company, Inc.*]

Roving   The sliver from the drawing machine is taken to the roving machine, where it is attenuated until it measures from $\frac{1}{4}$ to $\frac{1}{8}$ of its original diameter (Fig. 17.7). As the roving strand is ready to leave the roving frame, a slight twist is imparted to the strand, and it is then ready for the spinning frame. The fineness and intimacy of blending of the yarn depends to some degree on the number of times the slivers are doubled and redrawn during the roving operation.

Spinning   The final process in the manufacture of yarn is the spinning operation. In the spinning frame the *final draft* is given and the desired amount of twist is inserted. Final draft is the total amount of attenuation that the roving receives (Fig. 17.8). Both draft and twist are varied in accordance with the desired end product. Several methods are used for imparting twist during the spinning process. For cotton yarns the most common technique is *ring spinning* (Fig. 17.9).

In ring spinning, the drawn-out roving is guided in a downward direction through the *traveler*, which is a small inverted U-shaped device. The traveler moves around the ring at the rate of four to twelve thousand revolutions per minute. As the spindle revolves to wind the yarn, the latter has to pass through the traveler, which carries it around on the ring. This process imparts the desired twist to the yarn.

Other methods of spinning sometimes used include the flyer and cap systems. The flyer process was the original method of continuous spinning, but it is seldom used today because it is slow and production is low. However, yarns made by the flyer process are smooth and have a high luster.

In cap spinning the winding and twisting are accomplished by passing the end under a stationary cap while the bobbin revolves inside this cap. Yarns made by this technique are sometimes rather wiry,

right: **Figure 17.8**   Drafting the roving. [*Rieter Machine Works, Ltd.*]

far right: **Figure 17.9**   Spinning the yarn. [*Rieter Machine Works, Ltd.*]

harsh, and boardy; however, satisfactory fine-count yarns can be produced.

The yarn as it comes from the first spinning operation is a *single* yarn. This and other types of yarns will be discussed more fully in Chapter 18.

The Wool Systems

Wool and man-made fibers can be spun into yarns by the woolen or the worsted system. The woolen system is comparable to that used in spinning carded cotton yarns. These yarns are carded, drawn, and spun. Yarns made by the worsted system are similar to combed cotton yarns in that the fibers are combed after carding to produce smoother and finer-quality yarns (Figs. 17.10, 17.11).

Sorting   Each fleece is carefully opened, and an expert grader pulls the fleece apart and sorts the fibers according to fineness or width and length of fiber, and sometimes according to strength. The grade of fiber determines the type of product for which it will be used. Fine fibers that are relatively long are reserved for sheer wool fabrics and for worsteds; medium fibers of shorter length are suitable for woolens; coarse fibers, both long and short, are made into rough fabrics and carpets. The best-quality fibers come from the sides and shoulders of the sheep. Lamb's wool, sheared from animals about eight months old, is used in making very fine quality, soft-textured sweaters or similar products.

Scouring   After sorting, the wool is scoured. This involves washing in warm soapy water several times, followed by thorough rinsing and drying. Scouring is essential, for it removes the natural grease in the fiber, the suint or body excretions, dirt, and dust. Natural grease is recovered and purified and becomes lanolin, used in the cosmetic industry. Scouring is, of course, not required for man-made fibers.

Carding and Combing   Wool fibers are carded by passing them between cylinders faced with fine wire teeth. This procedure removes considerable vegetable matter, such as twigs and burrs that remain in the fiber after scouring, and begins to disentangle the fibers and straighten them. It can be compared with the carding of cotton fibers.

Before combing, the decision is made whether the fibers are to be used for woolen or worsted yarns and fabrics. Woolen yarns are carded only, and the fibers are quite random in arrangement. Considerable foreign matter is still present in woolen yarns, and if this is to be removed, it is eventually taken out by carbonization. The latter involves passing the fabric through a sulfuric acid bath and applying heat, which combines with the acid to burn out vegetable matter. A final rinse removes acid and carbonized matter.

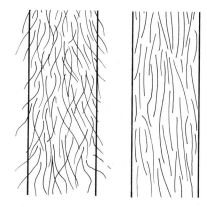

Figure 17.10   Woolen yarn (*left*), usually shows fibers lying in a random arrangement; worsted yarn (*right*) shows an ideal fiber lay.

Figure 17.11   Woolen (*left*) and worsted (*right*) yarns.

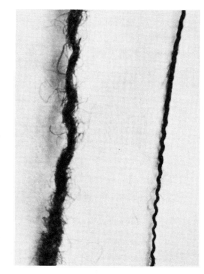

Fibers for worsted yarns are combed by passing through the combing machine, where the fibers are laid in a highly oriented arrangement. Short fibers are removed, and the remaining fibers are pulled into an untwisted strand called *top*. Wool tops can be dyed in this form, or the worsted yarns can be constructed and dyed later.

Spinning   The spinning of wool yarns usually is done by either the *ring* method or the *mule* method. Large amounts of wool are spun by the ring method today, because it is faster, it produces good-quality yarn, and the equipment occupies less space than the mule frame.

Mule spinning is classified as an intermittent system. It is still used for the best-quality soft yarns, so a brief description of the process is included. Unlike the systems previously mentioned, which are continuous-spinning techniques, the mule spinning frame accomplishes the drawing and twisting in one operation and the winding in a second operation. The carriage moves away from the spools of roving a distance of approximately 60 inches and during this outward move the yarn is drawn and twisted. As the carriage moves back to its original position, the yarn is wound onto the yarn package. Although this machinery occupies a considerably greater amount of space than other spinning frames, it does produce a soft yarn that is excellent for filling and for both warp and filling in blankets and other soft fabrics.

Staple Man-made Fibers

Staple man-made fibers can be processed by any of the spinning systems discussed, depending on the cut length of the fibers. Conversely, the yarn manufacturer orders the staple length he needs for the type of equipment available. If man-made fibers are to be used in blends with a natural fiber, the blending may be part of the preliminary blending process or it may occur at the drawing stage. Spun yarns of a single type of man-made fiber do not require blending, for they are relatively uniform as they come from the producer.

Yarns of staple fibers can be made with various diameters, from relatively fine to comparatively heavy and thick. All yarns composed of staple fibers must possess sufficient twist to hold the fibers in place, the amount depending upon the intended use. Yarns for warp, filling, or knitting have varying degrees of twist because of different requirements in fabric manufacturing. Additional twist is imparted to crepe yarns.[1]

---

[1] For more detailed discussions of yarn manufacturing and twist, see M. H. Gurley, *Man-made Textile Encyclopedia*, J. J. Press, ed. (New York: Textile Book Publishers, 1959), p. 229; and L. H. Hance, "Spinning," *American Cotton Handbook*, Vol. I, Dame S. Hamby, ed. (New York: Wiley-Interscience, 1965), p. 355.

Filament Yarns

The manufacture of yarns from filament fibers—whether natural or man-made—is a much more simple and direct process than that required for staple fibers. Silk is the only natural filament fiber, and the production of silk yarns has been discussed in Chapter 8. The size of the silk yarn depends upon the number of cocoons reeled off at one time.

Filament yarns of man-made fibers are made by either the continuous or the discontinuous process. In the continuous process the filaments are collected in groups (the number of filaments in each group determined by the desired number in the finished yarn) as they are extruded from the spinnerette. Then a specified amount of twist is added, and the yarn is ready for processing into fabric. In the discontinuous method the filaments are extruded, placed on cones or cakes, and shipped to yarn manufacturers, who combine the filaments into the desired arrangement, impart twist, and prepare the yarn for fabric construction.

Filament yarns are smooth and even unless they have been deliberately made irregular for novelty effects. They can be thick and heavy, gossamer sheer and light, or of any intermediate weight and diameter.

Simple yarns of filament fibers are lustrous and somewhat silklike in appearance. The luster can be reduced considerably by the addition of delusterants, but even delustered filaments tend to have more sheen than staple yarns.

Filament yarns have no protruding fiber ends, so they do not pill (unless the filaments are broken), and lint is not formed. Round filaments tend to shed soil, while multilobal filaments camouflage dirt.

The strength of filament yarns is determined by a combination of factors: fiber strength, number of filaments, denier of the actual yarn, and denier of each individual filament. Because the fibers are long, they receive equal pull, and fiber strength is maximized in the yarns.

The amount of twist in filament yarns is usually relatively low, but high-twist crepe yarns are made successfully from filament fibers.

New Methods of Yarn Preparation

During the past few years yarn manufacturers have developed new techniques in order to compete with rising operating costs and to develop some degree of automation in the textile industry. These new methods include continuous automatic spinning devices that

1. move the fiber directly from bale to sliver
2. move the sliver or tow directly to completed yarn
3. move the fiber directly to the yarn stage—open-end spinning—without intermediate stops

Fiber to Sliver   Automatic feeding of fibers from the bale to the sliver is accomplished by several types of equipment. The fiber is plucked automatically from the bale, fed to the blender, then moved to the carding equipment and, if desired, to combers. The resulting yarns are more uniform than regular yarns, and strength tends to be superior. Production speed is increased, and labor costs are reduced.

Sliver or Tow to Yarn   The sliver-to-yarn method utilizes a direct spinning frame. It eliminates separate drawing, roving, and twisting machines and provides for continuous movement from the sliver stage to the final yarn. Yarns tend to be coarser than those produced by older methods, but the new technique is less costly, more efficient, and faster. For end-uses where fineness is not important, particularly in woolen yarns, it is quite acceptable.

Tow-to-yarn processing is becoming very popular. Formerly, man-made fibers used in staple yarns had to be extruded as filaments, cut into short lengths, and then processed into yarns by either the cotton or the wool system. The tow-to-yarn procedure bypasses many of these steps. The tow is fed into a machine in which rollers, operating at different speeds, cause fiber breakage. The breakage occurs at various points, so no groups of fibers are broken at the same spot. Throughout this step parallel arrangement of fibers is maintained. The fibers then are fed to regular drawing and spinning equipment and converted into yarns.

Yarns made by this process are even, soft, uniform, and comparatively strong; they may have high bulk if desired. Considerable economy, reduced waste, and flexibility in the production of special effects are the virtues of this technique.

Fiber to Yarn   Fiber-to-yarn processing is frequently called *open-end spinning*. (It is also referred to as *break spinning* and *element spinning*.) This procedure converts fibers directly into yarns without separate carding, combing, drawing, and spinning operations. The major advantages include fewer knots in yarns, rapid yarn production, greater economy, and the ability to prepare yarn packages of any size.

Three basic fiber-to-yarn methods are in use: mechanical spinning, fluid spinning, and electrostatic spinning.

Mechanical spinning involves the contact of fibers with devices such as funnels, cones, baskets, sieves, or needled surfaces. The fibers are carried by gas or liquid current to the mechanical device, where they are collected. They are then taken off the collecting unit mechanically and delivered to the yarn-forming unit. Finally, twist is inserted. Fluid systems are very similar, except that in the final stage the fibers are delivered to spinning and twisting operations by fluid feeds instead of mechanical. Electrostatic systems utilize an electrostatic field to produce a parallel arrangement of fibers, transport and collect the fibers, and deliver the fibers to mechanical twisting equipment.

The machinery used for these new methods of yarn manufacture includes Rieter's Karousel, SACM's Flocomat, Robert's automatic spinning machines, Perlok's Turbostapler, the Pacific Converter, and Investa's BD200.

Split Film Yarns

Split or slit film yarns are produced from film or tape that is cut into narrow ribbons. These ribbons are fed by conveyor through a heat stretching or orientation zone to a stabilizing zone. They are then broken down into fibrils. The process is used on all thermoplastic fibers but is most commonly applied to polyethylene and polypropylene (olefins). Not only is it rapid and economical, but it can produce yarns of varying texture, denier, and appearance.

YARN PROPERTIES

Thread and Yarns

Thread and yarn are basically similar. *Yarn* is the term usually applied when the assemblage of fibers is employed in the manufacture of a fabric, while *thread* indicates a product used to join pieces of fabric together to create textile products. Thread is frequently of plied construction. It is fine, even, and strong. One authority explains that thread is made from yarn but that yarn is never made from thread.[2]

Thread must be so constructed that it can be adapted to either hand or machine sewing as well as embroidery or lacemaking. A satisfactory thread must have high strength and adequate elasticity, a smooth surface, dimensional stability, resistance to snarling, resistance to damage by friction, and attractive appearance. Several types of thread are available: simple-ply threads, cord threads, elastic threads, monofilament threads of man-made fibers, and multifilament threads.

Yarn Twist

As fibers, staple or filament, are formed into yarns, twist is added to hold the fibers together. The amount of twist is sometimes suggested broadly by such terms as low, medium, and high, but it is more accurately indicated by the number of turns per inch. The turns per inch (TPI) needed to form the best possible yarn varies with the yarn diameter. As the yarn becomes finer, it requires more twist. To deter-

---

[2]George E. Linton, *The Modern Textile Dictionary*, 2d ed. (New York: Duell, Sloan, and Pearce, Meredith Press, 1963), p. 946.

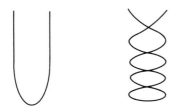

Figure 17.12 Balanced (*left*) and unbalanced (*right*) yarns.

mine the optimum degree of twist, a factor called the *twist multiplier* (TM) is used. The TM ranges from about three for soft yarns to six or more for hard-twist yarns. The twist multiplier is derived by dividing the turns per inch by the square root of the yarn count.

$$TM = \frac{TPI}{\sqrt{counts}}$$

To determine the number of turns per inch to use in making a desired yarn the manufacturer will multiply the twist multiplier by the square root of the counts. For example, if a manufacturer wishes to produce a hard-twist yarn with a count of 80s, he might establish a twist multiplier of 6. He would then multiply 6 by the square root of 80 to derive a TPI of about 53.

$$TPI = 6.0 \times \sqrt{80}$$
$$TPI = 6.0 \times 8.9$$
$$TPI = 53.4$$

The strength of yarns is due, in part, to the amount of twist that has been imparted. Strong yarns require considerable twist. However, beyond an optimum point additional twist will cause yarns to kink and finally to lose strength.

Balanced yarns are those in which the twist is such that the yarn will hang in a loop without kinking, doubling, or twisting upon itself. *Unbalanced yarns* have sufficient twist to set up a "torque" effect, and the yarn will untwist and retwist in the opposite direction (Fig. 17.12). Smooth fabrics require balanced yarns, but for crepe and textured effects, unbalanced yarns are frequently used. Crepe yarns are also produced by balanced yarns with enough twist to produce kinking.

The direction of twist is also important. Yarns can be twisted with either a right-hand twist (Z twist) or a left-hand twist (S twist). The direction of the twist conforms to the center bar of the letter (Fig. 17.13).

Various effects can be obtained by combining yarns of different twist direction, and durability may be increased by efficient plying of S and Z twist single yarns.

Yarn Number

Yarn number is a measure of linear density. *Direct yarn number* is the mass-per-unit length of yarn; *indirect yarn number* is the length-per-unit mass of yarn. Yarn number is frequently called *yarn count* in the indirect system. To some extent the yarn number is an indication of diameter when yarns of the same fiber content are compared.

Over the years various methods of determining yarn number have been developed. Cotton yarns are numbered by measuring the weight in pounds of one 840-yard hank; the count is then reported as the

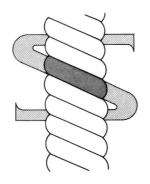

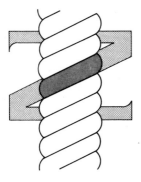

Figure 17.13 Diagrams of S and Z twist in yarn.

number of 840-yard hanks required to weigh 1 pound. For example, if 840 yards of cotton weigh 1 pound, the yarn number is 1s; if it requires 30 hanks to weigh 1 pound, the yarn is a 30s. A heavy yarn would be the ls, a medium yarn a 30s, and a very fine yarn a 160s.

Woolen yarn is measured by the number of 300-yard hanks per pound, while worsted yarn is measured by the number of 560-yard hanks per pound.

Silk and man-made fiber yarns are usually measured using the denier system. The denier is equal to the weight in grams of 9000 meters of yarn. This is an old system and dates back to early Roman times, when a coin, the *denier,* was the medium for buying and selling silk.

Some time ago the textile industry, or a segment thereof, proposed a universal system for yarn numbering or yarn count. This method, called *Tex,* would determine the yarn number by measuring the weight in grams of one kilometer of yarn. The result would be the tex yarn number. While the Tex system is favored by research groups, it has not met with commercial acceptance.

A universal system would make it easier to interpret research reports and quality-control studies. The reader is referred to the ASTM Standards and Kaswell's *Handbook of Industrial Textiles* for more detailed discussions of yarn numbering.

Simple
and Complex
Yarns

SIMPLE YARNS

Yarns that are even in size, have an equal number of turns per inch throughout, and are relatively smooth are called *simple yarns.* A simple, single yarn is the most basic assemblage of fibers suitable for operations such as weaving and knitting. These yarns can be made from any of the fibers and by any of the basic systems.

A simple-ply yarn is composed of two or more simple single yarns plied or twisted together. In naming a ply yarn the number of singles included usually precedes the word *ply.* For example, if two singles are used, the yarn is called a 2 ply; if four are used, a 4 ply (Fig. 18.1).

Simple cord or cable yarns are composed of two or more ply yarns twisted together (Figs. 18.2, 18.3). In identifying a cord it is necessary to indicate the number of plies in the cord. Thus, a "3,5-ply cord" indicates that each ply has five singles and that three of these 5-ply yarns are combined in making the cord.

Crepe yarns are a variation of simple yarns. However, a crepe yarn possesses a high degree of twist, so the yarn tends to kink. This kinkiness results in the rough texture characteristic of crepe fabrics. Nevertheless, the yarns are evenly twisted and even in size, so they are truly simple yarns rather than complex or novelty yarns.

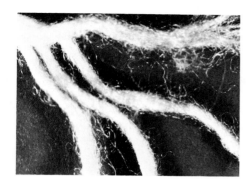

left: **Figure 18.1** A four-ply yarn, slightly magnified, with the fibers separated in one of the single yarns comprising the ply.

right below: **Figure 18.2** Sketch of a large cord yarn with three groups of ply yarns. Each ply has several single yarns.

Except for the textured effect achieved by crepe yarns, simple yarns, in themselves, do not create variation in fabric appearance. However, a combination of simple yarns of different size, different amounts of twist, and/or different fiber content can produce many interesting effects. The arrangement of yarns in groups (as in dimity) can also yield visual variety. Other changes in appearance depend on the number of warp and filling yarns per inch and the type and amount of twist. A large number of the simple yarns are used in fabrics where applied design through color or finish is the important thing.

Generally speaking, simple yarns produce smooth fabrics. Again, the arrangement of the yarns and combinations of various sizes in the fabric construction will influence the end product.

Simple yarns are usually considered to be durable, although the durability is affected by such factors as yarn number, amount of twist, and whether the yarn is single, a ply, or a cord. However, the uniformity of the yarn helps prevent snagging and tearing. Except for the highly twisted crepe yarns, which tend to shrink during care, simple yarns are most often easy to maintain. The arrangement of the yarns, the ply or cord construction, and the degree of yarn balance will dictate maintenance procedures to some degree. Even yarns, balanced yarns, and uniform arrangements of yarns produce fabrics that are comparatively easy to care for.

The statements made concerning the simple yarns are generalizations only and may be incalculably affected by differences in fibers, fabric structure, coloring methods, and finishing processes. It is, therefore, essential that the student of textiles be aware of the influence of other fabric "dimensions" on durability, maintenance, appearance, and comfort before establishing specific statements relating to any fabric.

COMPLEX YARNS

Complex or novelty yarns are made primarily for their appearance value. These yarns differ from simple yarns in that the structure is

Figure 18.3 Photograph of a cord yarn composed of four plies. Each ply has seven single yarns.

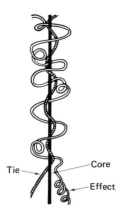

Tie — Core

— Effect

Figure 18.4 Diagram of the basic units in a three-ply complex yarn: the core or base yarn, the fancy or effect yarn, and the binder or tie yarn.

Figure 18.5 A slub yarn (magnified).

characterized by irregularities in size, twist, and effect. Complex yarns may be single or ply. Occasionally a cord or modified cord construction is used in complicated novelty yarns.

Complex ply yarns are usually composed of the following: a base, an effect, and tie or binder yarns (Fig. 18.4). The base yarn controls the length and stability of the end product. The effect yarn forms the design by the way in which it is applied to the base yarn. The tie or binder yarn is used to attach the effect yarn so it will remain in position.

Complex Single Yarns

Slub Yarns   A *slub yarn* may be either a single yarn or a 2 ply. In the single slub the yarn is left untwisted or slackly twisted at irregular intervals in order to produce soft, bulky sections (Fig. 18.5). In a 2-ply slub the soft and fluffy portion is held in place by a more tightly twisted single. The slubs in this latter yarn may alternate between the two singles, or all the slubs may be in one single while the second serves to hold the fibers in place. In any case, the lack of twist at irregular intervals causes the yarn to be fluffier and softer at those points. The resultant yarn is found in such fabrics as shantung, butcher rayon, and some linen.

Thick-and-thin yarns are similar in appearance to slub yarns (Fig. 18.6). However, the manufacturing process differs, as does the length of fiber used. Slub yarns are made from staple fibers, while thick-and-thin yarns are composed of filament fibers. As the filaments are extruded, the pressure forcing the spinning solution through the spinnerette is varied, so the filaments are thicker in some sections than in others. The resulting group of filaments forms a yarn of irregular size with thick-and-thin characteristics.

Flock Yarns   Flock yarns, frequently called *flake yarns*, are usually single yarns in which small tufts of fiber are inserted at irregular intervals and held in place by the twist of the base yarn. These tufts may be round or elongated. Flock yarn is used for fancy effects in suiting and dress fabrics. Tweeds, for example, usually contain flock yarns. One drawback to these yarns is that the flock nubs or tufts are easily pulled loose. However, since this occurs over a long period of time, it is not considered serious. Some authors classify flock or flake yarns as one variety of slub yarns.[1]

Complex Ply Yarns

There are several systems used in classifying complex ply yarns, and definitions of the various yarns differ.[2] The history of textile yarns

[1]M. Hollen and J. Saddler, *Modern Textiles* (New York: The Macmillan Company, 1968), p. 100.

and fabrics dates back thousands of years, so it is understandable that various descriptions for irregular and decorative yarns have evolved.

Bouclé Yarns   Bouclé yarns are characterized by tight loops projecting from the body of the yarn at fairly regular intervals. These yarns are of 3-ply construction. The effect yarn that forms the loops is wrapped around a base yarn, and then a binder or tie yarn holds the loops in position. Bouclé fabrics can be constructed by either knitting or weaving. The yarn is also available for hand knitting.

Ratiné and Gimp Yarns   Ratiné and gimp yarns are very similar to each other and, in addition, are rather like bouclé and loop yarns. The major difference between bouclé and ratiné or gimp yarns is that the loops are close together in ratiné or gimp, while in bouclé they are more widely spaced. The structure is similar, in that the yarn forming the loops is wrapped around the base yarn and then held in place by a binder or tie yarn. The ratiné yarn shows a taut, rough surface effect in overall appearance (Fig. 18.7). The manufacture requires two distinct twisting operations: after the yarn is first made, it is twisted in the opposite direction to establish the desired effect. The term *gimp* is used frequently as a synonym for ratiné. When a distinction is made, a yarn that has the loops formed by a very soft and slackly twisted yarn is referred to as a *gimp yarn,* while the loops on the *ratiné yarn* are of a soft but securely twisted yarn.

Loop or Curl Yarn   The loop yarn is of at least 3-ply construction. The base yarn is rather coarse and heavy. The effect yarn, which forms loops or curls, is made of either a single or a ply of two or more singles; it is relatively fine and has a small amount of twist (Fig. 18.7). The loops are held in place on the base yarn by the application of one or two single yarns that serve as ties. These binder yarns are fine and securely twisted. During the past decade loop yarns have become quite popular and are frequently used in mohair fabrics and in fabrics of other fibers that are (incorrectly) called mohair loops.

Nub, Knot or Knop, and Spot Yarns   The terms *nub, knot, knop,* and *spot* are often used interchangeably; however, there are minor differences between nub and knot yarns. A *nub yarn* (sometimes called a *spot*) is a ply yarn. It is made on a special machine that holds the base yarn almost stationary while the effect yarn is wrapped around it several times to build up a nub or enlarged segment. Sometimes the effect yarn is held in position by a tie or binder yarn, but in many

Figure 18.6 "Thick-and-thin" yarn within a fabric.

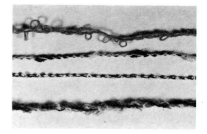

Figure 18.7 (*Top to bottom*) Loop, knot, ratiné, and loop yarns.

[2] For additional definitions, descriptions, and classifications, the reader is referred to K. P. Hess, *Textile Fibers and Their Uses,* 6th ed. (New York: Lippincott, 1958), p. 84; Hollen and Saddler, p. 101; and American Society for Testing and Materials, *Standards on Textile Materials,* Part 24 (1970), p. 50.

Figure 18.8   (*Top to bottom*) Loop, spiral, and chenille yarns.

cases the nub is so secure that no binder is needed. The *knot* or *knop yarn* (Fig. 18.7) is produced in much the same way as the nub yarn, except that some manufacturers add brightly colored fibers at the point of the nub.

Seed or Splash Yarns   Several authors classify seed and splash yarns as nub or knot yarns.[3] However, there is a tendency to differentiate between the types. According to several authorities, the splash yarn has an elongated nub that is very tightly twisted around the base. A seed yarn has a very tiny nub.[4]

Spiral or Corkscrew Yarns   Spiral or corkscrew yarns are complex yarns in which the desired effect is obtained either by twisting together yarns of different diameters, different sizes, or different fiber content, or by varying the rate of speed or the direction of twist. The *spiral yarn* is composed of two yarns of different size: one fine with a hard twist, the other bulky with a slack twist (Fig. 18.8). The heavy yarn is wound spirally around the fine yarn. A *corkscrew yarn* can be made by twisting together two yarns at an uneven rate, by twisting together two yarns of different size, or by twisting a fine yarn loosely around a heavy yarn so it gives the appearance of a corkscrew.

Spiral and corkscrew yarns may have more elongation than other types of yarns. The elongation of the corkscrew is governed by the amount of elongation of the central core.

The strength of all complex yarns is variable. Irregularities in amounts of twist and size produce an uneven stress distribution, which may result in yarns with some areas considerably weaker than others.

Chenille Yarn   Chenille yarns are used for special effects in fabrics and in the manufacture of chenille rugs. The yarn resembles a hairy caterpillar (Fig. 18.8)—*chenille* is French for caterpillar. A special doup-weave fabric (a leno-weave structure, see pp. 244–245) is constructed and then slit into narrow warp-wise strips that serve as yarn. This yarn, in turn, is used as filling in chenille fabrics (see Chap. 23). As the special fabric is slit, the loose ends of the crosswise yarns, which are soft and loosely twisted, form a pilelike surface. The yarn can be folded so the pile is all on one side of the final fabric, or it may be arranged so the loose ends form a raised surface on both sides.

Core and Metallic Yarns

Core and metallic yarns should be mentioned here. While not necessarily complex, they are certainly novelty yarns.

---

[3] George E. Linton, *The Modern Textile Dictionary*, 2d ed. (New York: Duell, Sloan, and Pearce, Meredith Press, 1963), p. 808; and *The American Fabrics Encyclopedia of Textiles* (New York: Doric Publications, 1960), p. 364.

[4] American Home Economics Association, *Handbook* (1970); and Isabel Wingate, *Textile Fabrics* (Englewood Cliffs, N.J.: Prentice-Hall, Inc., 1970).

Core Yarn   A core yarn is one in which a base or foundation yarn is completely encircled or wrapped by a second yarn. For example, the core may be rubber wrapped with cotton to give a comfortable, relatively durable, but highly elastic yarn. Or the core may be silk that is wrapped with gold or silver yarn.

Metallic Yarns   Metallic yarns are not new, but modern developments have produced varieties that are more durable and require less care. American-made metallic yarns are usually produced either by a sandwich-type construction or by a lamination process. In the sandwich construction, aluminum foil and pigment are "sandwiched" between layers of plastic—cellulose acetate, butyrate, cellophane, or polyester. The lamination process is similar; a metallized polyester (*Mylar*) is laminated or bonded to clear polyester. Color can be added to the adhesive used in the bonding process or to the polyester.

Metallic yarns are primarily decorative. Thanks to the plastic coating, they resist tarnishing, but care must be taken in pressing, for the plastic coating is easily softened or melted by high temperatures.[5]

Complex Yarns in Use

Complex or novelty yarns are valued primarily for their appearance. They add texture and design to a fabric. However, there may be problems in comfort, maintenance, and durability. Some complex yarns are rough and harsh, so they may actually be uncomfortable. On the other hand, loop yarns, such as those of mohair, are soft and pleasant to touch.

The rough surface of many novelty yarns and the irregular twist and loops that characterize many complex yarns may cause them to snag easily, and the flat abrasion resistance is reduced. These properties can contribute to inferior service when yarns of this type are used for upholstery. Fabrics of complex yarns usually require careful handling. However, because they are attractive, they are often selected for their appearance regardless of problems they might present.

[5]For detailed information concerning metallic yarns in use today, the reader is referred to American Home Economics Association, *Handbook* (1970), p. 33; *American Textile Reporter* (July 18, 1968), p. 104; and Leon E. Seidel, *Man-made Textile Encyclopedia*, J. J. Press, ed. (New York: Textile Book Publishers, 1959), p. 136.

Textured
Yarns

For many years the term *textured yarns* was just one of several expressions used in describing complex or novelty yarns. More recently, however, the term has been confined to a specific group of yarns and manufacturing processes.

Today's textured yarns are composed of filament or spun fibers. Most are processed directly from manufactured filament fibers. They may be regular or irregular in construction, so they often bear superficial resemblance to either simple yarns or complex yarns. However, textured yarns do possess individual properties of their own, and for this reason they require special attention.

ASTM defines *textured yarns* under the general heading of *bulk yarns:*

> *Bulk yarn:* A yarn that has been prepared in such a way as to have greater covering power, or apparent volume, than that of conventional yarn of equal linear density and of the same basic material with normal twist. Varieties include bulky yarn, textured yarn, and stretch yarn.
> *Bulky yarn:* A generic term for yarns formed from inherently bulky fibers such as man-made fibers that are hollow along part or all of their length, or for yarns formed from fibers that cannot be closely packed because of their cross-sectional shapes, fiber alignment, stiffness, resilience, or natural crimp, or both.

Textured yarn: A generic term for filament or spun yarns that have been given notably greater apparent volume than conventional yarn of similar fiber (filament) count and linear density. The yarns have a relative low elastic stretch. They are sufficiently stable to withstand normal yarn and fabric processing . . . and conditions of use by the ultimate consumer. The apparent increased volume is achieved through physical, chemical, or heat treatments or a combination of these.[1]

Some authorities have used the term *textured yarn* to include all three types called *bulk yarns* by ASTM. However, because of the growing popularity of stretch yarns and fabrics, a special chapter in this text is devoted to these products (see Chap. 28). The present chapter deals, primarily, with yarns defined as either bulky or textured.

Most textured yarns are manufactured from thermoplastic fibers. The ability to be *heat-set*—to be influenced in character by the application of controlled heat—is a necessary property for the production of many, but not all, textured yarns. In recent years several manufacturers have developed finishing and treatment techniques that impart stretch in completed yarns. However, in this chapter discussion is confined to those textured yarns formed as a result of fiber manipulation.

There are several processes used in the manufacture of textured yarns (Fig. 19.1). These include stuffer-box crimping, gear crimping, tunnel crimping, knit-deknit, false twisting, air-set, and edge crimping.

Despite the fact that textured yarns can be manufactured from staple fibers, filament fiber yarns are preferred for texturizing. These filament yarns are characterized by little or no pilling, yet they have the softness and bulkiness of staple fiber yarns (Fig. 19.2).

[1] American Society for Testing and Materials, *Standards on Textile Materials,* Part 24 (1970), p. 19.

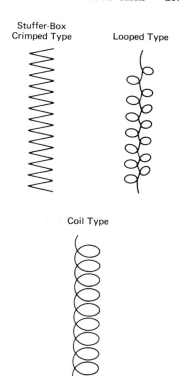

Figure 19.1 Diagram of three types of textured yarns.

Figure 19.2 Selected types of textured yarns. [*Allied Chemical Corporation, Fibers Division*]

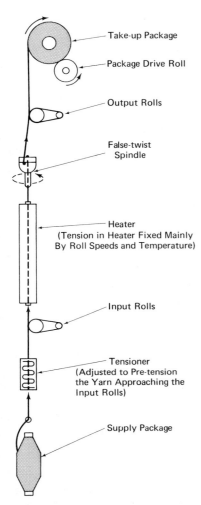

- Take-up Package
- Package Drive Roll
- Output Rolls
- False-twist Spindle
- Heater (Tension in Heater Fixed Mainly By Roll Speeds and Temperature)
- Input Rolls
- Tensioner (Adjusted to Pre-tension the Yarn Approaching the Input Rolls)
- Supply Package

Figure 19.3 Schematic drawing showing the yarn path through a typical false-twist machine. [© 1970 National Knitted Outerwear Association]

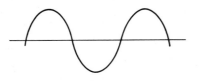

Figure 19.4 A sinusoidal curve.

TEXTURIZING PROCESSES

False-twist Methods

To create textured yarns by false-twist methods twist is inserted into simple filament yarns and set in place by the application of controlled heat. The yarn is then cooled and untwisted. As the twist is removed, the fibers kink or crimp sinusoidally because of the distortion resulting from the presence of the heat-set twist (Fig. 19.3). A *sinusoidal curve* is defined as a wavy line that has identical amplitude on each side of a central axis (Fig. 19.4). Simple yarns made by this process are characterized by either a left- or a right-hand torque, so they tend to be unbalanced and subject to distortion. To prevent the distortion, a right-hand and a left-hand twisted single are combined to form a balanced ply yarn. The best-known yarn made by this process is probably Helanca, manufactured by the Heberlein Corporation. Whitin ARTC yarns, produced by Deering-Milliken, are also in this category.

Helanca yarns were originally made of nylon fibers, but any thermoplastic fiber is suitable, and in recent years there has been a noticeable increase in the use of polyester fibers, as well as polyamides or nylons.

The twist method of manufacturing textured yarns produces some stretch in the final product, but this can be controlled by the amount of twist or turns per inch set into the yarn. Helanca and ARTC yarns are available in a wide range of stretch, from yarns that can be elongated up to 400 percent and still return to the original size to those with a small amount of stretch but with high bulk. It is possible to stabilize these yarns so the bulk remains but no noticeable stretch occurs.

Modifications of the preceding method are gaining popularity. One type of false-twist yarn is made on machines that combine the twist-set-untwist processes into a continuous operation. The yarn is highly twisted in one direction by a false-twist spindle on the supply side of the machine, and then the twist is removed on the take-up or out going side of the machine. Between the two steps, electrically heated elements set the twist.

Many false-twist machines are designed to process two yarns side by side imparting S twist to one and Z twist to the other (Fig. 19.5). These yarns are then plied together to produce balanced yarns, which possess many of the same properties as spun yarns manufactured on conventional yarn-spinning equipment.

Yarns made by the false-twist systems are often softer and more uniform than those made on other texturizing equipment. They have excellent stretch and/or high bulk. Trade names for some of the false-twist systems include Fluflon, Saaba, and Superloft.

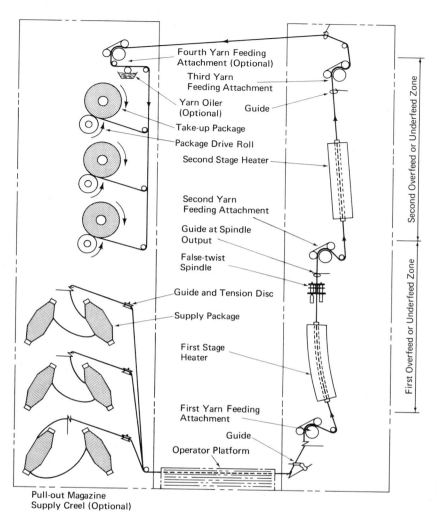

Figure 19.5 Diagrammatic representation of false-twist texturizing. [© *1970 National Knitted Outerwear Association*]

The Knife-edge Method

Textured yarns made by the knife-edge method possess a spiral-like curl or coil. The yarns are passed over a heated knife edge or heated and then passed over a knife edge while still hot. The student can easily visualize this texturing process if he will take a length of paper ribbon and draw it quickly over a sharp knife or scissor blade.

The best-known yarn of this kind is Agilon, a nontorque yarn manufactured by Deering-Milliken Research Corporation. The technique consists of drawing a thermoplastic yarn (usually nylon) over a hot sharp knife blade (Fig. 19.6). The edge of yarn in contact with the heat and knife is changed so that the resulting yarn has a bi-

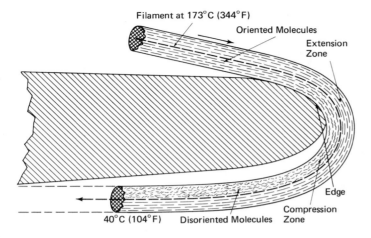

Filament at 173°C (344°F)
Oriented Molecules
Extension Zone
Edge
Compression Zone
40°C (104°F)   Disoriented Molecules

Figure 19.6 Sketch illustrating the principle of edge-crimping a heated monofilament yarn. Note the molecular orientation on the side of the filament away from the edge and the molecular disorientation on the side in contact with the edge. [© *1970 National Knitted Outerwear Association*]

component quality somewhat similar to wool. Agilon yarns have a high degree of elasticity but retain their sheerness, so they are desirable for stretch hosiery. Recent research has found the yarns to be satisfactory in knitted fabrics for sweaters and in certain types of upholstery and carpeting.

The Deering-Milliken Corporation states that the amount of stretch in Agilon yarns can be controlled. Because they are nontorque in nature, they can be used as either singles or plies. Moreover, the yarns accept dyes easily and uniformly. Agilon yarns produce comfortable items of apparel and can be combined with other types of yarns for novelty effects.

The Stuffer-box Method

In the stuffer-box method filament fibers are forced into a stuffing box or tube that causes them to develop a saw-toothed crimp. The crimp is heat-set, so when the filaments are removed from the tube, the crimp remains (Fig. 19.7). The greatest amount of bulk and a controlled

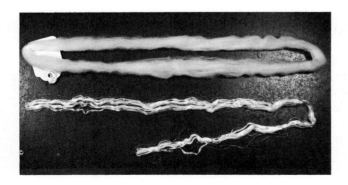

Figure 19.7 Stuffer-box textured yarns, showing filaments before texturizing (*below*) and after texturizing (*above*). Both skeins include the identical amount of fiber.

amount of stretch can be created by this method. Yarns are torque free and, therefore, produce satisfactory textured singles as well as plies. The best-known stuffer-box yarn is Ban-Lon. The Ban-Lon process is owned and controlled by the J. Bancroft & Sons Company of Wilmington, Delaware, and it can be applied to any thermoplastic fiber. Nylon is most frequently used, with the polyesters ranking second; other fibers have not gained sufficient popularity in Ban-Lon yarns to be economically practical at this time. The Ban-Lon process is adaptable to short staple fibers as well as filament fibers, but the most satisfactory product is obtained with filaments, because the degree of pilling is reduced. Therefore, most Ban-Lon is made from long filament fibers.

Ban-Lon fabrics are soft, strong, and easy to care for. They have a lively hand, excellent moisture absorption, controlled stretch, minimum pilling, dimensional stability, and adequate air circulation for comfort.

Another product that utilizes the stuffer-box method is Spunize. To make Spunize the technique is modified so that multiple ends of yarn can be handled at the same time. While Ban-Lon is generally used in wearing apparel, Spunize yarns are more common in carpets and upholstery fabrics and in industrial applications.

Air-jet Method (Looped Yarns)

Taslan, manufactured and licensed by DuPont, is the best example of a yarn textured by the introduction of loops into the filaments (Fig. 19.8). The process is a highly refined rewinding operation that provides for a brief exposure of multifilament yarn to a turbulent stream of compressed air. The air, in concentrated jets, blows the filaments apart and forms loops in the individual fibers. The resulting yarn is bulky, but it does not exhibit stretch. Since no heat is involved, filament yarns of any manufactured fiber can be bulked by the air-jet method. It has proved highly successful on glass fiber yarns for upholstery and drapery fabrics.

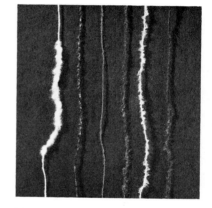

Figure 19.8   Taslan textured yarns that have been processed to simulate complex yarn structures. [*E. I. DuPont de Nemours & Company*]

Skyloft yarns, manufactured by the American Enka Corporation, use a similar air-jet principle.

Yarns that are bulked by the introduction of loops exhibit the following properties:

- a permanent change in the physical structure of the yarn
- a unique appearance, hand, and texture
- increased covering power
- subdued luster
- lower yarn strength and elongation

Gear Crimping

In gear crimping, yarns under controlled tension and temperature are carried between rotating, intermeshing gears, which give a saw-tooth configuration to the filaments. Novelty textured yarns can be formed by positioning the gear rolls out of mesh or by an intermittent crimping.

Tunnel Crimping

Tunnel-crimped yarn is fed into a tunnel in such a way that it arranges itself in a winding or sinuous form of desired amplitude. The filaments are then set by heat. Controlled vibrations facilitate proper shaping.

Knit-deknit

The knit-deknit process involves knitting filament yarns into fabric, heat-setting the fabric, then deknitting or unravelling the yarn. The unravelled yarns exhibit crimp and are ready for fabric manufacture.

Other Texturizing Processes

Bicomponent Texturizing   Two different formulas or modifications of a polymer fiber can be extruded simultaneously from side-by-side spinnerettes to produce a bicomponent fiber. If desired, the fiber can have built-in texturing potential. A crimp can be developed as the filament is formed or during later processing (latent crimp). Examples include Cantrece, Sayelle, Bi-loft, and Wintuk.

Spinnerette Modification   A spiral or helical crimp (texture) can be introduced into filaments by modifying the shape of the spinnerette, by changing the air flow at the spinnerette, or by vibrating the spinnerette.

Chemical Texturizing (Crimping)   Chemical texturizing can be applied to any man-made fiber. Immediate or latent crimp is possible, and several processes are in use or development.

Figure 19.9 Garment of Ban-Lon. [*Indian Head*]

TEXTURED YARNS IN USE

Textured yarns are found in a wide variety of fabrics (Fig. 19.9). They can be uniform in appearance or plied in special ways to resemble complex yarns. The major factors in the development and acceptance of textured yarns include comfort, appearance, and versatility.

Man-made filament fibers often prove to be uncomfortable apparel fabrics, because the filaments pack tightly together and prevent the movement of moisture or air through the fabric. However, texturizing creates bulk and space between the filaments, so the yarn itself will absorb moisture and provide a greater degree of comfort. In addition, fabrics of textured yarns may be considerably warmer than those of smooth, closely packed yarns, because the bulk acts as insulation.

The stretch property that is often a part of textured yarns has several advantages. For the manufacturer it means he can produce an item in a smaller range of sizes, since one size will fit a variety of figures. For the consumer it can mean a firmer and more comfortable fit with adequate freedom of movement. On the other hand, stretch can create problems. Some consumers find that stretch fabrics tend to bind and frequently blame stretch hosiery for foot problems. Many people wear stretch garments that are too tight, resulting in an unattractive appearance.

Maintenance of fabrics made from textured yarns is similar to the care given the same fiber in other forms. The added moisture retention increases the drying time for these fabrics, but washing procedures remain unchanged. Care must be taken to avoid snagging, for the crimp, curl, or loops present raised or rough surfaces that are easily caught. Durability is also similar to that of other fabrics made from the same fiber. However, the problem of snagging may lead to broken filaments or pilling and, perhaps, broken yarns that would shorten the life of the product.

The tremendous acceptance of textured stretch and bulky yarns has encouraged the development of other processes to create yarns with similar textured characteristics. Research has also been productive in relation to natural fibers. Special finishes have been devised for cotton and other cellulosic fibers to produce stretch yarns. According to the generally accepted definition, these are not truly texturizing techniques, but they result in end products that have many properties in common with textured yarns. Further discussion of these new techniques will be found in Chapter 28.

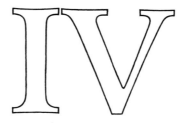

Fabric
Construction

Fibers, yarns, and single fabrics are combined in many different ways to produce the multitude of fabrics avilable to the modern consumer. Part IV outlines the techniques used in the manufacture of fabrics and the appearance, durability, maintenance, and comfort of these finished textiles.

Chapter 20 discusses the production of fabrics directly from fibers. Chapter 21 pertains to the construction of fabrics by knitting or interlooping of yarns and by stitch-knit methods. Woven fabrics, both basic and decorative or novelty, are treated in Chapter 22, while other methods of constructing fabrics—such as knotting, braiding, and multicomponent structures—are described in Chapter 23. Finally, Chapter 24 examines engineered yarns and fabrics, such as blends and combinations.

Figures IV.1 and IV.2 illustrate a number of time-honored methods of fabric construction. Fabrics can be knitted, knotted, and coiled from one thread, while plaiting, tapestry, and weaving employ two or more sets of threads. Many of these techniques are still in use today, but the repertoire of the fabric manufacturer has been enormously expanded by modern technology, as will be seen in the next chapters.

Fabrics from One Thread

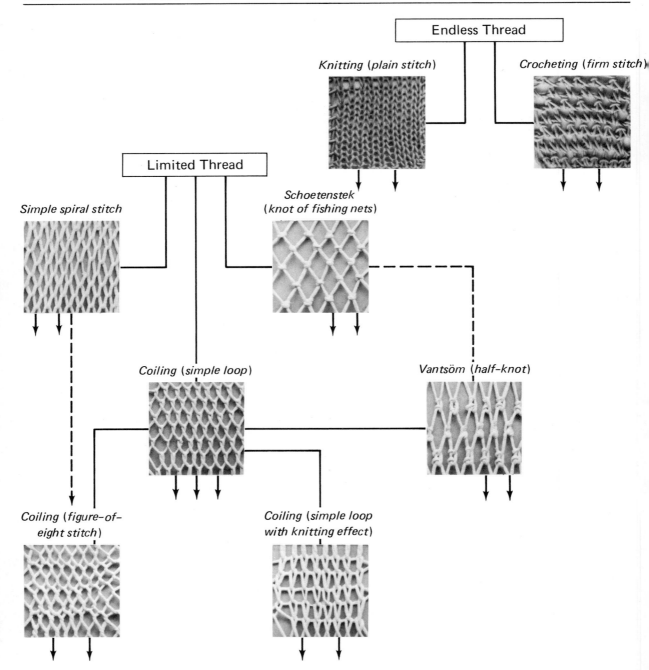

Figure IV.1 Examples of early fabrics made from one continuous thread, such as knitted and knotted fabrics. [*Ciba Review*]

Fabrics from Several Sets of Threads

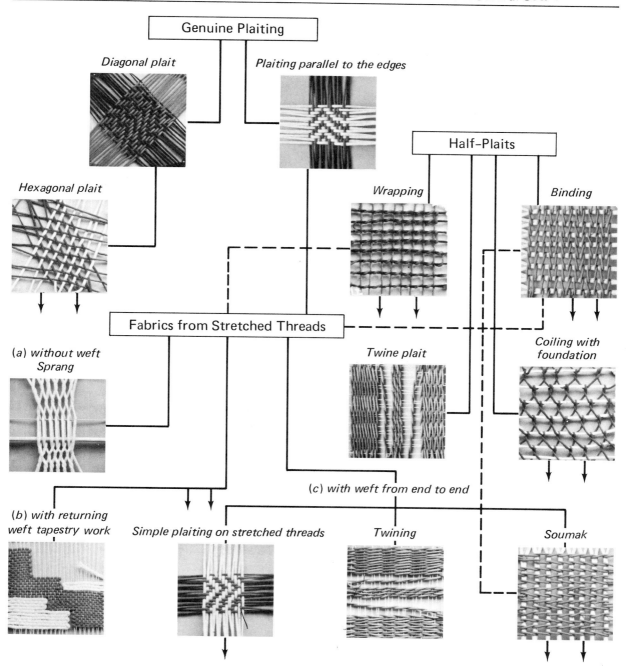

Figure IV.2 Examples of early fabrics made from several threads, including plaited and braided fabrics. [*Ciba Review*]

20

Felts
and Nonwoven
Fabrics

FELTS

The construction of fabric directly from fibers is both the oldest and the newest method of making cloth. Man's ingenuity in forming flexible covering materials first expressed itself in felts. The manufacture of felt is dependent upon the physical and, to some extent, chemical properties of wool, hair, or fur fibers. The most practical fiber for felt making is wool, but fur fibers are used in hat bodies and are sometimes blended with wool when certain properties are desired.

Felt made wholly or partly of wool is defined by ASTM as

> a structure built up by the interlocking of fibers by a suitable combination of mechanical work, chemical action, moisture and heat without spinning, weaving, or knitting. It may consist of one or more classes of fibers; wool, reprocessed wool or reused wool, with or without admixture with animal, vegetable, and synthetic fibers.[1]

A second definition of *felt* cited by ASTM is

[1] American Society for Testing and Materials, *Standards on Textile Materials*, Part 24 (1970), p. 27.

a textile (fabric) characterized by the densely matted condition of most or all of the fibers of which it is composed.

The ability of wool to coil upon itself, interlock, and shrink when subjected to heat, moisture, and pressure (including friction and agitation) is responsible for the felting of these fibers. In ancient times, long before recorded history, man found that he could take wool fibers, apply heat and water, and then pound the fibers with rocks to create a cloth that would hold together and conform to the general outlines of the body or add warmth to floors.

Today the manufacture of felt is highly mechanized. Wool fibers are cleaned, blended, and carded. After carding, two or more layers of fibers are arranged at right angles to one another. The number of layers depends on the planned ultimate thickness of the felt, but every layer alternates in fiber direction to the one immediately beneath it. The final thickness can vary from $\frac{1}{32}$ inch to 3 inches or occasionally more . Apparel felts are usually between $\frac{1}{16}$ inch and $\frac{1}{18}$ inch thick.

The layers or batts of carded fibers are passed through machines where they are trimmed and rolled. Moisture and heat are applied, and the batts are placed between heavy plates. The top plate vibrates, producing friction, agitation, and pressure, which cause the fibers to become entangled and pressed tightly together. The machinery is controlled automatically, so it stops when the desired thickness and hardness is attained. Fulling is the next step. This consists of shrinking the felt into a compact mass by the application of soap or sulfuric acid and then pounding with wooden hammers. Finally, felt is neutralized, scoured, rinsed, dried, and then stretched to the desired width.

Although wool, hair, and fur fibers can all be used in making felt, wool from sheep is most frequently employed because it possesses the best felting properties. Until fairly recently, felts were composed only of animal fibers; however, the increased popularity of felts has encouraged the blending of nonfelting fibers with wool to produce a lower-cost product. For example, rayon fibers are frequently blended with wool to make felts. Acceptable fabrics can be produced with up to 50 percent nonfelting fiber.

Felts have many industrial and domestic applications. Felt fabrics

- have good to excellent resilience
- are good shock absorbers
- are easy to shape
- will not ravel, so edges need no finish
- are sound absorbent
- have good insulating properties, with resultant warmth
- will not tear, though fibers may pull apart
- can be finished to be mothproof, water repellent, fireproof, and fungi-resistant

The breaking load of felts is low when compared with many woven or knitted fabrics. However, with intelligent selection of type and thickness, the consumer can obtain a felt that will be satisfactory for almost any end-use. The shopper seeking felts for apparel must remember that, because of low breaking elongation, felt garments should be loose to be durable. Other properties that may cause dissatisfaction include the fact that holes cannot be mended invisibly and that there is little or no elastic recovery. Felt will not return to shape after deformation caused by stretching or other forces.

Felt fabrics are used for wearing apparel, for home-furnishing items, for crafts and decorative accents, and for industrial purposes. The method of construction permits considerable variation in the thickness of the completed fabric and, therefore, enables the manufacturer to produce both flexible fabrics and comparatively stiff products. The flexible felts are desirable for apparel, such as skirts and jackets, and for tablecloths, pillow covers, and similar items. The thick fabrics are more appropriate to such products as rug pads and insulating materials.

Proper care procedures for felts are similar to those required of any wool fabric. However, because of the absence of yarn formation, the softer, thinner felts have comparatively low tensile strength; therefore, they should be handled carefully and never subjected to strenuous pulling or twisting. Dry cleaning is the recommended procedure for most felt products.

NEEDLE FELTS

Needle felts or needle-punched fabrics resemble felt in appearance, but they are made wholly or primarily from fibers other than wool. These fabrics are characterized by an intimate, three-dimensional fiber entanglement produced by the mechanical action of barbed needles, rather than by the application of heat, moisture, and pressure.

The fibers are first blended by the same techniques used for any other manufacturing process. The blended fibers are then arranged into a web or batt by mechanical or air-lay systems. This arrangement may be completely random, which gives the fabrics equal strength in all directions; or it may be parallel, so the resultant fabrics have increased strength in one direction. A layer of scrim or filament fibers may be added to the web for greater durability. Next the web (and scrim if used) is fed to the needle-punching equipment. In some cases the assemblage is tacked, using between thirty and sixty punches per square inch. The layer is then ready for actual needling.

While there are several types of needle-punching machines, the operation of all is similar. The needles, which have barbs protruding from the shaft, move through the layer of fibers, and the barbs push the fibers into distorted and tangled arrangements. The web is contained by metal plates above and below, so the fibers cannot be pulled or pushed beyond the web layer. As the web moves slowly through the machine, the needles punch as many times as desired for the end product. The number of punches per square inch varies from less than 800 to more than 2500. The higher figure is common for blanket fabrics, while low-cost carpet padding may use 800 or less.

The properties of needle-punched fabric depend on

1. the length and characteristics of the fibers used
2. the thickness, evenness, and weight of the fiber web
3. the arrangement of fibers in the web—parallel, criss-cross, or random
4. the density and pattern arrangement of needles in the needle board
5. the number of punches per second and the total number per square inch
6. the speed of movement of the web
7. the size of the needles and the number and arrangement of the barbs

Products commonly made by needle punching include indoor-outdoor carpeting, fiber-woven blankets, padding materials, insulation, and industrial fabrics.

BONDED FIBER FABRICS

Bonded fiber or nonwoven fabric employs one of the newest techniques of fabric construction. The term *bonded fabric* was formerly applied to consumer products, while *nonwoven* designated industrial materials. Currently there is much disagreement about these terms. While "nonwoven" is considered by many authorities to be preferred, this text will use "bonded fiber" and "nonwoven fabric" interchangeably, in order to distinguish between these structures and multicomponent fabrics that may be called *bonded* or *laminated.* Regardless of what the fabrics are called, the consuming public has accepted them for many uses, and production of these products increases each year.

A definition frequently cited for *nonwoven fabrics* states that they are materials made primarily of textile fibers held together by an applied bonding or adhesive agent or by the fusing of self-contained thermoplastic fibers. These fabrics are not processed on conventional spindles, looms, or knitting machines.

Nonwoven fabric dates back to the early 1930s. At that time a few textile companies began experimenting with bonded materials as one way of utilizing cotton waste. After World War II more firms became interested in nonwovens, and by 1960 approximately thirty-five were manufacturing products technically classified as nonwoven fabrics.

The sequence of steps in manufacturing these fabrics is fairly standard:

1. preparation of the fiber
2. web formation
3. web bonding
4. drying and curing

Most nonwovens or bonded fiber fabrics are sold in an unfinished state. If dyeing and finishing are involved, they become the fifth step.

Modern bonded fiber fabrics are no longer produced from waste fibers only. Although a small percentage of waste is still used, manufacturers have turned increasingly to good-quality fibers. Furthermore, while early bonded products were made of cotton, today's nonwoven fabrics utilize almost any type of fiber or combination of fibers. The length of fibers varies from $\frac{1}{2}$ inch to 2 inches, and some structures are actually fused filament fibers. Since there is no yarn spinning involved, fibers of different lengths and different chemical compositions can be successfully combined in a single bonded structure. The choice of the fibers depends upon the desired cost and performance of the end product. When the fibers have been selected, they are cleaned, bleached if necessary, and thoroughly blended.

The three techniques of web formation involve oriented, crosslaid, and random fiber arrangements. Oriented webs are those in which the fibers are parallel to the longitudinal axis. They can be formed on cotton or wool cards. Cotton cards result in uniform webs and fabrics with a texture similar to that of woven cloth. However, the production rate is low. Wool cards and garnetts produce webs at a rapid speed, but the web is less uniform. All webs of the oriented group are comparatively weak in the direction perpendicular to the lay of the fibers.

Cross-lay webs are made by combining two layers of fibers at right angles to each other. They are expensive to produce and no more satisfactory than random webs. Consequently, this technique is seldom used.

Random web formation is steadily gaining popularity with manufacturers, largely because special machines have been developed. The introduction of such units as the "Rando-Webber" (Figs. 20.1, 20.2), the Proctor Form Random Web Machine, and the Hunter Webformer has accelerated production of random-web nonwovens.[2] These machines employ an air-doffer principle. They are often described as *air-lay systems.*

The air-doffer principle involves the spreading and laying of fibers by controlled air currents (Fig. 20.1). Air suction pulls the fibers from the supply or feeder rolls or belts and deposits them in a random

---

[2]Francis M. Buresh, *Nonwoven Fabrics* (New York: Reinhold Publishing Corporation, 1962), p. 19.

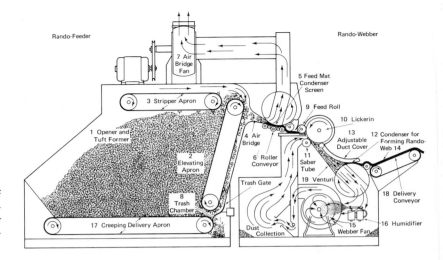

Figure 20.1 Schematic flow diagram of a combined Rando-Feeder® and Rando-Webber® machine, utilizing an air-doffer process for the manufacture of nonwoven fabrics.

arrangement on a condenser roll. The fiber mat is then fed into a compressor, which forms the fabric. Advantages of this method include uniformity in thickness, equal strength in all directions, and reasonable cost of manufacture.

Some investigators have worked with papermaking equipment for producing nonwoven bonded fiber fabrics. Fabrics made by this process have been used for disposable products such as cleaning and polishing cloths, "paper" garments, and interfacings. Consumers have welcomed the interfacings and polishing cloths, but acceptance of the apparel has been very slow, because the garments tend to be stiff and drape poorly. In addition, except for interfacings, the fabrics are usually considered disposable and, therefore, lack any durability. This method is a comparatively inexpensive technique for manufacturing. It is successful for selected types of low-cost fabrics.

Two basic techniques are employed in bonding fiber webs together:

1. An adhesive or bonding agent, either a dry powder or a liquid, is applied directly to the web in a separate step. The powder is usually a thermoplastic substance that is fused into the web by the application of dry heat. If a wet solution is used, it is spread uniformly over the web and then set by chemical action or heat.
2. Thermoplastic fibers are uniformly blended in the fiber mix and are evenly distributed within the web. Heat is applied, and the thermoplastic fibers soften and fuse over and around the other fibers. As the web cools, the fibers are all held firmly together.

The adhesive technique is most commonly used. These methods behave differently on different fibers or fiber blends, so the manufacturer must have considerable technical knowledge to make the right choice of bonding agent for the specific end-use of each fabric type.

Figure 20.2 One model of the Rando-Feeder® and Rando-Webber® machine used in the random fiber-lay technique. [*Curlator Corporation*]

The final stage in the manufacture of nonwovens is drying and curing. This step is extremely important when liquid binders are utilized. Drying devices include hot air ovens, heated cans, infrared lights, and high-frequency electrical equipment. The choice is dependent on the particular binding agent.

Nonwovens can be dyed or printed with standard techniques. However, at the present time, only a small percentage of these fabrics receive finishes, since most nonwoven yardage is sold in the same form that it leaves the drying and curing ranges.

Nonwoven fabrics appear in a number of products (Figs. 12.14, 20.3). One writer lists approximately 200 items that can be made from bonded fabrics.[3] Man's involvement with nonwoven fabrics may extend from birth to death: they are found in diapers, handkerchiefs, skirts, dresses, apparel interfacings, bandages, and shrouds. Other typical products include curtains, decontamination clothing, garment bags, industrial apparel, lampshades, map backing, napkins, place mats, ribbon, upholstery backing, and window shades.

Many nonwovens are manufactured for one-time use only, so care of the fabric is inconsequential. However, nonwoven fabrics can be laundered or dry cleaned if handled with care. The temperature of the laundry water should be that recommended for the specific fibers in the fabric, and agitation should be kept to a minimum. Twisting, wringing, or pulling must be avoided. Nonwovens in interfacings or underlining usually respond satisfactorily to the care demanded of the outer fabric.

Nonwoven fabrics available for consumer use include Pellon and Keyback interfacings. Of these, the Pellon products have attained the most widespread use, and well-planned advertising campaigns have made Pellon synonymous with nonwoven interfacings.

A major defect of nonwoven fabrics for apparel is their lack of good draping qualities. The fabrics tend to be firm and somewhat stiff. A further problem is their poor strength compared with woven or knitted fabrics of comparable weight.

In considering the future of the nonwoven fabric industry, a steady increase in industrial fabrics and in disposable items can be expected for some time. However, if bonded fiber fabrics are to be accepted for consumer goods, particularly apparel items, the industry must develop fabrics with good draping qualities and adequate strength. Furthermore, present-day finishing techniques and dye procedures will have to be adapted to these fabrics in order to create products for specific end-uses. Finally, nonwoven fabrics will require good styling and merchandising to be successful in a competitive fabric market. Present-day technology and research should cope with these problems quickly and satisfactorily.

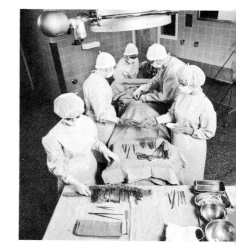

Figure 20.3 Nonwoven disposables in Purilon rayon for medical-surgical uses. [*FMC Corporation*]

[3]Buresh, p. 73.

Knitted
Fabrics

As this is written, knitted fabrics enjoy unprecedented consumer demand. Not only are traditional knitted products gaining popularity, but knits are making sizable inroads in areas heretofore dominated by woven fabrics. A recent market report states that the percentage of knits in apparel increased from 24 percent in 1963 to nearly 36 percent in 1969.[1] It is predicted that by 1975, more than 52 percent of all apparel fabrics will be of knit structures. The use of knits for other purposes will probably grow as much or more.

There are several reasons for this burgeoning acceptance of knitted fabrics. Knits can be made rapidly, so yarn-to-fabric expenses are much lower, and quality fabrics can be produced at comparatively low cost. The increase in travel, especially by air, has resulted in the need for lightweight, comfortable clothes that require little care and keep their neat appearance after sitting or packing. Knitted fabrics fit these qualifications. The tendency for knits to resist wrinkling is a factor in their acceptance for many end-uses. Fashion always plays a role in the selection of fabrics, and for several seasons leading fashion designers have included a wide range of knit fabrics in their creations.

[1] Bernard J. Obenski, "Potential for Knits," *Modern Textiles* (April 1970), p. 20.

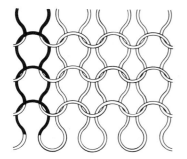

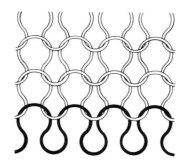

Figure 21.1 Diagram of a filling knit. The darkened vertical row indicates the *wale*.

Figure 21.2 Diagram of a filling knit. The darkened horizontal row illustrates the *course*.

In the construction and analysis of knits, two terms are used frequently: *wale* and *course*. *Wale* refers to a column of loops that are parallel to the loop axis (Fig. 21.1). The wales also run parallel to the long measurement of a knit fabric. A *course* is a series of successive loops lying crosswise of a knit fabric, that is, at right angles to a line passing through the open throat to the closed end of the loops (Fig. 21.2).

Machine knitting consists of forming loops of yarn with the aid of thin, pointed needles or shafts. As new loops are formed, they are drawn through those previously shaped. This interlooping and the continued formation of new loops produces knit fabrics. Two general methods are used: *filling* or *weft knitting* and *warp knitting*.

FILLING KNITTING

Filling or weft knitting, in simplified terms, involves the use of a single thread. The term *weft* is taken from weaving techniques, where it is used synonymously with *filling* to refer to the horizontal or cross-wise direction of a fabric. Fabrics can be manufactured by machine or by hand. Some knitting machines employ circular needles to produce tubular fabrics, which can be used in circular form or cut, shaped, and stitched (Pl. 1, following p. 244). In others the needles operate from a flat bed or a V-bed to create flat fabrics. The latter are fashioned or shaped during knitting, then stitched into such items as sweaters and hosiery. In both types the basic procedures are similar. A series of horizontal loops is formed on the needles, then for each succeeding row new loops are added as the needles hook into the thread, catch a new loop, and pull it through the preceding loop (Fig. 21.3). The old loop is forced off the needle, and the fabric is built. The resulting filling or weft knit fabric has slightly raised horizontal ridges (courses) and comparatively flat vertical rows of loops (wales) (see Fig. 21.1).

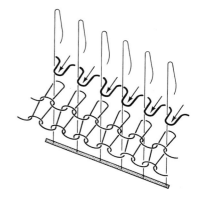

Figure 21.3 The loop-forming process in filling (weft) knitting. [© *1970 National Knitted Outerwear Association*]

Figure 21.4   The stockinette or jersey stitch.

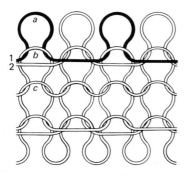

Figure 21.5 Diagram of a run-resist filling knit stitch used in some hosiery and similar fabrics. The numbers indicate the horizontal yarns, the letters the vertical loop rows.

There are four basic stitches used in manufacturing weft knits:

1. the flat or jersey stitch
2. the purl stitch
3. the rib stitch
4. the interlock stitch

Flat knit fabrics have distinct but flat vertical lines on the face and dominant horizontal ribs on the reverse side. Hand knitters recognize this effect as the *stockinette stitch* (Fig. 21.4). Flat stitches are used more frequently than any other stitches, because the process is rapid and inexpensive. Furthermore, flat knitting can be varied to provide run-resist and fancy patterned fabrics. Jersey stitch knits are used in making hosiery, sweaters, sportswear, and similar items.

A major disadvantage of regular flat knits is the ease with which they drop stitches if the yarn is broken. This results in vertical "runs" or "ladders," which destroy the appearance and usefulness of the fabric. Some flat knits, such as wool jersey, resist running because of the tendency for the wool fibers to cling together. A variation of the filling knit stitch (Fig. 21.5) will resist running.

The purl stitch (Fig. 21.6) produces fabrics that are similar to the reverse of the jersey stitch on both sides. This method is often used in the manufacture of bulky sweaters and in some children's wear. Many attractive designs and patterns can be created. Since the standard purl stitch forms a fabric identical in appearance on face and back, it presents no problems in construction and is considered reversible. The major disadvantages of this technique are that production is slow and machine maintenance is higher than for other types of equipment. The purl stitch is satisfactory in fabrics for such products as stoles and furniture throws, where stretch in both directions of the fabric is not required.

The rib stitch (Fig. 21.7) is usually made on a V-bed machine with two sets of needles that face each other. The stitches intermesh in opposite directions on a wale-wise basis, and the frequency of intermeshing determines the type of rib. If intermeshing occurs at every other wale, it is a 1×1 rib; if it occurs every two wales, the result is a 2×2 rib. Uneven ribs can be produced by interlacing in one direction for a certain number of wales and a different number in the other direction. Other rib fabrics can be knitted on circular machines equipped with two rows of needles.

The rib stitch is used whenever stretch is desired, and the resulting fabric has an excellent degree of elasticity. Rib knits are also warm. Their only disadvantage is costliness, because of the added fabric weight and the relatively low output of the machines. Ribbing is usually found at the lower edge of sweaters, on sleeves, and at necklines.

The interlock stitch (Fig. 21.8) is a variation of the rib stitch. In construction it resembles two separate 1×1 rib fabrics that are interknitted. Interlock stitch fabrics are thicker and heavier than regular

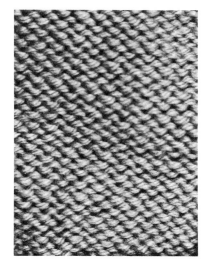

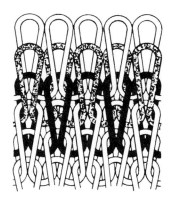

Figure 21.8 Diagram of the interlock stitch. [© *1967 National Knitted Outerwear Association*]

Figure 21.6 A purl stitch.

Figure 21.7 A 2 × 2 rib stitch.

rib stitch fabrics of the same gage. They are identical on both sides and are characterized by a soft hand, good moisture absorbency, and high dimensional stability. These fabrics cut easily, do not ravel, and do not curl at the edges. Even when fine yarns are used, the result is a firm, closely knit fabric.

Double Knits

Double knits are produced by the interlock stitch (Fig. 21.9) and by variations of that process (Pl. 2, following p. 244). Both surfaces of the fabric are somewhat riblike in appearance. Decorative effects can be achieved on double knits by the use of a Jacquard attachment for individual needle control on the knitting machine. This gives the designer much greater scope in developing interesting patterns. Double knits can also be made by knitting two distinct filling knit fabrics and providing for periodic binding stitches to hold the two layers together. From the basic double knit—the plain interlock—has come such variations as eightlock, single piqué, texi-piqué, pin tuck, and bourrelet. More complicated double knits include double piqué and Jacquard fabrics (Fig. 21.10).[2]

Double knits have good dimensional stability and resistance to runs. They are easy to cut and sew. Compared to single knits, they are relatively firm, heavier, less stretchable, and more resilient. Double knits are commonly made from polyester, triacetates, or wool. All varieties are very popular for apparel.

[2]Charles Reichman, *Double Knit Fabric Manual* (New York: National Knitted Outerwear Association, 1961), p. 11.

Figure 21.9 Double-knit interlock stitch.

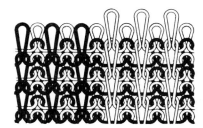

Figure 21.10 Diagram of one example of a Jacquard knit. [© *1969 National Knitted Outerwear Association*]

Figure 21.11 A pile fabric with a filling-knit base.

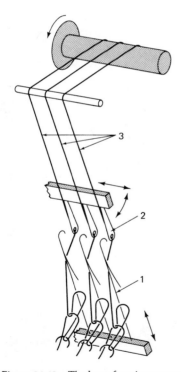

Figure 21.12 The loop-forming process in warp knitting. (1) Latch needles in the needle bar; (2) yarn guides in the guide bar; (3) warp yarn. [© 1970 *National Knitted Outerwear Association*]

Knitted Pile Fabrics

Knitted pile fabrics are usually made by filling knit procedures and resemble double-knit techniques (Fig. 21.11). To produce the pile, an extra set of yarns is drawn out in long loops and then cut or left uncut depending on the desired effect. The base fabric is generally a plain filling knit. Many pile coat fabrics are of knit pile construction.

A relatively new method for making pile knits involves the use of a soft yarn sliver for the pile yarn. This sliver is fluffy and full and creates a rich, luxurious pile. In fact, many of the "fake furs" available today are sliver knit fabrics.

WARP KNITTING

The term *warp knit* is also adopted from weaving techniques. In warp knitting the loops are formed in a vertical or warp-wise direction (Fig. 21.12). All the yarns required for the width of the fabric under construction are placed parallel to each other on a beam that resembles a warp beam for weaving (see p. 237). This beam is set into the warp knitting machine, and all yarns feed into the knitting area simultaneously. Each yarn is manipulated by one specific needle, and all the yarns form loops at the same time. Jacquard attachments can be added to warp knitting equipment to provide for individual control of each needle. The machine knits warp yarns into fabric by interlooping warp threads in adjacent wales. Because of the interlooping of these parallel warp yarns, warp knitting machines are frequently referred to as *knitting looms.*

Essentially, warp knitting is a system for producing fabric that is flat and has straight side edges. It can be manufactured rapidly and in great quantity.

Warp knits are classified according to the type of equipment employed. The common types are tricot, milanese, and raschel.

Tricot Knits

Tricot knits originated in England during the latter part of the eighteenth century. The first tricot machine was used in knitting silk hosiery cloth. Nearly a century later a large-scale mill was established in the United States. Current tricot knits are much like the products of that early mill.

Either plain or decorative tricot fabrics can be made on the same machine. This results in a saving to the consumer, because patterned tricot knits can be produced at approximately the same price as plain ones. The needles looping the adjacent yarns can be controlled in tricot knitting to create delicate, sheer, and attractive patterned fabrics (Fig. 21.13). Even lacelike fabrics are made easily.

The general characteristics of tricot knits are good air and water permeability, softness, crease resistance, good drapability, nonfray properties, run resistance, and elasticity. In addition, finishes are available for controlling the dimensional stability of the fabric and for providing good to excellent strength.

The thickness of tricot is influenced by the size of the yarns, the tightness or compactness of the stitches, the length of the guide bar movement, and finishing techniques, such as calendering (see p. 265). Fabrics of sheer and filmy appearance are made from fine yarns, while medium-size yarns produce firmer, more opaque textiles. In recent years tricot fabrics made of fine yarns have been calendered to decrease thickness and increase opacity.

Elongation is affected by the type of construction, the tightness of knit (the tighter the knit, the lower the elongation), finishing techniques, and gage (the finer the gage, the greater the elongation). *Gage* or *gauge* refers to the number of stitches in a specified unit width of a knitted fabric. It is comparable to yarn or thread count in woven fabric and is a measure of the closeness and compactness of the stitches or, in other words, a measure of fabric density.

Strength of knit fabrics is increased by using strong yarns, balanced construction, and fine gage. A combination of yarns of different fibers may increase or decrease the strength of the final fabric, depending upon the strength of the fibers used.

The cost of a tricot knit fabric is usually higher than that of a weft knit, and may occasionally be higher than a woven fabric of comparable weight and fiber content. Most often, however, warp knit fabrics are less expensive than woven fabrics, and they enjoy widespread acceptance because of their desirable inherent characteristics. Tricot knits are frequently employed for lingerie fabrics.

Milanese Knits

Milanese is a method of warp knitting similar to tricot. The resultant fabric is similar, but the machinery is quite different. Milanese equipment is not capable of producing as wide a variety of patterns.

Despite the fact that milanese knits are smoother, more regular in structure and elasticity, and higher in tear strength than tricot, they are disappearing from the market because the production rate is low and the pattern possibilities are severely limited. Tricot knits have a distinguishing faint horizontal line that is not observed in milanese fabrics. The latter, too, are used in lingerie.

Raschel Knits

Raschel knitting is a warp knit technique that needs special mention. It is one of the most versatile methods for constructing patterned knit

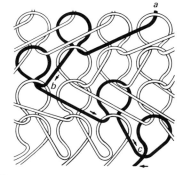

Figure 21.13 Diagram of a tricot warp-knit stitch.

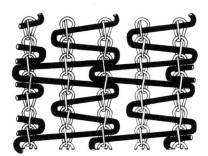

Figure 21.14 Diagram of a simple raschel crochet fabric. [© *1969 National Knitted Outerwear Association*]

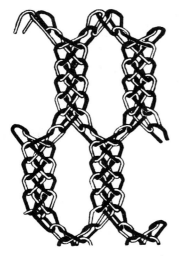

Figure 21.15 Diagram of a knitted net. [© 1970 National Knitted Outerwear Association]

fabrics. Among the many products of raschel knitting are crochet-effect dress materials (Fig. 21.14); compact, close-knit fabrics with designs produced by planned, "laid-in" yarns; laces of all types and in all widths (knit lace can be distinguished from true lace by the knit effect of the background net); powernets and similar elastic fabrics for foundation garments; curtain materials with lacelike designs; swimwear with elasticity and a knit pattern effect.

Unlike the tricot knit machines, which work best with comparatively fine yarns, raschel equipment uses any conceivable type of yarn. It is especially suitable to heavy yarns and coarse-gage designs.

Knitted net structures (Fig. 21.15) can be made on either raschel or tricot equipment, but the current emphasis is on raschel machines. These net fabrics result when, at planned intervals, no connection is made between two adjacent wales. The fabric spreads at those areas, leaving the openings characteristic of nets.

KNITTED FABRICS IN USE

Knitted fabrics possess several highly desirable properties. Probably the most important to the average consumer are excellent elongation and elastic recovery, plus good wrinkle and crush resistance (Fig. 21.16). Knits are preferred for traveling because they require little space in a suitcase and wrinkles hang out quickly when the clothes are unpacked.

The interlooping of yarns produces fabrics with considerable give or elongation. If not properly constructed or finished, and if the fiber choice is not appropriate, the fabric may stretch, sag, or shrink. This was a serious fault for many years, but recent improvements in manufacturing and finishing techniques, as well as the introduction of new fibers, have resulted in products with good size stability. Thanks to the interlocking of stitches, double knits exhibit excellent dimensional stability.

Knitted structures are porous and permit the free circulation of air, so they are comfortable. Most knitted fabrics used in apparel are characterized by high elongation plus good elasticity. They allow freedom of movement without permanent fabric deformation. These fabrics are soft and usually light in weight and still others need no ironing of any kind. Frequently knitted fabrics can merely be smoothed by hand after drying to restore their appearance. This is especially true of undergarments and cotton T-shirts.

It can be expected that knit fabrics will proliferate in both numbers and variety in the years to come.

Figure 21.16 Three-piece knit coordinate. Shirt of Arnel, sweater of Arnel plus Nylon, double-knit slacks of 100 percent Fortrel. [Celanese Fibers]

STITCH-KNIT STRUCTURES

It is difficult to isolate stitch-knit fabrics into a specific chapter or category. Although the word "knit" is included in the classification name frequently used, they are not truly knits. However, the equipment for stitch-knit resembles warp knitting machines used for tricot jersey fabrics, and the stich is similar to a chain stitch used in basic knitting processes. Therefore, the textile industry tends to group stitch knits with regular knits.

This group of fabrics has three basic subcategories: (1) comparatively flat fabrics composed of yarns; (2) pile fabrics; (3) fabrics utilizing fiber mats and yarns. The two major products are called *mali fabrics* and *arachne fabrics,* after the machines employed in their manufacture.

Fabrics Made of Yarns

Yarns are arranged in a planned pattern to produce the first type of fabric. They are then fed into a machine, where they are stitched in parallel rows across the width of the fabric to hold the yarns in position. The process utilizes chain stitch techniques and is very similar to the formation of knit loops (Fig. 21.17). Single bar tricot techniques are sometimes used.

The characteristics of the fabric depend on the relative weights of the base and stitching yarns. If the stitching yarns are considerably finer than the base yarns, the fabric resembles a woven structure on the face and a knitted structure on the back; if the stitching yarns dominate, the product has a distinct knitted appearance.

Fabrics in this category sometimes employ two sets of base yarns and one set for the stitch-knit operation. In this event the base yarns are arranged in a criss-cross manner, but not necessarily at a 90-degree angle as in the warp and filling of a woven fabric. If only one set of base yarns is used, the yarns are usually parallel to each other but perpendicular to the stitch-knit yarns.

Stitch-knit fabrics from yarn have pleasing drape and hand. They can be finished like woven or knit structures. The fabric properties are the result of the yarn arrangement, stitch-knit pattern, size of yarns, fibers selected, and fabric density.

Pile Fabrics

Pile structures can be made with a yarn base or a fiber mat base, but the latter is more common (Fig. 21.18). The base is fed into the stitching equipment, and, as the stitching occurs, the yarn is held in a loop to one side. When completed the loops can be cut or left uncut. For cut loops the procedure usually allows for the inclusion of spaced stitching yarns that conform to the base layer contours as a binding agent.

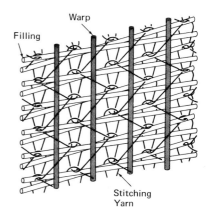

Figure 21.17 Diagram of the Malimo stitch-knit technique, utilizing three yarns. [*Crompton & Knowles Corporation, Textile Machinery Group*]

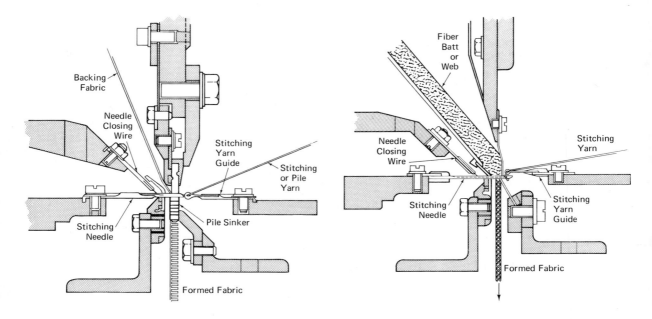

left: **Figure 21.18** Diagram of the Malipol pile stitch-knit process. [*Crompton & Knowles Corporation, Textile Machinery Group*]

right: **Figure 21.19** Diagram of the Maliwatt web stitch-knit process. [*Crompton & Knowles Corporation, Textile Machinery Group*]

The latest innovation in pile stitch-knit employs a base composed of a preformed fabric, as well as yarns or fiber mat. This results in a product similar to tufted fabrics.

Fabrics Utilizing Fiber Mats

Fabrics for special end-uses can be made by forming a fiber web and stitching it together for stability (Fig. 21.19). The web can be very thin (less than $\frac{1}{8}$ inch), or it may be $\frac{1}{2}$ inch or more in thickness. These fabrics resemble needle-punched materials to some degree.

Stitch-knit fabrics are economical to construct, for the process eliminates many steps involved in other techniques, and capital investment in space, materials, wages, and equipment is reduced. The design potential is great: fabrics that resemble woven structures, knit structures, and needle-punched structures can be made without all the different types of machinery. However, despite the many advantages, there has been little progress in the manufacture of stitch-knit fabrics in the United States. The reasons for this are not clear, but it must always be borne in mind that, despite a fabric's impressive list of qualities, the buying public may simply not like it. Perhaps this is the case with stitch-knits.

Because of their limited production, there is little information available about the care of stitch-knit fabrics. The consumer who might purchase such a product is advised to follow the maintenance instructions on the accompanying label.

Woven Fabrics

Weaving is one of the oldest arts known to man. While no actual looms from early civilizations survive, fabrics of fine quality have been found in the tombs of ancient Egypt, and designs on very old pottery provide indisputable evidence of man's skill in weaving. Painted pottery also gives us some idea of the type of loom that was in use in the ancient world. Figure 22.1 shows a red-figure Greek vase from the fifth century B.C. depicting the legend of Penelope, wife of Odysseus. Throughout the long voyage of Odysseus the constant Penelope refused to marry anyone else until she had completed a winding sheet for her father-in-law. By day she would sit at her loom and weave, and at night she would unravel everything she had done. At last, betrayed by her servant, she was compelled to finish the shroud and urged to choose from among her suitors. Just before the fateful decision was to be made, Odysseus returned, and the two were happily reunited.

Woven fabrics consist of sets of yarns interlaced at right angles in established sequences. The yarns that run parallel to the selvage or to the longer dimension of a bolt of fabric are called *warp* yarns or *ends;* those that run crosswise of the fabric are called *filling* yarns, *weft* yarns, *woof* yarns, or *picks.* Warp and filling are the terms in common use today. However, manufacturers often refer to ends and picks.

Figure 22.1 Vertical warp-weighted loom. Penelope at her loom, as depicted by Skyphos. Greek, 5th century B.C. [*Chiusi Museum*]

Early looms were very crude compared with modern mechanical weaving contrivances. Nonetheless, all looms, old and new alike, have a few basic principles in common: There is some system to hold the long threads under tension; there is a way to spread yarns apart so the crosswise thread can move through the opening or *shed;* and there is a device to pack the crosswise threads tightly together.

Until the early nineteenth century weaving was primarily a hand or manual process. In the late 1700s and early 1800s scientists and inventors such as Jean Marie Jacquard and Edmund Cartwright developed weaving looms that were partially machine powered. Later manufacturers produced looms that were entirely mechanical or power driven. Weavers were hostile to these first mechanical looms, for they feared the machines would take away their jobs. Consequently, the Industrial Revolution was well under way before its effects were felt in the weaving mills. During the late eighteenth and early nineteenth century, developments in weaving involved the addition of automatic features to existing looms to increase the speed of operation and reduce the frequency and amount of damage due to faulty functioning.

The parts of the basic loom are shown in Figure 22.2. The *warp beam* holds the lengthwise yarns. It is located at the back of the loom and is controlled so it releases yarn to the loom as it is needed. The *heddles* are wire or metal strips with an eye located in the center through which the warp ends are threaded. The *harness* is the frame that holds the heddles in position. Each loom has at least two harnesses and may have twenty or more. Harnesses can be raised or lowered in order to produce the *shed* through which the filling thread is passed and thus control the pattern of the weave. The *shuttle* moves back and forth in the shed, passing the filling threads between the warp threads. The *reed* is a comblike device, and the openings between wires in the reed are called *dents.* Warp threads pass through the heddles and then

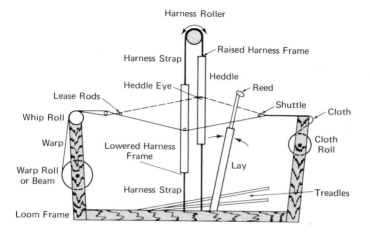

Figure 22.2 Diagram showing a simple loom.

through the dents. The reed keeps the warp ends from tangling and beats and packs the filling threads into their proper position. The reed is parallel to the harness. The *cloth beam* is located at the front of the loom and holds the completed fabric.

The basic weaving operation consists of four steps:

1. *Shedding* is the raising and lowering of the warp ends by means of the heddles and harnesses to form the shed of the loom, so the filling yarn can be passed from one side of the loom to the other (Fig. 22.3). In most looms the filling yarn is carried by a *shuttle,* a boat-shaped device that holds a *quill* upon which the filling yarn has been wrapped (Fig. 22.4).
2. *Picking* is the actual procedure of placing the filling yarn into the shed. The shuttle moves across the shed, laying the pick or filling as it goes.
3. *Battening,* sometimes called *beating* or *beating in,* consists of evenly packing the filling threads into position in the fabric.
4. *Taking up and letting off* involves taking up the newly manufactured

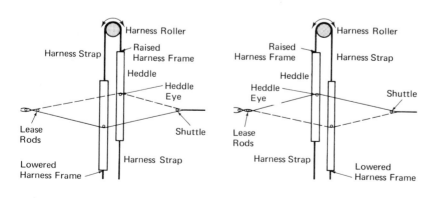

Figure 22.3 Diagram showing the movement of harnesses and warp thread to form the shed for a plain weave.

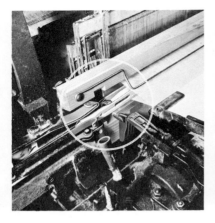

left: **Figure 22.4** Close-up view of a shuttle in position to move through the shed and deposit the filling yarn. [*Pepperell Manufacturing Company*]

center: **Figure 22.5** A large weaving room. [*Pepperell Manufacturing Company*]

right: **Figure 22.6** A power loom. [*Pepperell Manufacturing Company*]

fabric onto the cloth beam and letting off or releasing thread from the warp beam. The operation maintains uniform distance from harnesses to cloth.

Most fabrics are woven on a simple loom (Figs. 22.5, 22.6). For elaborate fabrics, modification of the basic loom, addition of special attachments to the loom, or specially designed looms are used. Looms for simple plain-weave fabrics have two harnesses, while from three to eight harnesses are required for modified plain, most twills, and most satin weaves. Weaving dobby patterns involves as many as thirty-two harnesses. The Jacquard attachment fits onto special looms and maintains individual control of every warp thread. It is used to manufacture elaborate weaves—often called *Jacquard weaves*—such as those found in brocades and damasks (see p. 243).

By the middle of the twentieth century the standard looms were nearly flawless in operation, so inventors turned to other types of improvements and changes. The current emphasis is on the perfection of shuttleless looms, which eliminate the movement of a shuttle back and forth across the warp threads at the shed. These new looms utilize either mechanical "fingers" that pick up and carry the yarn across or a jet of air or water to force the pick across.

Modern looms are made in different widths. Originally, cotton equipment produced fabric from 27 to 36 inches wide, and wool equipment usually resulted in fabrics from 50 to 72 inches wide. However, apparel manufacturers have found that wider fabrics are more economical for cutting, and weavers have discovered that wider fabric is less costly to produce per square yard, so most weaving mills now have some looms that weave fabric at least 60 inches wide.

PLAIN WEAVES

The plain weave is the simplest form of weaving. It consists of the alternate interlacing of warp and filling yarns, one warp up and one down, the entire width of the fabric (Fig. 22.7). This is referred to as a 1/1 weave. In Figure 22.8 the black squares indicate that the warp yarn is on the surface. The squares paralleling the vertical direction of the diagram can be visualized as warp yarns, while the horizontal rows represent the filling yarns. The term *tabby* is sometimes used as a synonym for the plain weave.

Many woven fabrics are constructed by plain-weave interlacing. Unless colors or finishes are added on one side only, plain-weave fabrics are usually reversible. The yarns can be packed loosely or compactly. The warp threads may equal the filling threads in number, but unequal arrangements can be produced by variations in yarn size and by unequal spacing of warp or filling yarns. When the number of warp yarns per inch is approximately the same as the number of filling yarns per inch, the thread count is considered balanced. Conversely, when the number of yarns per inch differs considerably between warp and filling, the fabric is unbalanced. The plain weave is comparatively inexpensive. Moreover, some of the most durable fabrics are manufactured by this technique.

Examples of plain weave include muslin, percale, print cloth, cheesecloth, chambray, gingham, batiste, nainsook, lawn, organdy, taffeta, linen toweling, handkerchief linen, dress linen, seersucker, chiffon, challis, shantung, china silk, some wool tweeds, and homespun.[1]

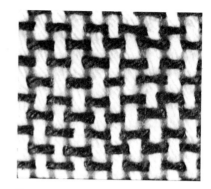

Figure 22.7   A plain-weave fabric.

Rib Variation of the Plain Weave

Interesting and attractive fabrics can be obtained from the plain weave by utilizing the rib variation. A diagram of this weave is identical to the regular plain weave. The rib appearance is produced by using heavy yarns in the warp or filling direction, by grouping yarns in specific areas of the warp or filling, or by having more warp yarns than filling.

Many rib-weave fabrics have heavy yarns inserted as picks. Examples of this construction include poplin, faille, bengaline, and ottoman. Dimity (Fig. 22.9) derives from alternation of fine and heavy yarns at planned intervals in the warp. Such alternation in both warp *and* filling is used for cross-bar dimity and certain tissue ginghams. Broadcloth results from a highly unbalanced yarn count in which there are many more warp yarns per inch than filling.

Comparative size of the ribs in some of the more commonly encountered rib weaves is one method of fabric identification. The

Figure 22.8   Diagram of a plain weave.

---

[1] Definitions and descriptions of some of these fabrics are provided in the glossary. Further definitions will be found in American Home Economics Association, *Handbook* (1970); George E. Linton, *The Modern Textile Dictionary*, 2d ed. (New York: Duell, Sloan, and Pearce, Meredith Press, 1963); *The American Fabrics Encyclopedia of Textiles* (New York: Doric Publications, 1960); and John Hoye, *Staple Cotton Fabrics* (New York: McGraw-Hill, Inc., 1942).

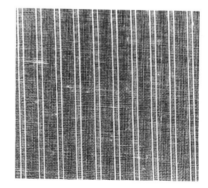

Figure 22.9   A rib-weave variation of the plain weave. The fabric is double-bar dimity.

Figure 22.10 A 2/2 basket-weave fabric.

Figure 22.11 A 4/4 basket-weave fabric.

below left: **Figure 22.12** Diagram of a 2/1 basket weave.

below center: **Figure 22.13** Diagram of a 2/2 basket weave.

below right: **Figure 22.14** Diagram of a 4/4 basket weave.

following is a selection of fabrics listed in order from fine rib to heavy: broadcloth, poplin, faille, grosgrain, bengaline, ottoman.

Basket Variation of the Plain Weave

The basket weave is generally defined as having two or more warp ends interlaced as a unit with one or more filling yarns. This construction is not so firm as regular plain weaves and frequently has lower strength, but basket weaves are attractive and have interesting surface effects (Figs. 22.10, 22.11). A basket variation in which two warps pass over and under one filling would be called a 2/1 weave (Fig. 22.12). Other fabrics of the basket-weave type can be found in 2/2, 2/4, and 3/2 constructions (Figs. 22.13, 22.14), but the 2/1 is common at the present time.

Oxford cloth is an example of a 2/1 basket weave. The two warp yarns equal in size the single filling yarn. The cloth is frequently used for shirts, because it is soft and comfortable. Some sources consider oxford cloth a separate weave variation of the plain construction. However, since it gives the impression of a basket weave, it will be classified as such in this text.

Monk's cloth is one of the best-known examples of basket weave. It is available in 2/2, 4/4, 8/8, and in uneven arrangements such as 4/3, and 2/3. Rather coarse yarns are used in monk's cloth. Yarn slippage may occur because of the loose weave.

Other examples of basket weave are found in selected examples of coat and suit fabrics, hopsacking, and flat duck.

Other Plain Weave Variations

In addition to the variations produced in plain-weave fabrics by modification of the weaving pattern, other design effects can be introduced without structural change. The use of complex yarns, at either

regular or irregular intervals, may create surface texture. The amount of twist in the yarns is the basis for diverse results: high-twist yarns are used in making crepe fabrics or voiles; low-twist yarns, plus special finishing techniques, produce napped fabrics (see Chap. 26).

Yarns of different fiber content (for example, metallic yarns) or of different colors (such as chambray and gingham) can be combined to develop interesting patterns. Spacing of the yarns produces a wide variety of plain-weave fabrics, from loose, sheer cheesecloth to firm, tightly woven taffetas. The selection and care of plain-weave fabrics is discussed in Part VI.

TWILL WEAVE

The second basic weave pattern used in the manufacture of fabrics is the twill weave. This technique is characterized by a diagonal line on the face, and often on the back, of the fabric. The face diagonal can vary from a low 14-degree angle (*reclining twill*) to a 75-degree angle (*steep twill*), with 45 degrees, which is considered a medium diagonal (*regular twill*), accepted as the most common. The angle of the diagonal is determined either by the closeness of the warp ends or by the number of yarns and the actual pattern of each repeat.

Twill-weave fabrics have a distinctive and attractive appearance. In general, fabrics made by the twill interlacing are strong and durable. Twills differ from plain weave in the number of filling picks and warp ends needed to complete a pattern. The simplest twill uses three picks and three ends.

The warp yarn goes over (*floats* over) two filling yarns and under one in the 2/1 twill (Fig. 22.15). In a regular twill each succeeding float begins one pick higher or lower than the adjacent float. In more complicated twills the progression may vary, but the diagonal effect will remain visible (Fig. 22.16). The number of pick yarns required to complete the twill pattern determines the number of harnesses needed on the loom. There must be at least three, and some twill patterns require as many as fifteen.

Twill fabrics have either a right-hand or a left-hand diagonal (Figs. 22.17, 22.18). If the diagonal moves from the upper right to the lower left of the fabric, it is referred to as a *right-hand twill*; if it moves from upper left to lower right, it is a *left-hand twill*. In even twill fabrics the filling threads pass over and under the same number of warps, while in uneven twills the pick goes over either more or fewer warps than it goes under. Uneven twill fabrics have a right and wrong side and, therefore, are not considered reversible materials. *Filling-faced* twills are those in which the picks predominate on the face of the fabric; *warp-faced* twills have a more evident warp on the surface. Filling-faced and warp-faced twills are never reversible. Even twills have the same number of warp and filling yarns on both face and back,

Figure 22.15 Diagram of a 2/1 right-hand twill.

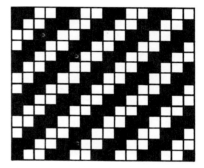

Figure 22.16 Diagram of a 2/2 right-hand twill.

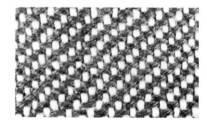

Figure 22.17 Right-hand twill fabric.

Figure 22.18 Left-hand twill fabric.

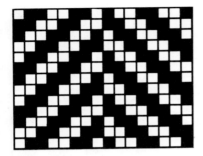

Figure 22.19 Diagram of a twill-weave variation—herringbone.

Figure 22.20 Herringbone fabric.

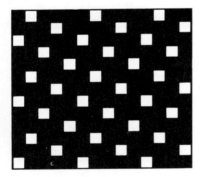

Figure 22.21 Diagram of a five-shaft satin weave.

Figure 22.22 Satin fabric.

and these fabrics can be reversible. Of course, finishes applied to one side would prevent reversing the fabric.

The yarns used in twill weaves frequently have comparatively tight twist and good strength. The twill weave permits packing yarns closer together because of fewer interlacings. This close packing can produce strong, durable fabrics, but if the yarns are packed too closely, the fabric will have reduced breaking and tear strength, abrasion resistance, and wrinkle recovery.

In addition to appearance and good strength, twills tend to show soil less quickly than plain weaves. However, twills are more expensive than plain weaves because of more complicated weaving techniques and looms required.

A frequent variation of the twill weave is the herringbone. In this design the twill is reversed at frequent intervals to form a series of inverted Vs (Figs. 22.19, 22.20).

Examples of fabrics made by the twill weave include denim, drill, jean, covert, gabardine, foulard, serge, surah, wool broadcloth, wool sharkskin, cavalry twill, elastique, flannel, and some tweeds.

SATIN WEAVE

Satin fabrics are characterized by long floats on the face (Figs. 22.21, 22.22). These floats are caught under cross threads at intersections as far apart as possible for the particular construction. At no time do parallel yarns come in contact with one another. This reduces the possibility of a diagonal effect occurring on the face of the fabric. In a satin fabric it is the warp ends that float on the surface. A variation of the satin weave, in which the filling picks float, is referred to as *sateen* (Figs. 22.23, 22.24). Filament fiber yarns are generally used for satins, while staple fiber yarns, often of cotton, are more common in sateens. There are, however, exceptions; cotton satins do exist and so do filament sateens.

The long floats of the satin weave create a shiny surface and tend to reflect light easily. This is accentuated if bright filament fiber yarns with a low twist of $\frac{1}{2}$ to 1 turn per inch comprise the floating yarns. The floats are snagged easily, and thus satin-weave fabrics are not as durable as plain or twill weaves. The length of the floats is governed by two factors: the number of harnesses, which determines the number of filling yarns over which the warp yarn floats, and the number of yarns per inch. A fabric with a high yarn count has shorter floats than one with a low yarn count when the same number of harnesses are used. When the yarn count and yarn size are held constant, the length of a float over, for example, four yarns obviously will be shorter than a float over seven yarns.

Satin-weave fabrics are lustrous. They are selected for their appearance and smoothness. Satins are frequently used as lining fabrics, because they are easy to slip on and off over other materials. There is a definite face and back to satin and sateen fabrics. Variations in satin can be produced by using highly twisted yarns in the filling to create a crepe effect on the back. This fabric is called *crepe-back satin* (or *satin-back crepe* when the crepe side is intended to be used as the right side).

Examples of satin weave include antique satin, slipper satin, crepe-back satin, satin-back crepe, faille satin, bridal satin, sateen, moleskin, and Venetian satin.

DECORATIVE WEAVES

Decorative weaves are called *fancy, figure,* and *design weaves.* They are formed by predetermined changes in the interlacing of the warp and filling yarns. This can be done by various attachments on or above the loom that increase its flexibility. Weave processes in this category include dobby, Jacquard, leno, pile, and double cloth weaves, as well as the use of extra warp or filling yarns.

Dobby Weaves

Dobby designs have small figures—such as dots, geometric designs and floral patterns—woven into the fabric (Fig. 22.25). These designs are produced by the combination of two or more basic weaves, and the loom (Pl. 3, following p. 244) may have up to thirty-two harnesses. The design is produced by a dobby pattern chain, which consists of a series of wood crossbars with metal pegs. Each crossbar controls a row of the pattern and mechanically determines which warp yarns will be raised and which lowered to produce the desired shed. Recent developments in pattern weaving include the double-cylinder dobby. This improvement vastly increases the number of potential designs that can be made by this method of weaving.

Examples of fabrics produced by dobby weaving are Bedford cord, piqué, waffle cloth, shirting madras, and huck toweling. The Bedford cord and piqué may utilize heavy yarns, called *stuffer yarns,* at the back to produce an accentuated cord.

Jacquard Designs

Fabrics with extremely complicated and decorative woven designs are manufactured using Jacquard loom attachments.

The Jacquard attachment was developed by Joseph Marie Jacquard. It was first exhibited at an industrial exposition in France in 1801, where the Emperor Napoleon saw the loom arrangement and met Jacquard. Fortunately, Napoleon visualized the importance of the development. The French government bought the idea in 1806. When

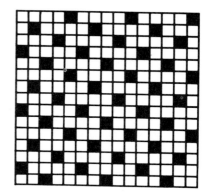

Figure 22.23   Diagram of a five-shaft sateen weave (filling satin).

Figure 22.24   Sateen fabric.

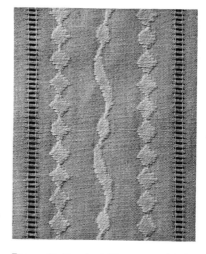

Figure 22.25   A dobby-weave fabric.

the loom was first introduced, many weavers believed it would replace them and cause widespread unemployment; considerable disorder and numerous riots resulted. However, as the transition to the factory system gained momentum, the importance of Jacquard's invention became evident, and instead of a danger it was considered a tremendous benefit to man.

The major advantage of the Jacquard machine is its ability to control each individual warp thread instead of a series as in regular harness looms. This separate yarn control provides great freedom for the fabric designer, and large, intricate motifs can be transferred to fabric. Within the last few years extremely elaborate patterns, including narrative scenes and reproductions of photographs, have been duplicated in fabric by the Jacquard device.

The pattern for the Jacquard loom is transferred to a series of perforated cards, one for each filling pick in the pattern. The card is punched to permit the lifting of knives on the machine that need to be raised to pass through the card; the others remain down. The shed is formed and the pick passes through. The punched cards, similar to a player-piano roll or an IBM card, are laced together and pass over a cylinder at the top of the machine. Each card stops on the cylinder for its particular pick, moves on, and a new card takes its place. This continues until all cards are used. When one repeat of the pattern or design has been completed, the cards start over.

The Jacquard loom is tall and requires a room with a high ceiling or openings into a second floor (Pl. 4, following). The machines are very complicated to operate and expensive to build, so the resultant fabrics are costly. As in the dobby weave, Jacquard designs are combinations of two or more of the basic weaves (Pl. 5, following). They are used for decorating fabrics, table coverings, and dressy apparel.

Examples of fabrics woven by Jacquard techniques are damask, tapestry, brocade, brocatelle, borders on terry-cloth towels, matelassé, and some bedspread fabrics (Pls. 6, 7, following). Jacquard mechanisms have been adapted for use on knitting machines in order to produce elaborate designs in knit fabrics (see p. 229).

Leno Weaves

Leno weave is also referred to as *doup weave* and sometimes as *gauze weave*. In the most correct usage doup is the name for the attachment on the loom that forms the leno weave. The doup attachment controls the warp threads and moves horizontally as well as vertically. Thus, the warp yarns can be interlaced and crossed between the picks.

When a distinction is made between leno and gauze, the term gauze is used to indicate only an open-mesh-type fabric, while leno is applied to all fabrics made by this special interlacing process. Another difference that may be encountered relates to the number of filling yarns

right: **Plate 1** Knitting machine. [*Dan River, Inc.*]

below: **Plate 2** Double-knit fabric, front (*left*) and back (*right*).

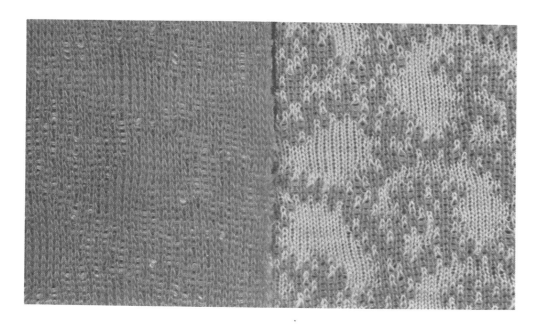

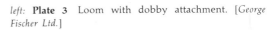

left: **Plate 3** Loom with dobby attachment. [*George Fischer Ltd.*]

below left: **Plate 4** Jacquard loom. [*Ruti Machinery Works Ltd.*]

below: **Plate 5** Fabrics woven on a Jacquard loom. [*Jack Lenor Larsen, Inc.*]

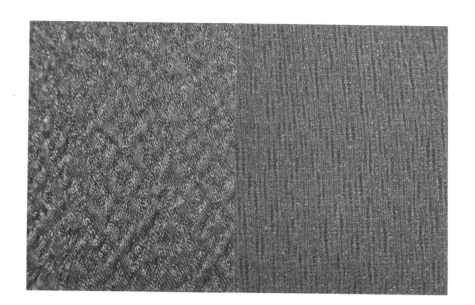

above: **Plate 6**   Matelassé fabric, front (*left*) and back (*right*).

left: **Plate 7**   Brocade (*left*) and brocatelle (*right*).

below: **Plate 8**   Lappet weave, front (*left*) and back (*right*).

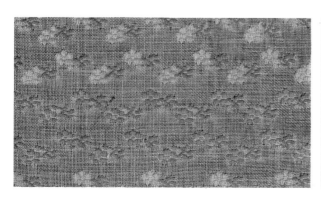

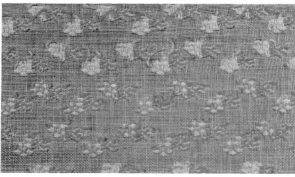

left: **Plate 9** Double cloth.

below left: **Plate 10** Tapa cloth, a nonwoven fabric from mulberry bark.

below: **Plate 11** Modern adaptation of primitive tapa cloth.

inserted between the crossing of the warps. If only one filling is involved in each crossing, the weave may be called *gauze;* if more than one, the term *leno* is applied. It is important to note that a "gauze" fabric is not always constructed by a gauze or leno weave. Gauze bandage, for example, is a plain-weave fabric.

The leno weave produces open-textured fabric that may be sheer or heavy (Figs. 22.26, 22.27). The unusual warp interlacing prevents slippage of the filling and reduces shrinkage. In addition, it adds strength to sheer fabrics. Some patterned fabrics combine one or more of the basic weaves with a leno construction.

Examples of the leno weave are found in curtain and dress marquisettes, some mosquito nets, and household bags, such as laundry and food bags.

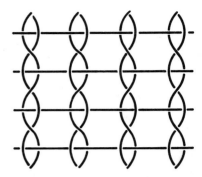

Figure 22.26 Diagram of a standard leno weave.

Surface Figure Weaves

Extra warp and filling yarns can be employed to produce many different designs. When extra warp yarns are used, they are wound on an additional warp beam and threaded into separate heddles so they can be controlled—depending upon the complexity of the pattern—either by the dobby attachment or by Jacquard mechanisms. Extra filling yarns are inserted by special shuttles, using either a box loom or a regular shuttle loom. The box loom is more satisfactory and permits considerable flexibility of design.

The three main design varieties that involve the use of extra warp or filling yarns are lappet, swivel, and spot or dot.

Lappet Weave   Lappet is a form of weaving in which extra warp threads are introduced in a manner that creates designs over predetermined portions of the base fabric (Pl. 8, following p. 244). Patterns are woven by means of an attachment to the loom, and the resultant designs resemble hand embroidery. The attachment consists of a frame or rack fastened to the loom near the reed. Long needles are carried in the frame, and the yarns to be used in making the design are threaded through these needles. When the rack is lowered, the needles are pressed to the bottom of the shed and held in position while the pick is laid. The rack is then raised, and the pick is beaten into the cloth. Next the rack is shifted sideways to a new location, and the same action is repeated. Each time the frame or rack moves sideways it carries the yarn in the needles across the surface of the fabric and creates the pattern. If long floats are formed on the back of the fabric, they are cut away; however, if the floats are short, they are usually left. This presents both an advantage and a disadvantage. If the floats are left, the pattern will be very durable, but the floats may be easily snagged. The lappet weave is considered strong, but it is comparatively expensive, so it rarely appears in fabric made in the United States.

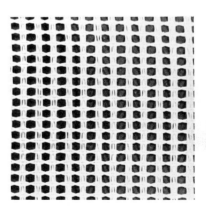

Figure 22.27 Modified leno-weave fabric.

Figure 22.28 Clipped spot fabric, face (*above*) and back (*below*).

Figure 22.29 Dotted swiss.

Figure 22.30 Uncut spot fabric.

Swivel Weave   The swivel weave differs from the lappet in that swivel designs are produced by extra filling threads. The yarns to be used in each pattern are wound on quills and placed in small shuttles. These shuttles are strategically located at each point where the design occurs. The pattern mechanism causes a shed to be made, and the shuttle carries the yarn through the shed the distance of the pattern. This is repeated for each row of the design. Between repeats the extra filling floats on the back of the fabric and is cut away after the weaving is completed.

The swivel process permits the weaving of different colors in the same row, because each figure has its own shuttle. In fabrics with small designs the swivel will save material and give figures a prominent, raised appearance. This method fastens the yarn securely as each figure is completed, and it cannot pull out. There is almost no fabric made by this process in the United States, but several Swiss manufacturers utilize it.

Swivel weaves can be recognized by the fact that the designing yarn is usually the same on face and back. It appears to go around a group of warp yarns several times.

Spot Weave   Spot or dot designs can be fabricated with extra warp or filling yarns. The yarns are inserted the entire length or width of the fabric in predetermined areas. If small, widely spaced dots or spots are made, the long floats on the reverse side are cut away, leaving dots that can be pulled out with little effort. The simplest designs use extra yarns that are different in weight and/or color from the background fabric. Designs in which the back floats have been clipped away are called *clipped spot patterns* (Fig. 22.28).

Filling floats are relatively easy to cut, but warp floats are rather difficult and require special apparatus. The durability of a spot design depends on the compactness of the background yarns that hold the design yarns in place. Spots in compact weaves are quite stable, while in loose weaves they can be pulled out of place rather easily, thus destroying the desired effect. Dotted swiss (Fig. 22.29), made in domestic fabric mills, is an example of a clipped spot weave. Despite the sheer background, this fabric is quite strong. "Eyelash" designs are also clipped spot, as are patterns with fringe effects.

Spot designs in which floating yarns are not cut are referred to as *uncut spot patterns* (Fig. 22.30). Frequently these are border designs, and the repeat patterns are often very close together. In some cases the patterns are reversible, with one side forming the mirror image of the other side. Uncut spot designs can be made with either a dobby or a Jacquard attachment on the loom to control the yarns in the design area. In both cut and uncut spot fabrics the yarns forming the design can be removed without disintegration of the background construction.

Pile Weaving

Woven-pile fabrics are those in which an extra set of warp or filling yarns is interlaced with the ground warp and filling in such a manner that loops or cut ends are produced on the surface of the fabric. The base or ground fabric may be either plain or twill weave.

Filling Pile   Filling pile fabrics are produced with two sets of filling yarns and one set of warp. Although the ground may be of twill or plain weave, twill is usually preferred for durability. The extra set of filling yarns floats over three or more warp yarns; frequently the floats are over five or seven yarns. After weaving is completed, the floats are cut and brushed up to form the pile. Velveteen and corduroy fabrics are manufactured by this method. Corduroy differs from velveteen in that the floats are interlaced in such a manner that rows are formed, and when the pile is cut, it produces a ribbed effect (Fig. 22.31).

Velveteen floats are interlaced to produce an allover effect when cut. The depth of the pile is controlled by the length of the floats: the longer the float, the deeper the pile. Pile yarns may interlace in either a V or a W form; the W is more durable, since the pile is held down by two ground yarns instead of one (Fig. 22.32).

Corduroys and velveteens are prepared for cutting in the same manner. The floats are treated to give them cutting surface, and the fabric is stiffened so it will remain smooth and firm. The pile may be cut by hand using a thin steel blade, but because this is tedious and time consuming, it is seldom used, except in countries where skilled labor is inexpensive. Most filling pile fabric is cut by machine. The fabric is held under tension and moves under sharp knife blades, which carefully cut the floats without damage to the base fabric. For corduroy, circular knives revolve and cut the rows of floats. There may be a knife for each row, or, for narrow-wale (rib) corduroy, there are knives for every other row and the fabric is put through the cutting machine twice, with the knives set to cut the alternate rows on the second run. Novel effects are obtained by cutting the floats so tufts of different lengths are produced or by cutting only certain sections to form interesting and intricate designs. Corduroy and velveteen are usually made of spun yarns.

Warp Pile   Warp-pile fabrics are those in which the pile is formed by extra warp yarns. Velvet, velour, rug velvet, and Wilton rugs are examples of warp pile. They are usually made from filament yarns. There are three general methods for constructing these fabrics: double cloth, wire-cut pile, and looped pile.

The *double cloth* technique is one of the most common methods used in manufacturing cut warp-pile fabrics. In this construction five sets of yarns are necessary: two sets of warp and two sets of filling form

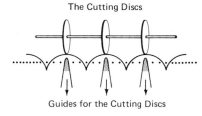

The Cutting Discs

Guides for the Cutting Discs

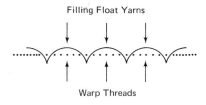

Filling Float Yarns

Warp Threads

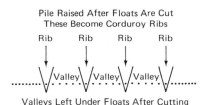

Pile Raised After Floats Are Cut
These Become Corduroy Ribs

Rib   Rib   Rib   Rib

Valley Valley Valley

Valleys Left Under Floats After Cutting
These Become Lines Between Ribs

Figure 22.31   Diagram illustrating the cutting of yarn to form corduroy. [*Cone Mills*]

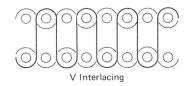

V Interlacing

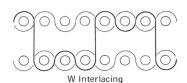

W Interlacing

Figure 22.32   Diagram of V and W interlacing of an extra set of yarns to form pile-weave fabrics, a method used when double fabric is to be cut through the center to form two distinct fabrics.

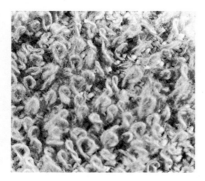

Figure 22.33 Uncut pile fabric—terry cloth.

the ground fabrics, and a third set of warp yarns is used to make the pile.

The pile yarns are interlaced with one set of ground ends and picks and then passed to the other ground set, where they interlace with those yarns. When the weaving is completed, a cut is made through the center of the pile yarns, and two separate fabrics are produced. When necessary, the pile is evened by a shearing process.

In the past, carpeting was often made by the *wire-cut* technique, but this has declined considerably. In the wire method two sets of warp yarns and one set of filling yarns are needed. The ground warp and pile warp operate independently. The pile is raised from the ground, a wire is inserted, and the pile is returned to the foundation cloth, where it is held in place by interlacing with filling yarns. At least three picks should interlace between two rows of pile loops. If the loop is to be uncut, the wires are smooth, so the loops are left intact. For cut warp pile a wire with a knife on one end cuts the loops as it is removed.

The manufacture of Wilton rugs depends on the wire technique with a Jacquard attachment. Furthermore, Wilton rugs usually employ yarns of different colors and sometimes of different structure. The colored yarns not showing on the surface at any one time are carried along below the pile, which increases the thickness of the base and produces a superior rug.

Terry-pile fabric (loop) construction is best known and most easily recognized in terry-cloth toweling. It is constructed with uncut loops of warp yarn on both sides of the cloth (Fig. 22.33). These loops are formed by holding the ground warp yarns taut and leaving the pile warp yarns slack. The shed is made, picks are placed, and this is repeated for a specified number of picks, usually three, without any beating in. After the picks have been placed, they are battened into position. This causes the slack warp yarns to be pushed into loops between the picks. Loops are formed usually on both sides of the fabric. The taut warp holds its position and remains smooth. Ply yarns are preferred for the ground warp because they are strong. The warp yarns that form the loops are soft, fluffy, and absorbent. Two low-twist yarns treated as one are frequently used for the pile to provide these desirable characteristics. Loop or uncut pile is found in such items as turkish toweling, terry cloth for robes, selected types of carpeting, and some upholstery fabrics.

Double Weaves

Double weaves are most accurately defined as those in which at least two sets of filling and at least two sets of warp yarns are interlaced. The most common types of double cloth have two sets of warp and two sets of filling, with or without a binder set. Warp-pile weaves that employ five sets of yarn are considered double cloth during the weaving, but, since they are cut apart, they are no longer double cloth when placed on the market. The fabrics considered in this unit are not severed. They are characterized by good strength, a variety of design detail, and extra weight.

At times, the term *double weave* is used to indicate fabrics in which two sets of filling and one set of warp yarns, or two sets of warp and one set of filling yarns have been interlaced. These are more accurately called *backed fabrics.*

Double cloths woven from four or five sets of yarns are found in heavy or unusual fabrics. They are frequently designed to be reversible, with compatible colors or patterns on the two sides (Pl. 9, following p. 244). For example, a coat fabric of double cloth might have a plaid face and a plain back, or one color for the face and a different but harmonizing color for the back. Matelassé and brocatelle are types of double cloth in which extra sets of yarns are used to create puffed effects in the design on the face side of the fabric.

Fabrics requiring additional bulk or those with unusual design effects can be produced with three sets of yarn, such as one warp and two filling. The warp yarn is not easily visible, for the filling threads predominate on both face and back. A common fabric of this type is found in blankets that have a different color on each side.

Yarns with varied amounts of twist and different fibers, combined by the several weaving techniques, permit the manufacture of many elaborate and attractive fabrics. Double or backed fabrics are used for design, weight, strength, and warmth. The most important double fabrics appear in coatings, blankets, and elaborate creations such as matelassé, double brocade, and brocatelle.

Other
Fabric Construction
Processes

The fabrics discussed in this chapter are primarily of unusual design or manufacture. They include new products about which little has been written and those confined to rather specific uses.

BRAIDED FABRICS

Braided fabrics are characterized by a diagonal surface effect. They are made by plaiting three or more yarns that originate from a single location and lie parallel before the interlacing occurs (Fig. 23.1). The yarns intercross from one side to the other, resulting in a column of horizontal V's. Narrow braids can be joined together to form wide fabrics or large articles, such as rugs. They are made either in flat rectangular or in circular formations.

Circular braids appear in such everyday items as shoelaces and insulation for wires. Flat braids are used for trimming. All braided fabrics have considerable stretch in the length and some in the width. This property causes problems in the application of braided trim, because it is difficult to keep even. However, the stretchiness enables a skilled person to apply braid so it lies smooth at corners and curved areas as well as on straight edges.

Fabrics made by braiding are not limited to yarn forms. Cut strips of fabric, leaves, leather, straw, or any other flexible product can be braided to create attractive and unusual fabrics.

NETS

Nets are open-mesh fabrics. They have large geometric interstices between the yarns (Fig. 23.2). Early nets were made by knotting the yarns at each point of intersection, and to some extent this is still done. Knotted nets have a comparatively large mesh and will not slip or spread. All net fabrics made before 1800 were done by hand. In 1809 a machine was developed that duplicated the net construction so accurately that only an expert could distinguish the handmade from the machine-made product.

In recent years nets have been constructed on tricot and raschel knitting machines. They still have the open-mesh effect, but since knitting only interloops the yarns, these fabrics are not as durable as those formed by the knotting technique.

Nets are used for such items as evening apparel, curtains, millinery veils, screening, and hammocks.

LACES

Lace has been defined in many ways, and authorities differ about what really constitutes lace. However, it is safe to say that lace is an open-work fabric consisting of a network of threads or yarns formed into intricate designs (Fig. 23.3).

It is not known when lace was first made, but specimens in museums have been dated as early as 2500 B.C. However, regardless of its origin, we can assume that lace was developed for beauty and adornment.

Lace is truly the aristocrat of textile fabrics. No other material is so difficult to make yet so delicate, requires so much proficiency in manufacturing, or is so demanding of creative ability. Everything about lace is different. It is a product of yarn twisting, and the machinery for lacemaking is among the most complicated known.

The development of lacemaking equipment involves some interesting history. Several early inventors devoted considerable time and energy to finding mechanical devices that would reproduce the intricate designs of handmade lace. In 1808 John Heathcote completed a machine that successfully made bobbinet. This was followed in 1813 by the first Leavers machine. John Leavers and his brothers continued improving the machine and in 1837 incorporated the motion of the Jacquard loom attachment. The resulting product, the Leavers lacemaking machine, remains essentially the same today as it was in 1837.

The Leavers machine is the most complicated piece of textile

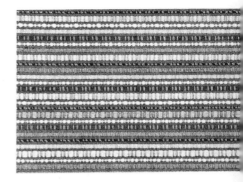

Figure 23.1   Braided yarn in filling of fabric. [*Jack Lenor Larsen, Inc.*]

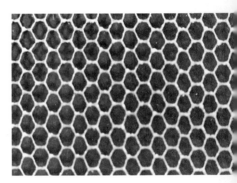

Figure 23.2   Net fabric.

Figure 23.3   Lace fabric.

equipment in the world. Each machine has approximately *40,000* parts, most of them movable and built to a tolerance measured in thousandths of an inch. A single machine weighs more than 30,000 pounds and requires 500 square feet of floor space. Despite its size, it produces some of the finest, most delicate fabrics known anywhere. In the Leavers lace machine, rather than one shuttle that carries the filling picks back and forth, there are about 4000 bobbins, each of minute size, spaced approximately nineteen to the inch. These bobbins are mechanically controlled to move in all directions and will intertwist with the warp threads according to a planned pattern.

Yarns used in making lace are stronger and more firmly twisted than those for other types of fabrics. The adaptation of man-made fibers, such as nylon with its inherent high strength, has added durability to lace.

Lace can be manufactured in many widths and shapes. The variety of patterns is practically limitless. Specific characteristics are associated with specific laces, and the following terms will help the consumer identify them.

Alençon lace is a delicate and durable lace of solid design outlined with cord on a sheer net background. It cannot be called "Alençon" unless it is actually manufactured in that French city.

Allover lace is a fabric approximately 36 inches wide with a pattern repeated over the entire surface. The pattern may be large or small.

Antique lace is handmade bobbin lace of heavy thread with a large, often irregular, square-knotted net base on which designs are darned.

Binche lace is Flemish bobbin lace that has a scroll floral pattern and a six–pointed star ground, sprinkled with snowflake-type figures.

Chantilly lace is bobbin lace with a fine ground and pattern. The designs are outlined by cordonnet of thick, silky threads. (*Cordonnet* refers to a three-ply yarn of nylon or silk. Singles are Z twist, ply is S twist.) The fabric is popular for hat and bridal veils. Originally made of silk, chantilly is now created from several man-made fibers.

Galloon lace is narrow lace with scallops on parallel edges.

Insertion lace is a narrow strip of lace with the long parallel edges smooth, so it can be attached between two pieces of fabric.

Needlepoint lace is made entirely with a sewing needle, rather than with bobbins. It is worked with buttonhole and blanket stitches over a design on paper.

Rose-point lace is Venetian needlepoint lace with a delicate and full design of flowers, foliage, and scrolls connected by string cordonnet.

Tatting lace is knotted lace worked with the fingers and a shuttle.

Valenciennes lace (*val lace*) is flat bobbin lace worked in one piece with the same thread forming both the ground and the design. The ground openings may be round or diamond-shaped. It is usually narrow in width.

Venice lace (*venise*) is a needlepoint lace decorated with floral motifs and designs connected with irregularly placed picot edges.

Most lace fabrics are fragile and require careful handling, but those made of strong fibers such as nylon, are durable and give long life. Regardless of its properties, lace is treasured for its decorative appearance. Whether selected for a single occasion or for long-term use, its function is to adorn.

Laces made on Leavers machines are expensive. In order to provide a variety of "lace" fabrics at competitive prices, manufacturers now produce a relatively high proportion on raschel knitting machines (see p. 231). These fabrics have the same general appearance as true laces, and it is difficult for the consumer to distinguish one from the other. However, a true lace has knotted yarns, while raschel laces are made by interlooping patterns. Both types give good service that belies their fragile and delicate appearance.

TAPA CLOTH

One of the earliest fabrics known to man was created from the bark of the paper mulberry tree by natives in the South Sea Islands. This fabric, called *tapa* (Pl. 10, following p. 244), is still made by hand in the South Pacific. The bark is cut into thin layers, soaked for a period of time, and then pounded into a thin, filmy layer. The layers are combined to produce the desired weight of the final product. Tapa cloth is usually printed by hand with native designs, using colors in the tan and brown range with touches of black. The natural color of the tapa is a light tan. The fabric is used for clothing and for indoor matting by the natives. It is a frequent purchase of tourists in Polynesia.

Recently a series of cotton fabrics inspired by the original tapa designs have been placed on the market (Pl. 11, following p. 244).

FILM FABRICS

Films are not true textiles in that they are not composed of fibers. However, because they are used today for such a wide variety of products, it seems advisable to mention them (Fig. 23.4). Films may be clear and transparent, colored and transparent, translucent, or opaque. A frequent application is protective clothing, such as rainwear. These films are made from the same chemicals as some of the man-made fibers, but they are extruded in sheets instead of filaments. This difference can be visualized if one compares Saran Wrap®, a film, with saran fiber, which is used in rugs and furniture webbing.

The films vary in thickness, from thin layers such as those used in rainwear to heavy vinyl films for upholstery. Many films are supported or laminated; that is, they are sealed onto a fabric base that has been

Figure 23.4 A clear vinyl printed with opaque enamel, for drapery or wall panel. [*Jack Lenor Larsen, Inc.*]

knitted or woven. This lamination adds resistance to tearing and increases durability. Nonsupported films are not considered "long-life" fabrics.

MULTICOMPONENT FABRICS

Multicomponents include *bonded fabrics, laminated fabrics,* and *foam-backed fabrics.* Some authorities group all these products under the heading of bonded fabrics, while others classify them as laminates. Inasmuch as there are differences among them, they will be discussed as subtypes of multicomponent fabrics.

A multicomponent fabric is one in which at least two layers of material are sealed together by some effective adhesive. The adhesive may be a thermosetting resin, a thermoplastic material, or heat-treated polyurethane foams.

The most common multicomponent is made of two fabrics securely bonded or sealed together by an adhesive. These structures frequently use tricot knit as the backing fabric, in order to provide a self lining and to stabilize the fabric. Bonded fabrics do have good stability, resistance to stretch or deformation, and a low incidence of raveling. However, the bonding exhibits widely varying properties, and consumers may encounter difficulty, though recent products of this type have shown considerable improvement.

Another type of multicomponent has a layer of fabric bonded to a layer of foam, which adds warmth and stability. In sandwich construction a layer of fabric is bonded to a second fabric by means of a thin layer of foam in the center. Besides providing stability and warmth, the sandwich laminate has a finished appearance on both sides.

Among the other multicomponent products are the following:

1.   film laminated to fabric. This creates an attractive plastic-effect fabric with the durability of a woven or knit structure.
2.   film laminated to foam. These fabrics are relatively durable and have good insulative properties.
3.   two fabrics sealed together with heat and/or adhesives by the chemstitch method. This technique provides for the crinkling or rippling of one of the fabrics to achieve a quilted effect.
4.   a layer of fabric bonded to a layer of fibers. This results in an interesting material for special end-uses, with the added advantages of warmth, insulation, and bulk.

Multicomponent fabrics are gaining popularity. In addition to the built-in lining effect, they offer easy construction into end-use items by both manufacturers and home sewers; greater comfort than single-layer fabrics, because of the smooth inner surface; resistance to wrinkling; and easy-care properties if the fabrics are adaptable to laundering.

The consumer should examine multicomponent fabrics to determine that

- the fabric layers are sealed together firmly
- the grain lines are held in their proper relationship
- adequate care instructions are included
- there is no unusual odor
- there is no stiffness
- the adhesive is not visible
- the foam, if used, is not discolored

TUFTED FABRICS

Handmade tufted fabrics originated in the American Colonial period. At that time hand tufting was practiced as an art and was limited to making fabrics for special uses. About 1900 the craft was revived, and machines were developed to produce tufted fabrics at rapid speeds.

Tufting is a process of manufacturing pile fabrics by inserting loops into an already woven ground fabric (Figs. 23.5, 23.6). This ground fabric may be of any type and composed of any fiber, but most tufted fabrics use a base of cotton, linen, or jute in a close or heavy weave. The tuft yarns may also be of any fiber. Cotton, rayon, wool, acrylic, polyester, nylon, and acetate fibers are popular in tufted fabrics.

Pile loops, cut or uncut on the face, are inserted with needles and held in place either by a special coating, applied to the back or by untwisting the tufted yarn and shrinking the base fabric. Tufted fabrics are less costly than their woven counterparts.

Major growth in the production of machine-made tufted fabrics has occurred since 1940. In that year the first tufting machine wide enough to handle a bedspread was built by Joseph Cobble. Continued research resulted in modern equipment capable of producing as many as 300 bedspreads in an eight-hour period.

Tufted carpets and rugs made their appearance in 1950 with the development of special machinery. The success and popularity of these

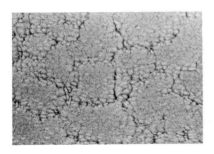
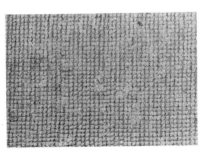

left: **Figure 23.5** Tufted carpeting, face.

right: **Figure 23.6** Tufted carpeting, back. The photograph shows the tuft yarns lying against the base fabric. Like most carpeting at the present time, this piece was backed with jute fabric to improve dimensional stability.

carpets has been phenomenal. Approximately 90 percent of all carpets produced in the United States are made by the tufting process. (Another 5 to 6 percent are needle punched leaving a very small amount produced by weaving methods.) This predominance of tufted carpets derives from the fact that a great variety of fibers can be used successfully, and durable carpets can be made for lower cost than by other construction methods. A few firms are marketing tufted blankets, but in this area needle-punched fabrics appear to have the advantage.

By controlling the amount of yarn being fed to each needle, the size of the loop is determined, and this in turn, produces variations in texture and design. Tufts can be cut or uncut, and combinations of both are used to create interesting effects.

Identification of tufted fabrics is comparatively simple if there is no secondary backing. Tufts run in the lengthwise direction, and the underside shows the back loop of the tufts as a separate yarn, not an integral part of the weave. However, most tufted carpeting is either sealed to another fabric to increase dimensional stability and prevent damage, or is coated with a heavy layer of adhesive or sealing compound. These backings make it difficult to see the characteristic tufted structure.

Color depends upon the arrangement of colored yarns or the use of space-dyed yarns. Other design effects can be achieved by the spacing and height of the tufts.

Textile manufacturers agree that tufted fabric production will continue to expand, especially in home-furnishing fabrics.

POROMERIC STRUCTURES

DuPont introduced a poromeric structure to the public in the early 1960s. It is trademarked Corfam® and is used primarily in shoe uppers. Corfam resembles leather, but it is a truly synthetic product. DuPont describes it as a microporous and coriaceous material of multilayer construction. It is said to be a microporous urethane polymer reinforced with polyester fiber.

Unlike leather, Corfam permits the free passage of air and moisture, so the shoes are more comfortable. The product wears very well and is easy to clean. It does not scuff readily.

Several other manufacturers are producing materials similar to Corfam, but as yet the output has been limited, and characteristics have not been evaluated.

In late 1971 DuPont sold the rights to Corfam. It is to be produced, but there is no information on quantity or marketing plans.

Engineered Yarns
and Fabrics

Engineered yarns and fabrics are those in which two or more textile fibers are used. The fibers can be blended or combined in various yarn and fabric structures:

1. Different fiber types can be blended in single yarns.
2. Yarns of different fiber content (that is, single yarns each of one specific fiber) can be woven into fabric.
3. Single yarns of different fibers can be plied together.

FIBER BLENDS

The accepted definition of a *blend*, as stated by ASTM, is

> a single yarn spun from a blend or mixture of different fiber species.

According to this definition, only the first type of fiber mixture listed above would qualify as a true blend. Unfortunately, many consumers have come to associate the term *blend* with any fabric containing two or more fibers, regardless of how they are used. As a rule, woven fabrics designated as blends by fabric producers employ blended yarns throughout the construction. Exceptions to this might occur when the warp yarn is a blend and the filling a single yarn, or vice versa. Fabrics in which blended yarns are used throughout will be more likely to give desired characteristics. Knit fabrics that are true blends generally have blended yarns throughout the construction. Nonwoven fabrics are blends when two or more species of fiber are used in constructing the fiber mat.

Blends are composed of various percentages of the fibers involved. Thus, a blend of 65 percent polyester and 35 percent cotton utilizes yarns in which each single yarn strand has approximately two polyester fibers for each cotton fiber.

For most blends on the textile market, optimum percentages have been established for at least one of the fibers involved. For example, it has been fairly well agreed among textile manufacturers that in blends of polyester and cotton, the percentage of polyester should range between 50 and 65 percent. A blend of 55 percent acrylic with 45 percent wool results in a fabric with satisfactory washability, and 15 percent nylon fiber is frequently added to wool that has been finished by chlorination to produce "washable" wools. The nylon increases abrasion resistance and helps prevent damage to the wool from the chlorination process. In some cases nylon may be added to low-grade wool even when there has been no chlorination treatment.

In manufacturing blend fabrics the fibers may be intimately mixed before yarn manufacturing, or the blending may occur during the drawing operations. For nonwovens the fibers are intimately mixed before entering the felting or bonding machinery. Unless a thorough blending occurs, the end product will not be uniform.

A factor of major importance in building blend fabrics is that all man-made fibers must be cut to a length comparable to that of the natural fiber involved in order to facilitate yarn production. If all fibers in the blend are man-made, they must be cut to the length that is used effectively on the particular yarn-making machinery employed. For example, if a blend of rayon and polyester is to be processed into yarn on cotton machinery, the fibers are cut to the size of an average cotton fiber; if processed on wool equipment, the fiber length should be comparable to average wool fibers.

Engineered textile fabrics are the result of considerable research, development, and testing. Instead of developing new fibers, manufacturers are concentrating on combining existing fibers in various ways to produce yarns and fabrics with specific qualities. A blend that is properly engineered exhibits the most desirable properties of all fibers used.

Blends can be developed to provide the consumer with special performance qualities or to meet predetermined end-use requirements (Fig. 24.1). They may be designed strictly for appearance; to combine appearance and performance; to include small quantities of a luxury

Figure 24.1 Canvas jeans of 50 percent Dacron polyester and 50 percent cotton by Avondale. Pullover of stretch nylon combined with Lycra spandex. [*E. I. DuPont de Nemours & Company*]

fiber for prestige effects; or as a means of reducing cost. Blends of polyester and cotton are usually sought for their performance. Rayon and acetate may be blended for appearance, particularly if a subtle cross-dye effect is desired, and for performance, especially if a fabric with appealing drape and hand is the goal. Silk, vicuña, or cashmere are sometimes blended with a less costly fiber to lend prestige to the fabric. These fibers do have many desirable properties and, if used in sufficient amounts, contribute pleasing characteristics to the final fabric. Blends that contain relatively low-cost fibers—such as rayon, acetate, or cotton—with more expensive fibers—acrylics, polyesters, or nylons—will be less costly than fabrics composed entirely of the higher-cost fiber.

COMBINATION FIBER FABRICS

Combination fabrics are also composed of two or more fibers. However, instead of each single strand of yarn being an intimate blend of the fibers involved, each yarn is made from a single species of fiber, and the combination is developed by using some yarns of one fiber, and some of another. Many combination fabrics have yarns of one fiber in the warp and yarns of a second fiber in the filling. When more than two fiber species are included, the yarns are arranged in a manner that creates special design effects, special color effects by cross dyeing or yarn dyeing, or desired performance properties. Checks can be produced in combination fiber fabrics by skillful arrangement of the yarns so they can be cross dyed to create the desired pattern. Some stripes are obtained in the same manner.

Strength is introduced into combination fiber fabrics by using yarns of high breaking load in the direction that requires extra resistance to force or by spacing strong yarns among yarns of low-tenacity fibers, so the strong yarns increase resistance to breaking force. Nylon yarns in the warp direction, with a low-tenacity fiber in yarns for the filling, would produce such a product.

Fabrics are manufactured that are neither true blends nor true combinations. For example, a fabric recently used for lingerie has a nylon yarn in the warp and an intimate blend yarn of cotton and polyester in the filling. Other fabrics that use fibers in this manner occasionally appear on the market.

Engineered fabrics are usually developed for predetermined end-uses, and they perform best under these circumstances. However, the product manufacturer may select engineered fabrics for end-use items to which they are not suitable. In such instances the consumer may not be satisfied with the item.

It is particularly important for the fiber content of a blend or combination fabric to be labeled properly. Unfortunately, the law does not require identification regarding fiber arrangement. For example, if an item is labeled 50 percent rayon and 50 percent polyester, the consumer cannot be sure the fibers have been intimately blended. Should the fabric be a combination, with warp threads of rayon and filling of polyester, it would not exhibit the wrinkle resistance of a blend of these two fibers.

Combinations of this type can often be verified by applying the burning test to the yarns; however, not all combination fabrics respond with adequate individuality to provide a discerning test result. Moreover, few consumers have sufficient knowledge to benefit from the test if it did apply. Fortunately, the better-known and reliable manufacturers select end-use fibers with consumer satisfaction uppermost in their minds. They choose fibers, yarns, or fabrics they believe will give the most satisfactory performance.

Finish
and Color
Application

"The finishing of a fabric marks the occasion of its birth."[1] This statement is not meant to be amusing; it is a fundamental fact. A modern fabric, as the average consumer knows it, does not exist until it has been subjected to various finishing procedures.

Most fabrics that reach the consumer market have received one or more finishing treatments, and, except for white fabrics, color in some form has been applied. The textile industry tends to consider the application of dyestuffs as a finishing step, but for convenience, dyestuffs are discussed in separate chapters in this text.

The history of finishes (excluding color) is sketchy. Smoothing fabrics on flat stone surfaces was the forerunner of calendering; application of white clay was a rudimentary sizing. The first written information concerning finishes dates from the mid-nineteenth century. We assume that, before that time, routine finishes were anything but routine. Mercerization is one of the oldest finishes. John Mercer discovered the effect of caustic soda on cotton in 1853, but it was H. A. Lowe who perfected the process in 1889. Shrinkage control by finishes such as Sanforization was developed, for the most part, in the twentieth century.

This text has divided finishes into two major groups: (1) routine or general, and (2) functional (including those that influence final

[1]"Modern Finishes," *American Fabrics*, No. 28 (Reporter Publications, Inc.), p. 56.

appearance). Each of these can be either mechanical or chemical and durable or renewable.

Stretch fabrics are, in a sense, a summation of textile science. They could, with some justification, be included in any of the main sub-divisions of this book, for they can be created by fiber technology, by methods of yarn production, by fabric-construction techniques, or by the application of special finishes. Because these fabrics do encompass the entire range of textile technology, they have been placed in Part V at the end of our discussion of finishes.

Routine
or General
Finishes

The application of finishes is the province of the converting industry. Manufacturers in this group devote their research and production to finishes that will change, improve, or develop the appearance or desired behavior characteristics of a fabric. Fabrics are said to be in the *greige* (or gray) before finishing. This term does not indicate a specific color, for the unfinished fabric may be tan, gray, white, or already colored through yarn or fiber dyeing techniques.

Finishes can be classified into various groups. There are mechanical or chemical finishes, permanent or nonpermanent finishes (preferably called *durable* or *renewable*), general or functional finishes, and finishes that do or do not alter appearance.

Mechanical finishes are applied to fabrics by equipment such as copper plates, perforated cylinders, or tentering frames. Chemical finishes are those in which acids, alkalies, bleaches, detergents, resins, or other chemical substances are used to produce the end product. Scientific development in chemical finishes during the past two decades has been extensive. Such end-products as minimum-care cottons and shrink-resistant wools are examples.

Finishes are said to be durable if they can withstand a "normal" amount of wear. Normal is, obviously, a relative term, and one person's interpretation may be very different from another's. The industry measures the durability or permanence of a finish by its ability to withstand tests designed to simulate average use and care. Renewable finishes are those that rub off or are removed easily by washing or dry cleaning They can sometimes be replaced.

Routine or general finishes are discussed in this chapter in alphabetical order and are identified as mechanical or chemical. Durability or permanence of finish and of the fibers and fabrics that usually receive the treatment are also considered.

Routine finishes may or may not influence the type of care a fabric should receive. Since most of them are used to produce a standard fabric type, care procedures are not explained except in special instances.

BEETLING

Beetling is a mechanical finish applied to cotton and linen fabrics. It increases the luster of fabrics by flattening the yarns to provide more area for light reflection. The fabric is fed over rolls that rotate in a machine where large hammers rise and fall on the surface of the fabric. The continued pounding flattens the yarns and closes the weave. The beetled finish will withstand wear and maintenance if the fabric is laundered carefully and ironed with pressure to restore the flat appearance. Fabrics for table coverings are often beetled to add luster and to make them lie flat.

BLEACHING

Fabrics, yarns, or fibers can be bleached to make them white or to prepare them for dyeing or printing. Bleaching is a chemical finish. It is relatively durable when the bleaching method is appropriate to the fiber or fibers involved. Whiteness retention of textile products is important to the consumer and may require frequent bleaching during the life of the article.

The public seldom considers the effect of preliminary bleaching on colored fabrics. However, if a fiber or fabric has not been properly bleached during routine finishing, it may return to its natural color, thus causing a change in the color applied to the fabric. For example, if wool is bleached by a reduction process, it will reoxidize in the presence of oxygen in the air and return to its natural yellowish color. When this happens, a light blue wool could become green-blue.

The particular chemical used for bleaching depends on the textile fiber. Cellulosic fibers, such as cotton, are bleached with chlorine compounds (sodium hypochlorite is one of the chemicals used) or hydrogen peroxide. Perborate bleaches are available for home use, but they are seldom employed commercially, since hydrogen peroxide reacts in much the same way and is more effective.

Chlorine compounds for bleaching protein fibers result in loss of fiber strength and eventual deterioration of the fabric. Silk responds well to hydrogen peroxide, and wool can be bleached by hydrogen peroxide or by sulfur dioxide gas in moisture, forming sulfurous acid. Hydrogen peroxide is an oxidation type of bleach, while the sulfur dioxide process results in a reduction bleach. The latter, called *stoving*, is less durable on wool, because the wool may reoxidize in air and return to its original color. Man-made fibers sometimes require bleaching, but if they are properly processed during manufacture, many fibers are naturally white.

The bleaching process involves several steps. First the fabric is saturated in the solution of bleach, then the temperature is raised to that recommended for the bleach and fiber involved. The bleaching action is allowed to continue for a prescribed length of time. Finally, the fabric is thoroughly rinsed and dried.

A current trend in bleaching is the use of optical brighteners. These substances react rather like dyes and are applied from a solution. They alter the reflectance characteristics of the surface of the fabric, producing a visual effect of whiteness. (See p. 280 for a full discussion of optical brighteners.)

BRUSHING

Brushing is a mechanical finish. It involves the removal of short, loose fibers from the surface of the fabric. Cylinders covered with fine bristles rotate over the fabric, pick up loose fibers, and pull them away by either gravity or a vacuum. This finish is usually applied to fabrics of staple fiber content. Filament yarns do not have loose fiber ends unless breakage of the filaments has occurred during previous processing stages.

CALENDERING (PRESSING)

Calendering is applied to cottons, linens, and silk, as well as to rayons and other man-made fiber fabrics. *Pressing* is the term used for wool fabrics. Basic calendering and pressing are mechanical processes and must be renewed after each laundering or cleaning.

The finish is similar to ironing but is done with much greater pressure (Figs. 25.1, 25.2). It gives a smooth surface to fabrics. More complicated calendering processes include moiréing, embossing, and schreinerizing. Because these add design to fabrics and can be combined with other finishes, they are discussed with functional finishes in Chapter 26.

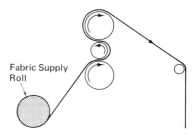

Fabric Supply Roll

Figure 25.1 Diagram of calender rolls, showing the method by which fabric passes through and around the rolls.

Figure 25.2 A calender machine. [*Pepperell Manufacturing Company*]

CARBONIZING

Carbonizing is a chemical finish applied to wool fabrics. Wool yarns and fabrics frequently contain vegetable matter that was not completely removed during carding. To eliminate this, the wool fabric is immersed in a solution of sulfuric acid; it is then subjected to high temperatures for a brief period of time. The acid and heat react to convert the vegetable matter to carbon, which is easily removed by a final scouring and, if necessary, brushing. The process must be carefully controlled to prevent fiber damage, which would result in weakened fabrics.

CRABBING

A mechanical finish applied to wool fabrics, crabbing permanently sets the weave. The fabric is immersed in first hot then cold water, and passed between rollers. If properly fed into the rollers, the warp and filling yarns are set at a true 90-degree angle to each other. Improper crabbing contributes to "off-grain" fabric.

DECATING

Decating is a mechanical finish. It is used on wool, silk, rayon, and blends that include these fibers and can be done wet or dry. The finish helps to set the luster, and in wool especially, develops a permanent sheen. On rayons, silks, and blends it softens the hand, delays the appearance of breaks and cracks, reduces shine, and helps even the set and grain of the fabric.

The decating process involves pressure and moisture. In dry decating, steam and then cold air are forced through the fabric; in wet decating, first hot and then cold water is forced through the fabric. This is followed, in both instances, by a final pressing.

FULLING

Fulling is a mechanical finish applied to wool to produce a compact fabric. When wool is removed from the loom, it bears little resemblance to the fabric the public associates with wool. It is loose and hard in texture. To make the fabric compact and soft it is fulled by applying the proper amount of moisture, heat, and friction. The fabric yarns shrink together and the fabric softens to the desired texture. Worsteds are usually fulled less than woolens.

HEAT SETTING

Thermoplastic (heat sensitive) fibers are generally given a heat-setting finish to produce a special shape or to ensure a stable fiber. While the process is mechanical, the heat changes the physical characteristics of the polymer.

These fibers have both a melting temperature and a glass transition temperature (T_g). The glass transition temperature is the point at which the amorphous regions of the fiber develop flow or "melts." At this temperature fibers can be shaped. Only the interior of the fiber need reach the T_g temperature.

In addition to the flow of the amorphous regions, there are some changes in the crystalline areas of the polymer. Small crystallites may disappear and large crystallites may grow larger and assume the configuration in which the fiber is held—smooth, bent, folded, or uneven.

The rules for heat setting are that (a) fiber temperature must reach or exceed the T_g before the fiber is allowed to cool; and (b) any subsequent resetting or reshaping must be done at a temperature greater than that used in the first setting.

A major reason for heat setting is to introduce dimensional stability, and the degree of dimensional stability is determined by the temperature, the period of exposure, and the amount of force used to hold the fabric in the desired shape and size during setting. Other characteristics introduced by heat setting include resiliency, which contributes to wrinkle resistance; elastic recovery, which aids in size retention; and relatively permanent design details, such as pleats, planned creases, or surface embossing. Fabrics can be heat-set in a smooth, flat shape or with pleats and creases pressed in. They can even be made to assume an end-use shape, such as nylon hosiery.

If at any later time the fabric is exposed to temperatures higher than the heat-setting temperature, or if it is subjected to heat for an extended period of time (longer than the heat-setting period), it will take on a new shape. For example, if the fabric is wrinkled or folded where it should not be, these undesirable conditions are permanently set into the material. To prevent this, certain maintenance requirements must be satisfied. Laundering, drying, and smoothing temperatures must be safely below the glass transition temperature of the fiber.

INSPECTION

Fabric inspection involves three possible steps. Originally only wool fabrics received detailed inspection, but today nearly every fabric is examined before it leaves the manufacturing plant. The process is, obviously, mechanical and somewhat subjective.

Perching

Perching is a visual inspection. The name derives from the frame, called a *perch*, of frosted glass with lights behind and above it. The fabric passes over the perch and is inspected visually (Fig. 25.3). Flaws, stains or spots, yarn knots, and other imperfections are marked.

Figure 25.3 Perching—a visual inspection of fabric. [*Pepperell Manufacturing Company*]

Burling

While burling is usually applied to wool, it, too, is being used now in relation to other fibers. Burling is the removal of yarn knots or other imperfections that can be repaired without producing inferior fabrics.

Mending

Mending is, obviously, the actual repair of imperfections. It may leave marks that result in fabric being classified as "second quality," or it may be done so the repair is not visible.

MERCERIZATION

Mercerization is a chemical finish applied to cellulosic fibers, especially cotton. It adds a luster to fabric, improves dyeing characteristics, and increases strength. In mercerizing, yarns or fabrics are immersed in a solution of 18 to 27 percent sodium hydroxide (caustic soda). For conventional mercerization the yarns or fabrics are held under tension during the finishing procedures. Slack mercerization is one method of introducing stretch properties into fabrics (see Chap. 28).

The finish swells the fibers, giving them a round cross section that reflects light to create a gloss or sheen. The natural twist of cotton fiber is largely removed. Mercerization under tension produces fibers with increased strength and increased affinity for dyestuffs.

Although chemicals are used in the mercerization process, the change in the fiber is actually physical, for the cellulose fibers experience some molecular rearrangement. The crystallinity is decreased, and the arrangement of the molecules is such that they share stresses more equally than before, so the fiber has increased tenacity. After exposure to the sodium hydroxide, the yarn or fabric is rinsed several times and then given a cold acid sour bath to neutralize any remaining alkali. This is followed by additional rinses to assure complete removal of the acid from the fabric. The yarn or fabric is held under tension until the last rinse.

SCOURING

Scouring procedures vary depending upon the fibers involved. Some fabrics are sold as they come from the loom, while others are scoured to remove foreign materials that might be present. The latter include natural waxes, dirt, processing oils, and sizing compounds used on yarns during weaving. Fugitive colors introduced for yarn identification are removed by scouring.

Scouring and bleaching are sometimes carried out together in large kettles called *kiers*. Until recently most fabrics were scoured in rope form, but today many converting plants use a continuous scouring and bleaching technique, in which the fabric is processed in full width in a J box. This new method generally uses hydrogen peroxide as a bleaching agent on all fibers, and minimum damage is done to the fabrics.

Soaps or synthetic detergents with alkaline builders are the common scouring agents. For protein fibers a neutral or slightly acidic synthetic detergent is often used. Fibers with natural impurities or fabrics with noticeable amounts of sizing are the most frequent candidates for scouring. Despite the use of chemicals, no molecular change occurs.

SHEARING

Shearing is a mechanical process applied to fabrics of most natural fibers and to many fabrics constructed from staple-length man-made fibers. It involves cutting or shearing off undesirable surface fibers or evening nap or pile.

After singeing and subsequent processing, fiber ends or loose fibers may protrude from the fabric surface. Shearing cuts off these ends and permits a clear view of the weave. For pile or napped fabrics, shearing evens the surface to give a uniform appearance. By manipulating the shearing it is possible, also, to cut designs into pile fabrics.

The shearing machine has a wide, spiral cylinder to which cutting blades are attached. It resembles a lawn mower in action. The fabric passes over brushes that raise the fiber ends or the fabric nap, then it moves over the cutting blades.

SINGEING

Singeing consists of burning off the fuzz of fiber ends on fabric in order to obtain a smooth surface. Fabrics of natural fibers and staple-length man-made fibers can be singed to produce a clear, smooth appearance.

Before singeing, the cloth is brushed to remove loose fibers, lint, and dust. The fabrics are singed in full width, and they are kept flat and free of wrinkles, creases, and curled selvages. In one method the fabric passes over two or more heated plates or rollers; the second plate or roller is red hot and singes the surface. Another system is to pass the fabric directly over open gas flames. The fabrics are moving rapidly, 100 to 300 yards per minute, to prevent damage. After singeing, the fabric is immersed in water to extinguish any sparks or afterglow.

Singeing is basically a chemical finish, since the reaction is one of oxidation.

SIZING

Sizing is the application of various materials to a fabric to produce stiffness or firmness. It is a chemical process in that substances are added to the fabric. Cellulose fabrics are sized with starch or resins. (The resins are usually a part of other finishing processes, so they are considered with functional finishes in Chapter 26.)

Starch is applied to cellulosics, particularly cotton, to create luster and to improve the body of the fabric (Fig. 25.4). It adds weight and can make inferior fabrics look attractive until laundered. Starch also prevents fabrics from soiling quickly.

In one procedure the fabric passes through the starch solution and then between rollers that pad the starch into the fabric and remove excess solution. Another technique passes the fabric over rollers that revolve in starch solution. The roller deposits the starch and pads it into the fabric. Starch is generally considered a nonpermanent finish. Gelatin and dextrins also may be used.

TENTERING

Tentering is the mechanical straightening and drying of fabrics. A clip or pin tenter frame is used (Fig. 25.5). The fabric is held between two parallel chains, either by clips or by curved pins. The chains spread apart to the desired fabric width, move with the fabric through the drying unit, and release the fabric to be rolled or folded onto bolts.

If the fabric is picked up by the tenter chains in such a manner that the filling yarns are not absolutely perpendicular to the warp yarns, the fabric is finished off grain and exhibits *skew*. This poses problems

left: **Figure 25.4** Application of starch sizing. [*Pepperell Manufacturing Company*]

right: **Figure 25.5** Tentering of fabric. The photograph shows a tentering machine with the fabric held by clips. [*Pepperell Manufacturing Company*]

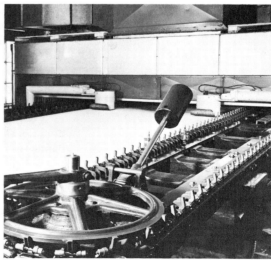

in use. When resin finishes have been applied, or if fibers are thermoplastic and have been heat-set, the grain cannot be corrected.

There are a number of devices that help to set the fabric on true grain. A tenter with a variable chain drive enables the operator to slow down one chain and keep the filling threads in their proper location. The same variable mechanism can be controlled by "electric-eye" sensors that adjust the speed of the individual chains. The controls stop the tenter if the grain becomes too crooked and provide for readjustment of the fabric.

The marks of the clips or pins used to hold the fabric are frequently seen on the selvage edge of fabrics. Some fabrics are made with heavy selvages to prevent or reduce tenter damage to the fabric.

Heat setting of man-made fiber fabrics is frequently combined with tentering. If the feed rolls operate more rapidly than the fabric take-up rolls, slack occurs in the width and permits the fabric to shrink to new stable dimensions. If heat-set fabrics are stretched during tentering, shrinkage may occur in laundering, tumble drying, or ironing at temperatures considerably less than those used for heat setting.

WEIGHTING

Weighting is a sizing technique applied to silk fabrics. After complete degumming, silk fibers are very soft. To make heavy or stiff materials, manufacturers resort to weighting the fabrics with metallic salts, such as stannous chloride. The absorbency characteristics of silk protein make this a feasible procedure. However, weighting, if overdone, causes silk fabrics to crack and split.

The Federal Trade Commission established rulings in 1938 that require silk fabrics with more than 10-percent weighting (except black, which may have 15 percent) to be labeled with the proper informative data. Silk may be labeled as "pure dye" if the weighting is less than the established 10 or 15 percent. Some manufacturers are using sericin and other organic gums to stiffen the fabric. This is satisfactory as long as products are dry cleaned, but washing will gradually remove some types of gum. Sericin particularly accentuates the tendency for silk to water spot.

Weighted silk has body and density, but the fabrics are not as durable, since they are more sensitive to sunlight, air, and perspiration damage.

26

Special Finishes

Despite their name, many "special" finishes have come to be regarded as essential by the consumer. They can be divided into two major categories: finishes that change or modify the *appearance* and/or *hand* of the completed fabric, and *functional finishes,* which alter, improve, or change the *behavior* or service characteristics of the fabric and produce certain properties. An example of the first is a plissé crepe; of the second, a durable-press fabric.

In the discussion that follows no attempt has been made to establish the relative significance of finishes, and their order in the text should not be construed as indicative of importance. Finishes that affect appearance are grouped by similarity of process; functional finishes are discussed alphabetically, except for related groups.

FINISHES THAT ALTER APPEARANCE OR HAND

Special Calendering

The preparation of some fabrics involves special calendering—smoothing under pressure—which imparts design to the fabric surface. The actual finishing process is mechanical. Permanence or durability

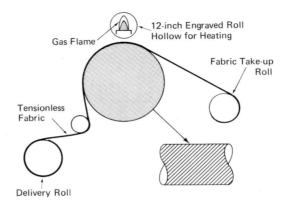

right: **Figure 26.1** Diagram of a Schreiner calender. The insert shows the presence of angle lines on the face of the calender roll.

below: **Figure 26.2** Diagram of embossing rolls.

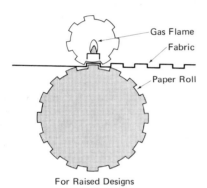

For Raised Designs

of appearance depends on several factors: if fibers are thermoplastic, the calendering can actually soften them and impart a permanent design effect; if a resin is applied to the fabric, a durable calender design can be produced; if pressure alone is used on nonthermoplastic fibers, the design will probably be lost during the first laundering.

Schreinering Schreinering is produced on a schreiner calender (Fig. 26.1). The metal roll has a series of fine lines, about 250 per inch, engraved so they form an angle of roughly 20 degrees to the construction of the cloth. The angle is usually such that the lines are parallel to the twist in the yarns This finish produces a soft luster and is used frequently on cellulosic fibers such as cotton and linen. In addition, the rolls flatten the yarns and create a smooth and compact fabric. During the last several years schreinering has been used on tricot-knit lingerie fabric of nylon and polyester fibers. It flattens the yarns and results in a more opaque fabric.

Moiré In the days before man-made fibers moiré was known as "watered silk," and the finish was applied primarily to silk fabrics. Today, moiré is used on many fibers, such as acetate and nylon. A moiré finish is characterized by a soft luster and a design created by differences in light reflection. Rib fabrics, such as failles, taffetas, and bengalines, work best in producing moiré.

The ribbed fabric is doubled and fed between rollers that exert pressure and add heat. Two rollers are involved—a large one covered with cloth and a smaller one that is heated and often includes a design. The ribs in one thickness impress images on the other thickness by flattening the ribs. If a pattern has been "scratched" in the metal roll, the design is transferred to the fabric. When there is no definite design on the metal roll, a bar (irregular or broken) effect is created. For quality moiré fabrics the cloth must be even and the ribs must be uniform.

A moiré finish on nonthermoplastic fibers, such as rayon or cotton, can be made relatively durable by impregnating the fabric with resin finishes before moiré calendering.

Embossed Surfaces   Embossed fabrics are characterized by three-dimensional designs. Before the introduction of resin finishes, embossing lacked durability, but now resins aid in making embossed patterns relatively permanent. Thermoplastic fibers produce fabrics that hold embossed designs.

The calenders used for embossing may consist of two or three rolls (Fig. 26.2). In the two-roll method, one roll is of cotton or paper, and the second roll, which is metal, has the engraved design. The metal roll is half the diameter of the cotton or paper roll. In the three-roll method, the center roll is engraved metal, while the outer two rolls are cotton or paper.

After the design is engraved on the steel roll, the large roll is dampened, and the machine is run without fabric passing through. The pattern on the steel roll is deeply impressed in the soft roll. When the impression is sufficiently deep, the machine is run until the soft roll is dry. The fabric is then fed through the calender, and the design is transferred to the fabric. Any type of design can be adapted to embossing (Fig. 26.3).

Polished Surfaces   Modified calenders, with special chemicals produce fabric with a degree of permanent polish.

Glazed Surfaces   A friction calender is used to produce glazed surfaces such as those found on glazed chintz or polished cottons. Three rolls are used. The center roll is cotton or paper, and the other two are metal. The third roll (metal) operates at a high speed, developing a polish on the fabric by friction. To make the glazed finish permanent, the fabric is impregnated with resins before calendering. Nonpermanent glazed finishes are accented by the use of wax, glue, starch, and/or shellac.

Ciré   Ciré is a high polish frequently applied to silk or blends including silk. It is accomplished by impregnating the fabric with wax or with a thermoplastic substance and passing it through a friction calender. This finish is not considered permanent, although it may be rather durable if handled carefully. Ciré is a fashion finish, and, therefore, it is available only when the current styles dictate its appearance. Sometimes called "the wet look," ciré is presently used on a wide variety of fabrics.

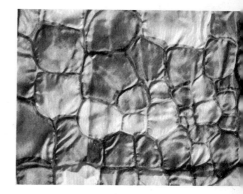

Figure 26.3 Embossed fabric.

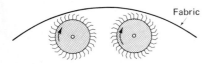

Figure 26.4 Diagram of napping rolls. The bent wires nap the surface of the fabric.

Raised Surfaces

Gigging and napping are the two principal methods used to raise fiber ends to the surface. Staple or short fibers in spun yarns are essential in fabrics that are to be napped. The action is mechanical, and the result is comparatively durable. The nap hides the yarns and weave and produces a soft, hairy appearance. Flocking creates a raised effect by adding short fibers to the surface of the fabric.

Gigging   Gigging is a napping process used on wool, rayon, and other fibers where a short lustrous nap is desired. Teasels obtained from a special variety of thistle plant are attached to a cylinder. The fabric is then fed into the machine, and the teasels gently tease or pull the fiber ends to the surface to produce the nap. Teasels are strong and durable.

The nap obtained by gigging is soft. The process is gentle and does little damage to the fabric. The nap is sometimes pressed flat, as for wool broadcloth, or it may be left full and fluffy as in soft blankets.

Napping   Napping is applied to cotton, rayon, wool, and any other staple fiber yarns when a deep nap is desired or when fibers will not respond to teasels. The process utilizes cylinders on which there are fine metal wires with small hooks (Fig. 26.4). These hooks pull fiber ends to the surface and produce the nap. Napping can be done on one or both sides of the fabric (Fig. 26.5).

Fabrics used for napping should contain soft-spun yarns with low twist and comparatively loose fibers. Plain-weave soft-filled sheeting fabrics and soft-filled twill-weave fabrics are preferred for napping or gigging. In these fabrics the warp yarns are strong enough to provide adequate strength to the fabric, and the soft filling yarns are easily roughened, so they permit pulling the fiber ends to the surface.

Fabrics with napped surfaces include flannels, flannelettes, blankets, and some coating and suiting materials. Suede cloth and duvetyns are made by napping the fabric and shearing the nap to produce a smooth, compact, and uniform surface. Napped fabrics should not be confused with pile surfaces produced by fabric construction.

Flocking   Flocking is sometimes considered a printing method, but because of its similarity to raised surface finishes, it is included with finishes in this text.

Flocking consists of attaching very short fibers to the surface of the fabric by means of an adhesive. The result is a textured appearance. Frequently, some areas of the fabric are flocked while others are left smooth to produce a pattern.

The flock is usually of rayon fibers, because they are inexpensive, but cotton, wool, or other short fibers can be used. The best fibers for flocking are those that can be cut so they are square on the ends.

Figure 26.5 A napped fabric. Note that the weave is hidden. [*Collins & Aickman*]

The adhesive is printed onto the fabric in the desired pattern, then the flock is applied by one of two methods. The vibration or mechanical method is used to apply flock to one or both sides of the fabric. The flock is circulated in a container through which the fabric passes (Fig. 26.6). As the fabric moves, it vibrates and builds up static that attracts the flock. The flock adheres to the areas where adhesive has been applied. The fabric then moves into a drying chamber where the adhesive dries with the flock firmly embedded. Finally it is brushed to remove flock in areas where there is no adhesive.

The second method, the electrostatic or electrocoating technique, depends on electrical charge of fibers and the presence of an electrical field above and below the fabric (Fig. 26.7). The fabric, printed with the adhesive, passes over an electric field, which establishes an atmosphere that forces the loose fibers in the area away from one of the electrical fields and toward the second. With the fabric moving, the loose fibers strike the adhesive, and the electrical field orients the fibers and pulls them into the adhesive. The fabric moves into a drying area where the adhesive is dried to hold the flock fibers in place.

Flocking is comparatively permanent to laundering as long as high temperatures are avoided. Dry cleaning, however, may cause damage by softening or dissolving the adhesive.

Recently, flocking has been applied to a wide variety of fabric types and end-uses, including apparel and home furnishings. Interesting designs can be formed on the base fabric. Solid flocked surfaces that resemble napped or pile materials are available. Among the fabric types duplicated by flocking are suede and velour.

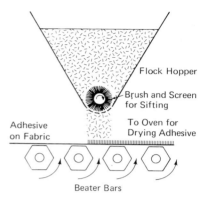

Figure 26.6 Mechanical flocking.

Acid Finishes

To produce transparent or parchmentlike cottons with permanent stiffness, such as organdy, cotton fabric is treated with sulfuric acid. The fabric is immersed in the acid under carefully controlled conditions for a very brief time and then quickly neutralized. One finish, developed in Switzerland, is called the *Heberlein process*. *Bellmanized* and *Ice organdy* are the terms used in the United States.

This type of acid finish can be applied to the entire fabric to produce a clear organdy. By printing an acid-resistant substance on the fabric before treatment, designs can be developed with both opaque (frosted) areas and transparent areas. The finish works effectively on all cotton, but mercerized cotton responds most satisfactorily.

The durability of stiffness and transparency characteristic of organdy depends on the quality of the finish. A well–applied finish will be long lasting and not weaken the cloth. Fabrics with this finish wrinkle badly during laundering and require considerable ironing.

A second acid finish produces "burned-out" designs. It employs a fabric composed of two properly selected fibers—one that is easily

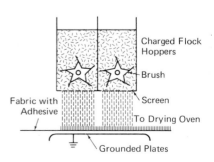

Figure 26.7 Electrostatic flocking.

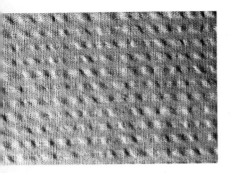

Figure 26.8 Plissé—a fabric design produced by a finish.

destroyed by acid, such as rayon or cotton, and another that is acid resistant, such as wool, acrylic, or polyester. The fabric is exposed to an acid, usually sulfuric, which burns away the first fiber to leave sheer areas. If carefully controlled, the acid can be printed onto a fabric to produce interesting and intricate designs. Careful planning of the fiber content and arrangement is essential for satisfactory results.

Burned-out designs can also be developed by using other chemicals (not necessarily acids) that will dissolve away one fiber in specified areas. Acetone is used on fabrics composed of acetate and a second fiber not affected by acetone. The acetate is destroyed in the treated areas, leaving the second fiber and an interesting design. Phenol is effective on fabrics made partly of nylon.

Basic Finishes

The application of chemical bases or alkalies—frequently called caustics because of their corrosive action—produces certain finishes.

Plissé crepe, a crinkled or crepelike cotton, results from the action of sodium hydroxide on cotton fabric in selected areas. Caustic soda in a paste form is printed onto the fabric in a predetermined pattern. This causes the coated areas to shrink and the untreated areas to pucker because of fiber shrinkage around the puckered areas (Fig. 26.8). After the caustic is removed, the crinkled effect is comparatively durable. However, heavy or prolonged ironing will stretch out the fabric and result in dimensional change, so ironing is not recommended. In any case, the natural appearance of plissé is quite pleasing, and ironing is usually not needed.

Crinkled, embossed, or plissé effects in irregular designs on nylon fabrics can be obtained by using phenol. The organic base accomplishes the same results on nylon fabrics as caustic soda (sodium hydroxide) accomplishes on cotton fabrics. Both processes must be carefully controlled.

Not all plissé effects on nylon are the result of treatment with phenol. Nylon fabrics that resemble seersucker (Fig. 26.9), in which the crinkle follows specific yarns, or crinkled check effects where yarns appear to be the dividing line, are often accomplished by using nylons with different shrinkage characteristics. One group of yarns shrinks while the second group does not, and, thus, the desired crinkled effect is achieved. The shrinkage occurs under hot-wet processing or finishing. The purchased fabric will have good stability.

Stiffening Finishes

Sizings discussed in the previous chapter contribute stiffness to fabrics, but they are relatively temporary. Recently, thermosetting resins or plastic compounds of various types have been used with success in

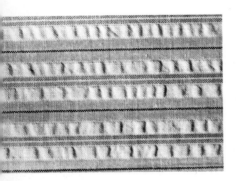

Figure 26.9 Seersucker—a fabric design created in the weaving process.

producing fabrics with quite durable stiffness. They keep sheer fabrics crisp and attractive, prevent sagging and slipping of yarns and wilting of fabric, reduce the formation of lint, and help maintain a smooth surface that is resistant to snags and abrasion. Resin finishes contribute to dimensional stability.

In addition to thermosetting resins, chemicals used for these finishes include various cellulose compounds, such as ethyl cellulose, and thermoplastic substances, such as vinyl acetate and polyacrylates.

Softening Finishes

Softening finishes improve the hand and drape of a fabric. They may also add body to fabric, facilitate application of other finishes, subdue the coarseness imparted during processing, and increase the life and utility of the fabric. Batiste is an example of a fabric treated with a softener.

Softening agents are more common than any other finish. They include a variety of products, such as oil, fat, and wax emulsions; soaps and synthetic detergents; substituted ammonium compounds; and silicone compounds. The silicone compounds produce relatively durable softening effects.

Softening finishes have gained new importance with the acceptance of durable-press fabrics. Durable press (see p. 298) stiffens fabrics considerably, and, to maintain a pleasant hand, softeners are included in the finishing procedure. A wide selection of fabric softeners are available to the consumer for home care of fabrics. These quaternary ammonium salts (substituted ammonium compounds) maintain or restore fabric softness and are also satisfactory antistatic agents.

Optical Finishes

Delusterants   Man-made fibers, such as rayon, acetate, and nylon, often have a high degree of luster, because their relative transparency or their shape reflects light. A few long light rays reflected from a surface give more luster than many short rays, because the latter are diffused as they cross and blend. Luster can be diminished by intelligent selection of fibers or yarn-construction techniques. (Conversely, if desirable, luster can be enhanced by the same methods.) However, when this is not feasible, supplementary substances can be added to the fiber solution or special finishing procedures can be used to reduce excessive luster. A common system is to introduce pigments into the spinning solution. Titanium dioxide, a white pigment, reduces luster by the breakup of light reflection, thus creating an opaque fiber. The finishing techniques include special heat treatments that soften yarn and fabric surface to change light reflection. A recent innovation from France is a finish applied to tricot knits, called *Chavacete*, which softens the luster of acetate fabric and produces a silklike appearance. It also develops a pleasing hand.

External delusterants are frequently barium salts. Barium sulphate is the most common and is applied in a two-bath procedure that results in the formation of an insoluble deposit on and in the fiber. Other external delusterants are applied in a single bath and deposited on the fiber. These include china clay, aluminum oxide, zinc oxide, and methylene urea. Nyshield, used to soften the luster of nylon hosiery, is an example of an external delusterant.

Optical Brighteners   Many fabrics lose their brightness, whiteness, and clearness during processing and maintenance. In an attempt to prevent this and to maintain white and bright fabrics, optical brighteners have been introduced. These are sometimes called "optical bleaches," but the term in this case is inappropriate, since no bleaching occurs. Optical brighteners are used by fabric converters in finishing, and they are added to many home laundering agents, so the consumer can restore brightness each time a product is laundered. The substances attach themselves to the fabric and create an appearance of whiteness and/or brightness by the way in which they reflect light. In fact, they actually reflect more light than is visible under normal conditions. One writer explains the behavior of these optical brighteners thus: "When exposed to sunlight, they absorb invisible ultraviolet light and reflect it as visible blue light. They act . . . like a bluing agent but add the fluorescent effect."[1] There are many brands available, but they are all generally called *optical* or *fabric brighteners.*

Other Finishes That Affect Appearance and Hand

Several finishes discussed as routine or general techniques produce changes in appearance. These include calendering, mercerizing, fulling, singeing, and beetling. Finishes such as sizing, fulling, and heat setting influence the hand of fabrics. (See Chap. 25 for a discussion of these techniques.)

If properly applied by reliable converters, finishes resulting in changed appearance or hand do not reduce fabric durability in the end-uses for which the fabrics are usually selected. Nondurable finishes will be destroyed by laundry or cleaning. Some, such as optical brighteners, are easily replaced. However, if such finishes as glazing, ciré, plissé, and flocking are not permanent, they are lost during maintenance, and the fabrics will never look as attractive as when new. Permanent or durable finishes will retain their appearance with proper care.

[1]H. C. Speel and E. W. K. Schwarz, eds., *Textile Chemicals and Auxiliaries,* 2d ed. (New York: Reinhold Publishing Corporation, 1957), p. 83.

The appearance of finishes should be considered in fabric selection as a part of planned end-use. This is discussed further in Part VI.

FUNCTIONAL FINISHES

To provide the public with fabrics that have special service qualities, the finishing industry has been called upon to develop many new techniques. Finishes in this category are of two basic types: (1) external finishes; (2) internal finishes or chemical modifiers.

External finishes are those that are applied to the surface of fibers, yarns, and/or fabrics and that do not combine chemically with the fiber. They include softening agents; film-forming finishes, such as starches, selected thermosetting resins, and thermoplastic resins; surface deposits, such as delusterants, slip-resistant finishes, and hygroscopic agents; and corrective finishes, such as water repellents, fire inhibitors, moth repellents, or bacteriostatic agents. External finishes sometimes alter the appearance and hand of fabric.

Internal finishes are deposited within the fiber. They combine chemically with the fibers and modify or inhibit some inherent fault or weakness in the actual fiber structure. In some cases internal finishes produce cross-linking of fiber molecules. This can be compared to parallel chains of paper clips that are held together at intervals by clips perpendicular to the long chains.

Internal finishes are applied to fibers with porous surfaces. They include thermosetting resins, such as urea and melamine formaldehyde, used for dimensional stabilization, crush resistance, and durable press; weighting agents, such as metallic salts; and chemical modifiers that alter the fiber chemistry, such as cyanoethylation and acetylation of cotton. Internal finishes do not alter the appearance of fabric, but they occasionally modify the hand to some degree.

Abrasion-resistant Finishes

Many of the newer manufactured fibers, particularly nylons, have inherent resistance to abrasion. However, natural fibers and some man-made fibers (including rayon) may be damaged by rubbing. To reduce this type of fabric damage, manufacturers do one of two things. They can blend fibers of high abrasion resistance with those of low resistance, or they can apply soft thermoplastic resins, which appear to increase the fabric's resistance to abrasion damage. According to one source the addition of acrylic resins improves abrasion resistance.[2] The acrylic resin must be selected with care, for some are too soft to be effective, while others are too hard and introduce brittleness.

[2] J. E. Lynn and J. J. Press, *Advances in Textile Processing* (New York: Textile Book Publishers, 1961), p. 314.

The problem of abrasion is extremely complicated, but it is believed that a substantial part of the resistance produced by these resins results from the fact that the resin binds the fibers more firmly into the yarns and, thus, increases the time and amount of abrasion required to roughen the surface by fiber breakage.

Recent evidence indicates that abrasion-resistant finishes may increase the wet soiling of fabrics. Therefore, their use is lessening. They are still popular for trouser pockets, carpet backings, and hat bands.

Absorbent Finishes

Absorbent finishes increase the moisture-holding power and speed up the drying action of fibers, yarns, and fabrics. Fabrics treated with these finishes are capable of absorbing more moisture than they normally would, and they also tend to dry easily. One type of absorbent finish causes the absorbed moisture to break up into small micelles that evaporate readily. A second type holds the moisture by absorption and disperses it into the yarns. The drying of fabrics with this type of finish is delayed an amount of time commensurate with the additional moisture that must be given off.

Evidence that absorbent finishes are effective is meager. The substances are combinations of wetting agents and humectants or hygroscopic agents.[3] The use of these substances has decreased since the early 1960s, but some are still applied to synthesized fiber fabrics to increase moisture absorbency.

Absorbent finishes have been applied to towels, diapers, underwear, sport shirts, and other items where moisture absorbency is desirable. They may aid in the application of dyestuffs. These finishes are external coatings on yarn and fiber; they are applied from aqueous baths and have fair durability. Trade names encountered for absorbent finishes include Hysorb, Nylonized, Telezorbant, and Sorbtex.

Antislip Finishes

Finishes applied to a fabric to reduce or eliminate yarn slippage are called antislip, slip resistant, or nonslip finishes. They help keep yarns in their proper position in the fabric and reduce seam fraying. Three types of products are used. Rosins—hard, waxy substances left after distillation of volatile turpentine—have been widely applied, but they have poor washfastness unless they are treated with a heavy metallic salt, such as zinc acetate. Colloidal dispersions of silica decrease slippage by reducing surface slickness; these finishes are not durable. Urea and melamine formaldehyde resins, which produce other service qualities, are effective agents in reducing yarn slippage and are the most durable of the three types.

[3]"Products/71," *Textile Chemist and Colorist* (September 1970), p. 201.

Antistatic Finishes

Static buildup in fabrics has long been recognized as a problem by textile scientists. Static charge or static electricity is controlled by introduction of humidity into the air and by the use of various weaving lubricants in the processing of natural fibers and rayon. With the introduction of acetate, nylons, polyesters, acrylics, and other new man-made fibers, the problem of static electricity gained magnitude. It was difficult to manufacture and process these fibers even in humid rooms without some type of coating to carry the electrons away and prevent their buildup on the yarns and fabrics.

In addition to difficulties encountered in the production of fabrics, static buildup caused soiling of the fabric. Thus, finishes to reduce or eliminate this buildup have become highly desirable.

Antistatic finishes work by one or more of three basic methods. First, the finish may improve the surface conductivity and thereby help the electrons to move either to the ground or to the atmosphere. Second, the finish may attract molecules of water to the surface, which, in turn, increase the conductance and carry away the static charges. Third, chemical finishes may develop an electric charge opposite that of the fiber, which neutralizes the electrostatic charges. The most effective finishes work in all three ways. However, because fibers differ in the type of static charge they generate (some are positive, some negative), in order for the third method to be effective, there must be different finishing agents for different fibers.

The presence of static buildup on fabrics is evident to the consumer when garments cling to the body or to other fabrics; sparks with sufficient force to be seen or felt jump from the wearer to metal after the person has walked across floor coverings or slid across upholstery; crackling sounds are heard as a person walks or takes off a coat or sweater; a visible spark is produced by rubbing the fabric. Static is annoying, and it is a major problem to the yarn and fabric manufacturers.

Most antistatic finishes are cationic surface active agents based on quaternary ammonium compounds. Other active substances include those that are cationic or positively charged, and a few nonionic or neutral surface active agents are employed.[4] As yet, most of these finishes are not durable and must be replaced after each laundering. The addition of fabric softeners such as Nusoft, Sta-Puf, Downy, Diasof, and Felsoft to the final laundry rinse helps to control static. Negastat is a product designed specifically for static control.

The finishing industry continues to work on the development of durable antistatic agents. The most successful of these are incorporated into the fiber, including Fybrite, a component of the formula for a polyester, and 22N, a static-free nylon. There are a few durable antistatic

---

[4] Speel and Schwarz, p. 136.

finishes, which are frequently based on polyethylene glycol or poly-alkylene glycol ethers. The trade names include Stanax, Permastat, Aston, Nopcostat, and Valstat.

Bacteriostats

Bacteriostatic agents or antiseptic finishes are applied to fabrics for three reasons. They may control the spread of disease and reduce the danger of infection following injury; they help to inhibit the development of unpleasant odors from perspiration and other soil on fibrous structures; and they reduce damage to fabrics from mildew-producing fungi and rot-producing bacteria.

Evidence indicates that substances to prevent fabric deterioration from microorganisms were known in Ancient Egypt. The fabrics used to wrap mummies were preserved by applying spices and herbs, which protected them from rot. The current usage of antimicrobial finishes dates from about 1900. However, during World War II the importance of these finishes was emphasized. The German army treated soldiers' uniforms with quaternary ammonium compounds, and records indicate that men wearing the treated fabrics suffered considerably less infection from wounds. At the present time antimicrobial finishes are becoming very popular.

Bacteriostatic finishes may be durable or renewable. The renewable ones are external finishes that produce a "climate" unfavorable to the microorganisms. Some durable finishes are surface coatings that have been made insoluble so they remain on the yarns and fabrics during care, while others are internal finishes that are insolubilized within the fiber structure. It is possible to include bacteriostats with other finishes, such as water repellents.

Finishes to prevent the growth of microorganisms appear on fabrics for a wide variety of apparel, home-furnishing, commercial, and industrial products. Apparel items include socks, shoe linings, foundation garments, sportswear, and babies' clothing, especially diapers. Sheets, pillowcases, mattress padding and covering, carpet underpadding, carpeting, blankets, and towels are among the many items in the home that are treated. Fabrics for tents, tarpaulins, and auto convertible tops have a longer life when treated with finishes that reduce rot and mildew damage.

The importance of this group of finishes can be observed in the fact that merchandising organizations and product manufacturers have trademarked finishes for their own use. For example, Sears Roebuck uses the name *Kenisan* to indicate bacteriostat finish on their products; Chatham Blanket Company refers to their products as *Hygienated*.

A number of research projects have proved the value of bacterio-static finishes. A lower incidence of reinfection from athlete's foot was noted when shoe linings were treated. A reduction in diaper rash was evident in a group of babies who were clothed in treated diapers. Homemakers have remarked on the absence of musty odors that accompany mildew in hot, humid climates when floor coverings are protected with mildew-resistant finishes.

There is some evidence that free formaldehyde, present in fabrics after treatment with various finishing resins, produces a temporary bacteriostat. However, when the formaldehyde gradually dissipates, any bacteriostatic property disappears.

Renewable bacteriostat finishes are recognized by such trademarks as Sanitone, Sanitized, Dowicide, and Borateem. "Durable" finishes applied during the converting process include such active ingredients as triethanolamine, plus acetic acid, plus a metal salt; zirconium compounds with copper or mercury compounds; neomycin; silver salts; and bacitracin. In addition, cyanoethylation of cotton aids in producing mildew- and rot-resistant fabrics. Despite the trade terminology, these finishes are "durable" only to a point. Tests have found that most will withstand a maximum of fifteen launderings, and many are removed before the fifth laundering. Furthermore, bleaching with chlorine compounds removes the finishes completely. A few trademarks for these "durable" finishes are Guardsan, Hyamine, Marcocide, Nuodex, Permacide, and Vancide.

Flame Inhibitors

Finishes that reduce the flaming, charring, or afterglow of fibers and fabrics are important for safety. Most finishes in this group produce fabrics that *will* burn in the direct path of flame but that self-extinguish when the source of flame is removed. A truly fireproof fabric will not burn even in the path of direct flame, but actually, only asbestos and fiber glass have this property. Finishes cannot provide completely safe products. They can, however, reduce the danger of complete destruction of treated fabrics (though they do not eliminate damage altogether) and provide a margin of safety that may prevent serious harm to individuals.

Flame inhibitors are not new. According to one author's research, the history of these finishes can be traced back at least 300 years.[5] In 1821 J. L. Gay-Lussac produced some flame retardants for Louis XVIII of France. Versmann and Oppenheim did comparative studies of flame inhibitors in 1859 and found ammonium phosphate and tungsten salts to be effective. These are still used in renewable finishes. In 1922 Kling and Florentin studied the use of borax and boric acid and arrived at a recommended mixture of these substances. Ramsbottom and Snoad tested the same compounds in 1947. Renewable finishes and home applications for cellulosic fibers still employ the latter two chemicals.

[5]McQuade, cited in Speel and Schwarz, p. 466.

There are many recorded incidents of serious fires caused by the flash burning of brushed rayon negligees or sweaters ignited by cigarette ashes. More notorious were the group disasters—the Coconut Grove nightclub fire in Boston and the circus-tent fire in Hartford. As a result of these tragedies, the Federal Government enacted a bill to control the interstate sale of highly flammable fabrics. This legislation, passed in 1953, applied to all textile fabrics intended or sold for use in wearing apparel, and all such fabrics contained in articles of wearing apparel. In 1967 the law was amended to prohibit the introduction or movement in interstate commerce of articles of wearing apparel and fabrics that are so highly flammable as to be dangerous when worn by individuals or used for other purposes. The definitions now include fabrics that may *reasonably be expected* to be used in wearing apparel or interior furnishing. These amendments strengthen the law at a time when public concern has been directed toward the hazards of flammable textile products. In addition to Federal legislation, there are many local and state laws that place restrictions on fabrics for public buildings.

Flame-resistant finishes are of two general types. One group is water soluble and must be replaced after each laundering or, in many cases, dry cleaning. The second group is considered durable. The latter will withstand dry cleaning, laundering, and weather. However, the durability of these finishes is quite variable, so it is important to have a label that indicates what the finish will withstand and the approximate life of the finish.

The finish should be linked or bound to the fibers in such a way that it will release flame-retardant chemicals at the time flaming temperatures are reached. It is essential that the release does not occur under ordinary conditions. Improperly applied finishes or improperly chosen chemicals can result in fabric degradation. The most desirable finishes do not change the hand or appearance of the fabric. In practice, however, many finishes used to inhibit flaming do affect certain colors and tend to produce harsh and somewhat stiff fabrics. Where safety is important, these side effects are overlooked.

Flame-retardant finishes are classified in one of the following groups:

1. water-soluble compounds that must be reapplied after exposure to excess moisture. These include borax, boric acid, ammonium phosphate, ammonium sulfate, and various mixtures of these chemicals.
2. insoluble salts applied by dissolving in a suitable solvent, applying to the fabric, and either reacting with a second compound to render them insoluble, or evaporating the solvent. These include metal oxides, such as ferric oxide, stannic oxide, and manganese dioxide.
3. oils, waxes, or resins that incorporate chlorinated substances,

bromine compounds, phosphorous compounds, antimony compounds, or other similar flame retardants.
4. substances that react with the fiber to produce molecular change. These are usually confined to cellulose fibers and produce cellulose ethers or esters[6] with flameproof properties.

The compounds used to react with cellulose molecules are complex organic substances. They are usually characterized by the presence of phosphorous in their formulation, although antimony and titanium compounds and borate compounds are used in some processes. One of the most important, commercially, is tetrakis hydroxymethyl phosphonium chloride (THPC). Other flame retardants combine a phosphorous product with ammonia, urea or melamine formaldehyde, and the halogens.[7]

Not all types of finishes are suitable for all fabrics. The selection of finish for the particular fiber and fabric is determined by such factors as anticipated end-use, appearance, and desired hand.

Flame-retardant finishes act in various ways, including

1. release of gases or foams that provide a flame-smothering atmosphere.
2. the formation of Lewis acids (acceptors of electrons) at flaming temperatures. These acids catalyze the dehydration of cellulose to carbon and water. This dehydration forms larger carbonaceous aggregates that do not oxidize as easily, and flaming is inhibited.
3. release of substances that accelerate the degradation of cellulose with the release of volatile products that inhibit flaming.

The individual who wishes to treat fabrics at home can apply a rinse of borax and boric acid (2 quarts of water, 7 ounces of borax, and 3 ounces of boric acid). Other finish formulas can be obtained by requesting USDA Bulletin L454 from the Superintendent of Documents, Washington, D.C.

Flame inhibitors are often applied in conjunction with other finishes, such as water repellants, crease resistants, and soil resistants. Trade names of flame inhibitors that may be encountered in consumer goods include Banfire, CM Flame Retardant, Fi-Retard, Firegard, Firemaster, Pyropel, Pyroset, Pyrovatex, and X-12. The durability of fire retardants is variable, and the consumer should read labels carefully. The application of fire-retardant finishes usually increases the cost of textiles, so the consumer must decide whether the finish is important enough to warrant the extra expense.

[6] Speel and Schwarz, p. 479.
[7] For an excellent reference to fire retardants, the reader is referred to John W. Lyons, *The Chemistry and Uses of Fire Retardants* (New York: Wiley-Interscience, 1970).

Fume-fading Resistant Finishes

Certain dyestuffs on certain fibers are subject to color loss or change caused by atmospheric fumes. These gases in the atmosphere are oxides of nitrogen compounds.

Dyes applied to acetate fibers are particularly susceptible to fume fading, and the same dyestuffs may cause problems on nylon or polyester fabrics.

A major step in efforts to reduce color breakdown was taken when pigments were introduced into the solution of the fiber polymer before extrusion. However, not all fabrics subject to fume fading can be dyed economically by solution coloration. For items colored after the yarns or fabrics have been made, finishes can be added to reduce or prevent fume damage.

Simple alkaline substances, such as borax, are sometimes used, but these are not permanent and require renewing after laundering. Comparatively durable finishes are produced by various tertiary amines. Trademarks for finishes that reduce or eliminate fume fading include Antifume, Crestofume, Emkafume, Neufume, Permafume, and Protex. In the case of acetate fibers, the preferred technique is solution dyeing (see Chap. 29).

Metallic and Plastic Coatings

In an effort to produce fabrics that reflect heat, fabric converters have developed a finish in which aluminum coating can increase warmth or coolness, depending upon the situation.

An important use of aluminum-coated fabrics has been for lining coats or jackets. It was originally hoped that the metallic finish would help retain body heat and reflect heat lost from the body by radiation. However, several research studies have determined that these finishes are not effective and do not keep the body warm. The insulative value is related to fabric construction rather than finish and to a psychological feeling of protection.

A popular use for reflective fabrics is in drapery linings. The fabric helps maintain a constant room temperature: it reflects sunlight in summer and retains heat in winter. It is especially helpful when entire walls of glass have these drapery covers.

Aluminum finishes with adhesives that resist solvents will dry clean fairly well, but others are lost during cleaning. In general, they have low resistance to washing. A metallic coating will not compensate for inferior fabric construction, nor will it make a loose-weave fabric really warmer.

Satins and taffetas of good quality are excellent candidates for these finishes. Trademarks frequently observed include Milium, Temp-Resisto, and Therm-O-Ray.

Plastic coatings decrease heat loss by reducing air circulation. Fabrics coated with plastics (similar to coated fabrics like oil cloth) are used as drapery linings. They not only help prevent temperature change but reduce the amount of soil that can penetrate the draperies. In garment manufacturing, imitation leather fabrics made by applying a coating of plastic to a fabric base have been successful. Because they have low air permeability, they resist wind penetration. Therefore, the fabrics will be warmer on a windy day than a fabric without the plastic coating that is comparable in weight and thickness. Some plastic-coated fabrics tend to stiffen at low temperatures and may become uncomfortable to wear.

Mothproofing Finishes

Fibers containing protein, such as keratin, are especially susceptible to damage by moths and carpet beetles. The cystine amino acid residues in keratin are believed to be what the moths prefer, and this explains why moths eat hair fibers containing keratin and not fibroin of silk. Carpet beetles are likely to eat keratin, fibroin, and some other protein substances.

Fibers other than the protein type may be damaged by insects trying to escape confinement and reach desirable food. The same situation arises when other fibers are blended with keratinaceous fibers. Wool is the most susceptible and most frequently damaged fiber. Recent reports have stated that damage to wool by moths and carpet beetles results in an annual loss valued between $200 million and $500 million.

While the homemaker may blame the holes she finds in wool or wool blends upon the "clothes moth," the damage may have been caused by any one of thirteen species of moths and carpet beetles. Most of these insects live on keratin, but they will destroy other products in order to obtain it. The term "moth" will be used to indicate the entire group. The larva of the moth is the culprit in fiber damage. During the eating period the larva increases its weight approximately 300 times.

Finishes to reduce or prevent damage by moths can be renewable or durable. The substances may be effective because

1. they give off an odor that repels the mature moth and prevents the deposit of eggs. Naphthalene crystals or moth balls are examples. These are comparatively unsuccessful in preventing damage.
2. the agent gives off a gas (which may or may not be noticeable to humans) that is toxic to the mature moth and the destructive larvae.

Products such as Crestocide, Hartocide, Mitin, Neocid, and Repel-O-Tac are usually added to the fiber during dyeing or fulling operations. These are considered durable. Most mothproof agents that have fluorine as the active ingredient are water soluble; any exposure to water will remove the finish, and it will have to be renewed. Larvex and Eulan are in this category.

Good practices in the care of wool or wool-blend fabrics should be observed by the consumer regardless of mothproofing finishes present. Soiled wool should never be stored, for it is highly subject to attack by larvae. Closets should be kept clean, and spraying of the closet is an extra precaution. Carpets must also be kept clean.

The yearly damage to wool is a constant stimulus to research workers to develop effective mothproofing methods. Work is being done in two areas. One technique involves modifying the wool keratin by breaking the cystine linkage and making it unpalatable to the larvae. A second area of research seeks to produce wool that is toxic or unpalatable to the larvae because of substances fed to the wool-producing animal.

A worsted yarn, sold by Deering Millikin as "Dimension Three," is reported to be durably mothproof, soil resistant, and pill resistant. The trend to include mothproofing with other finishes is important, and more and more wool products will be offered with several functional finishes as an integral part of the fabric.

Stabilization Finishes

A question frequently asked by the consumer when purchasing a textile item is, "Will it shrink?" The problem of fabric shrinkage is not new, and some solutions were developed years ago. However, before the 1930s fabric shrinkage was generally left to the consumer, who preshrunk fabric herself or bought garments large so they would fit after laundering. Stabilization also controls stretch, and it is not unusual for a fabric to shrink in one direction and stretch in the other. Any variation in size and shape will cause dissatisfaction.

Fabric converters and processors recognize two distinct types of shrinkage: relaxation or residual shrinkage and felting shrinkage. *Residual shrinkage* is relaxation shrinkage remaining in the fabric when it is purchased. *Relaxation shrinkage* occurs when some factor causes a release of stress imposed during fabric manufacturing and finishing. *Felting shrinkage* is caused by certain fiber characteristics and may continue over long periods of time. The term *progressive shrinkage* is frequently used to indicate either residual or felting shrinkage.

Relaxation shrinkage is rather complex. It is the most common type and occurs when some operation such as laundering releases the tensions imposed during fabric manufacture, so the yarns return to their original length. This type of shrinkage is sometimes progressive, in that all the potential shrinkage may not take place during the first laundering. The delay may be caused by the presence of various finishing agents, and, as they are gradually removed, the additional relaxation shrinkage occurs. Dry cleaning can also cause relaxation shrinkage, especially if a wet cleaning process is used.

A second problem with relaxation shrinkage is that, while shrinkage occurs during laundering, ironing will often restretch or strain the fabric. This may continue for the life of the garment, so the size will vary with each laundering period.

Some fabrics shrink or stretch with changes in humidity. This is a fiber property, but it is reflected in the fabrics. It occurs when a fiber is more easily stretched or elongated when humidity is high, and the weight of the fabric causes some fibers to extend. As humidity decreases, the fiber returns to position. This type of shrinkage is visible in some drapery fabrics, where an actual variation in length can be noticed on damp and dry days. The chain weighting in draperies increases the effect.

Fabrics occasionally exhibit shrinkage in one direction and stretch in the other if they have been held under high tension during the drying period following other finishing processes.

Nearly all fabrics composed of natural fibers and many of man-made fibers exhibit relaxation shrinkage. The amount will vary, but if left uncontrolled, it will result in dissatisfied consumers.

Residual shrinkage occurs over a long period of time. It can result from changes in yarn or fiber shape or from specific fiber properties. In this discussion the term *progressive shrinkage* will be used synonymously with *residual shrinkage.*

Felting shrinkage is primarily a characteristic of hair fibers. It occurs when fibers entangle as a result of heat, moisture and pressure, or agitation. Wool fabrics are particularly susceptible to felting shrinkage.

The techniques used to control shrinkage vary with fiber content. In recent years fabric stabilization has been included with other finishes, such as wrinkle resistance and water repellency. Durable-press or minimum-care finishes are discussed in more detail later in the chapter, but they, too, help to control fabric dimensions.

Cotton and Linen   The primary methods for eliminating residual or relaxation shrinkage in cotton and linen fabrics are mechanical. The simple method, frequently employed by the consumer as well as the fabric converter, is to wet the fabric thoroughly, dry it in a tensionless state (*slack drying*), then smooth it out by calendering or ironing. A second method involves feeding the cloth into the tenter frame in a slack condition and applying stretch to the filling. The third method is probably the most common and will produce fabrics that have less than 1- or 2-percent shrinkage. This system is called *compressive shrinkage.* Fabric is fed over a feed roll that stretches the surface and then under a heated drum, where the stretch surface is compressed

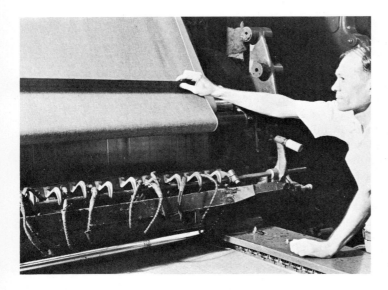

Figure 26.10 Equipment used in the Sanforizing finish. [*Pepperell Manufacturing Company*]

as it is dried. Fabrics carrying trade names such as Sanforized and Rigmel are examples of compressive shrinkage (Fig. 26.10).

Recently, chemical methods for fabric stabilization have gained popularity. Mercerization of cotton reduces residual shrinkage. It can remove up to 50 percent of the potential shrinkage.

Other chemical treatments to control shrinkage include resin impregnation, acetylation of cellulose, and application of special cellulose finishes. Of these, only resin impregnation is commercially important.

Resin impregnation will control dimensional stability of fabric as long as the resin remains on the fabric. One authority states that there is a simple and continuous relationship among shrinkage control, wash-and-wear, and durable press.[8] Any good cross-linking resin will produce shrinkage resistance at a 2- to 3-percent add on (that is, a 3-percent increase in weight); wash-and-wear requires 4 to 5 percent, and durable press 6 percent or more. The stabilization of the fiber dimensions and configuration gradually increase as the amount of cross-linking increases. The disadvantage of resin treatments is that the hand and behavior of the fabric may be altered. There is a tendency toward a harsher, stiffer fabric.

Acetylation of cellulose changes the molecular structure to control dimensional stability. The use of cellulose solutions bonds the fibers together to a greater or lesser degree. Neither of these techniques is in commercial use today.

Rayon   Rayon fabrics, unless treated, are subject to both relaxation and residual shrinkage. Resin impregnation has been the most fre-

[8]J. W. Weaver, in a personal communication with the author.

quently used method for controlling this shrinkage. The technique forms the resin within the fiber, not on the surface. While certain changes in fabric hand may occur, the control of shrinkage over-shadows any disadvantages. New resins have fewer undesirable side effects and produce better fabric stabilization.

An effective way to prevent rayon shrinkage is by chemical reaction (cross-linking) with acetals. The Avcoset process is of this type. This process forms an acetal cross-link of cellulose molecules by reaction between a cellulose ether and formaldehyde. While this reaction is effective in reducing shrinkage, special additives are required to pre-vent excessive fabric tendering and weakening and to retain the desired hand and appearance.

One method that produces dimensional stability and also prevents discoloration during laundering utilizes the dialdehyde glyoxal. The glyoxal reacts with urea to form dihydroxyethylene urea. The latter is then reacted with a small amount of formaldehyde to create dimethylol-dihydroxyethylene urea. This cross-links the cellulose to form stable fibers and thus stable, self-smoothing, nonyellowing fabric. The reaction is as follows:

Urea   +   Glyoxal   ⟶   Dihydroxyethylene urea

Dihydroxyethylene + Formaldehyde ⟶   Dimethylol dihydroxyethylene urea

Nylon and Polyesters   Man-made fibers of the thermoplastic vari-ety, such as nylons and polyesters, are usually heat-set during finishing. This process stabilizes fabrics of these fibers.

Wool   Wool has always posed many problems in relation to shrink-age, In addition to relaxation shrinkage, wool also has a high degree of felting shrinkage. Felting shrinkage is a result of the behavior of wool fiber. Sponging or steaming will eliminate much of the relaxation shrinkage of wool, but felting shrinkage calls for much more drastic treatments.

Resin impregnation of wool fibers by the same general methods used on other fibers controls shrinkage. The main difference is that a higher concentration of resin is required for wool. Furthermore, there tends to be an unfavorable effect on the yarns and fabric, and the finish is not as durable to laundering as it is on other fibers. Lanoset and Resloom are two resin finishes that have been used on wool. Because of the undesirable side effects, and because new treatments for wool fabrics have been developed, this process has been substantially removed from the market.

Other methods used to control the shrinkage of wool include treatments with enzymes that attack the fiber scales; silicone finishes that add water repellency as well as stability; and the new interfacial polymerization methods. The last of these was developed by the Western Regional Research Laboratory of the USDA and has been named *Wurlan*. In this treatment, the wool fibers are coated with a microscopically thin layer of a polymer, in this case a polyamide substance. The interfacial polymerization technique has proved effective, and other substances have been employed successfully. Companies using the Wurlan process usually have their own trademarks; the products may be referred to as *Wurlanized*, or the company name could be combined with the term *Wurlan*. When other chemicals are substituted, the fabrics are given different names. One of the best known is Stevens "H_2O" Wool. Interfacial polymerization processes exhibit durable stabilization with little or no change in fabric appearance or hand.

An old method still used to produce nonfelting and, thus, shrink-resistant wool fabric is chlorination. The application of alkaline hypochlorite modifies the scale structure of the wool fiber and prevents felting. Early chlorination processes tended to weaken wool fibers. However, recent research developments, which involve the addition of potassium salts (such as potassium chloride) to the chlorination bath, reduce fiber damage.

Dylanize is one of the better-known chlorination processes. It produces stabilization by an oxidation process involving chlorine compounds. Other trademarks for wool chlorination treatments include Schollerized, Protonized, and WB-7.

The advantages of shrink-resistant finishes are obvious. A product that shrinks or stretches and changes size results in poorly fitting, uncomfortable items and general discontent with the product.

The consumer should be cautioned about shrink-resistant finishes, particularly on wool. They often pose problems, especially for home sewing, in that it is difficult to shrink out fullness during construction. Loose weaves, in which yarns can be packed more closely, present no serious difficulty, but firmly woven fabrics may require pattern modification to obtain neat, even seams.

According to the Federal Trade Commission, textile products that are labeled in respect to shrinkage should give the percentage of maximum shrinkage. However, there are no regulations that require such statements concerning shrinkage to appear at all. Recent unpublished research indicates that many fabrics, especially cotton, shrink excessively though the fabrics are labeled as preshrunk. It might be a wise precaution in buying yardage to test a sample of fabric before using it in constructed products. This is even more important if fabrics are to be tumbled dry. For example, the Sanforized guarantee is based on flat drying. Tumble drying, a much more common method today, causes more shrinkage than flat drying and results in a greater change than label information would indicate.

Many fabrics of man-made fibers are considered highly dimensionally stable. However, it is still generally advisable to preshrink these fabrics before cutting a pattern, particularly when working on polyester knit fabrics.

Water-repellent and Waterproof Finishes

Waterproof finishes are those that coat or seal a fabric so water does not pass through it. Such fabrics are nonpermeable to air and, thus, are not comfortable in wearing apparel. Water-repellent finishes result in a fabric that *resists* wetting and is relatively porous.

Early methods used to produce waterproof fabrics coated the fabric with rubber, oxidized oil, or varnish. While they prevented water from passing through, some were heavy, bulky, and uncomfortable. Oiled silks were light in weight, but they were not so durable as rubber or varnish-coated fabrics. Some materials are still waterproofed by these early methods, but modern fabrics are coated with synthetic polymers that repel water.

Water-resistant or water-repellent finishes are popular in consumer goods, because the fabrics are comfortable and the finish does not alter the original appearance. Water repellency is determined to a large extent by fabric construction as well as finish. The fabric should be made so the largest interstices between yarns or fibers are smaller than raindrops. The yarns should be soft and the yarn count high. The fabric still has adequate air permeability.

To be effective, the finishing agent must coat each individual fiber and it must adhere tenaciously. Many of the finishes used to produce water repellency are also good stain repellents, because the compounds used are similar. Logically, then, most fabrics that have good water repellency possess good resistance to water-borne stains.

Early water-repellent coatings were removed by dry cleaning and/or laundering. The first "durable" water-repellent finish was introduced in the United States in 1942. It was used by the Armed Forces for protective apparel and was based on quaternary ammonium compounds. At the present time there are durable, semidurable, and renewable water-repellent finishes on consumer goods.

Durable finishes are based on silicone compounds, pyridinium compounds, fluorochemicals, methylol stearamide, and the ammonium compounds. Trademarks include Hydro-Pruf, Sylmer, Cravanette 330, Ranedare S, Aquagard, and Impregnole for silicone finishes; Norane and Ranedare for pyridinium finishes; Zepel and Scotchgard for fluoro-chemical finishes; Tanpel S for methylol stearamide finishes; and Zelan and Cravenatte for quaternary ammonium compounds.

Semidurable finishes are based on two kinds of compounds that are mixed just before application. One of these is a type of soap, while the second is a zirconium or rare earth-metal salt. Durane and some Cravanette finishes are examples of this group.

Renewable water repellents are generally composed of aluminum compounds or wax emulsions. As the term implies, renewable finishes must be reapplied after cleaning. Trade names of this group include Cyanatex, Aridex, and certain selected types of Norane and Cravanette finishes. The fluorochemical finish Scotchgard is available in aerosol spray cans and can be applied to fabrics in the home. The sprayed product has effective water repellency.

Water-repellent fabrics are common in raincoats, all-weather coats, hats, capes, umbrellas, and shower curtains.

Stain- and Soil-resistant Finishes

Removal of stains from fabrics has been, and is, a constant problem for consumers, so finishes that reduce staining and soil are always welcome.

Soiling or staining of fabrics occurs in three different ways: redeposition of soil during laundering or dry cleaning, deposit of dry soil from the air or by contact, and spot soiling by contact with foreign matter. Soil is redeposited when fabrics are incorrectly laundered. Materials such as carboxy methyl cellulose (CMC) aid in preventing redeposition and are frequently included in detergents.

Substances used to produce fabrics with resistance to water-borne stains are similar, in some cases identical, to those for water repellency. Silicone chemicals are effective for this as well as for other end results. Resistance to both oil- and water-borne stains is produced by fluoro-chemical finishes.

Soil-resistant finishes reduce the rate of soil deposition on a fabric either by creating an electric charge that repels the soil or by producing a smooth surface to which soil will not adhere. Fabrics treated with soil-resistant finishes are, therefore, easily cleaned.

Finishes used to reduce soiling act as direct barriers. Apparently, the finish occupies places on the fiber that would otherwise be filled by the soil.[9] They create a smooth surface and the soil falls off.

Spot staining causes the most problems in the care of fabrics, so finishes that prevent stains are a real aid to the consumer. Stain-repellent finishes are made from such compounds as silicones, fluorochemicals, waxlike derivatives, triazine compounds, and pyridinium compounds. The presence of the finish prevents stains from penetrating into the fiber by developing a high degree of surface tension on the fabric; by producing a substrate on the fabric, so oil and water actually "float"; or by presenting a surface with a very low free-energy factor, which prevents reaction between the stain and the surface.

The fluorochemical finishes are among the newest in this group (Fig. 26.11). They create a surface with a low free-energy factor and are considered extremely successful. Zepel and Scotchgard are in this category and are becoming important for home furnishings.

Sylmer is a silicone finish that has proven quite satisfactory. Other silicone finishes listed as water repellents increase resistance to staining caused by water-soluble substances.

Many of the other compounds used in producing water-repellent fabrics will reduce possible fabric damage from water-borne stains.

One problem has become apparent with the increased use of stain-repellent finishes. If dry soil becomes embedded in the fabric, it is nearly impossible to remove, although frequent cleaning will help. Liquid stains are relatively easy to remove, but any stain should be cleaned as soon as possible if optimum use of the finish is to be achieved.

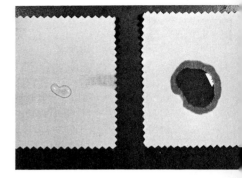

Figure 26.11   Fabric with a fluorochemical finish (*left*) and fabric without the finish (*right*). Oil does not penetrate the treated sample, but it soaks and spreads rapidly into the untreated fabric.

Soil-release Finishes

With the increased use of synthetic fibers and durable-press finishes, the difficulty of removing soil has increased. In an effort to alleviate the problem, manufacturers have incorporated soil-release properties with durable-press finishes or have applied soil release alone. These finishes operate on one of two principles: they provide a hydrophilic surface that attracts the water and permits it to lift off soil, or they coat the fibers so the soil never penetrates.

Many soil-release finishes are nearly identical to soil-repellent finishes. They often provide several side benefits, such as preventing soil redeposition, introducing antistatic qualities, and improving the softness and hand of fabrics. Chemicals used in soil release include

[9]J. M. Salsbury, T. F. Cooke, E. S. Pierce, and P. B. Roth, "Soil Resistant Treatment of Fabrics," *American Dyestuff Reporter* (March 26, 1956), p. 190.

fluorocarbons, acrylate emulsions, and hydrophilic copolymers. Trade names for the better-known processes include Dual-action Scotchgard, Fybrite, Visa, Cirrasol, and X-it.

Despite the fact that soil-release finishes have been on the market since 1967, and that they have been constantly improved since that time, soil release still poses a few problems to the consumer. It may cause some color loss or reduce the durable-press qualities. However, if properly applied and if the product is cared for according to directions, the promise of soil release will be justified.

In an effort to provide a product that gives the consumer what he really wants, one manufacturer specified that soil-release systems must

- remove all types of common soil in home laundering, with normal detergents
- remove all common or kitchen stains
- remove stains stemming from hair tonics, vaseline, deodorants, and the like, removing to some degree, lipstick, liquid eye makeup, and hair dye
- remove all oily stains, grease, motor oil
- be durable in continued home launderings
- be nonyellowing or nondulling
- maintain the same hand as untreated fabrics
- permit absorbency
- be durable for commercial laundering
- *and* maintain every permanent-press quality[10]

Durable-press and Minimum-care Finishes

Wrinkle recovery is an expression used in technical literature to indicate the ability of a fabric to recover from folding deformation while the fabric is dry. *Crush resistance* is similar, but this term usually describes the recovery from crushing of a pile fabric. *Durable press* refers to the ability of a fabric to retain an attractive appearance during wear and to return to its original smooth surface and shape after laundering. These characteristics can be imparted to fabrics by finishing processes and by fiber choice. The following discussion is limited to finishes.

A recognized defect of cellulose fibers is their tendency to wrinkle badly during wear and maintenance. This results in unattractive products that require considerable ironing to restore a neat appearance. Before the 1920s the only method known to minimize wrinkling of cellulosic fiber fabrics was to apply starch, and this was only a temporary solution. In 1919 textile scientists made the first measurements of fabric creasing and recovery. In the following decade the first finishes to reduce wrinkling were developed, and they were applied to cotton and linen. These early finishes had one undesirable side effect: they

---

[10]"As You Like It," *American Fabrics*, No. 81 (Winter 1968), p. 91.

caused considerable loss of strength and fabric deterioration. Linen fabrics had sufficient strength to spare, and the finish made the fabric look so much better that the processing was accepted by consumers with little awareness that a strength loss had occurred. Cotton fabrics, on the other hand, did not stand up well under processing, and in 1948 less than 1 percent of all cotton was treated. The consumer, however, was impressed with the few examples he saw and did not seem to realize that such fabrics were less durable. In fact, one of the best-kept secrets in the textile industry was the loss of strength and flex abrasion resistance in cellulose when finished for easy-care characteristics.

The introduction of man-made thermoplastic fibers, which could be heat-set to build in optimum appearance values, made the easy-care concept a major goal of the textile industry. Manufacturers of cellulosic fiber fabrics, as well as the U. S. Department of Agriculture, expended considerable time and money in research to find techniques that would produce satisfactory easy-care products. Rapid strides were made during the 1950s, when "wash-and-wear" or "minimum-care" fabrics became the usual, not the exception. In 1964 the introduction of durable- or permanent-press fabrics presented consumers with truly easy-care textiles.

Chemicals used in easy-care finishing often cause reduced abrasion resistance, breaking strength, and tearing strength of cellulosic fibers. To help alleviate these negative aspects, most processed fabrics now combine a fiber such as polyester or nylon with the cellulose. Original research, however, was performed on 100 percent cotton fabrics, and the first durable-press fabrics introduced to the public were 100 percent cotton.

There are two major theories about why the finishing process improves wrinkle resistance and reduces the care required. One theory states that the finishing resins build a memory into the fiber, so it returns to the size and shape it had during the finishing operation. The second theory, the cross-linking theory, holds that the finishing agents (resins) react chemically with cellulose molecules lying parallel to one another so they become cross-linked. This causes the molecules to experience a strain whenever they are moved into a new position, and when strain is removed, they return to the original arrangement and flatten the fabric. If the finish is applied when the fabric is creased, the product returns to its creased state after deformation. Many durable-press finishes are probably a combination of cross-linking and memory.

In simplified terms the durable-press operation consists of padding the fabric continuously as it moves through the solution of finishing chemicals, drying on a frame to desired dimensions, curing at an elevated temperature, scouring, and redrying. Durable-press finishes can be cured at either of two points in processing. Precured durable press follows the procedure outlined above. Delayed-cure or deferred-cure involves additional steps or a different order of processing. The Koratron patent specifies no cure until the final product is completed. In other words, the resin remains unpolymerized until the final oven cure has been performed. Other delayed-cure methods require that a partial cure occur at the time of finish application and that the final cure and polymerization be undertaken after the manufacture of the textile product.

The chemical generally used in post-cured finishes is dimethylol-dihydroxyethylene urea. The precured processes employ a variety of resins, including dimethylol-dihydroxyethylene urea, dimethylol-ethylene urea, and methylated hexamethylolmelamine. For precured white fabrics carbamates and dimethylolpropylene urea are most effective. A few manufacturers are trying a modified glyoxal-type resin, which shows promise of replacing some of the rather expensive carbamates.

Durable-press finishing utilizes one of these cross-linking resins, plus a catalyst, a softening agent, a fulling agent, and a wetting agent. The cross-linking chemicals have various advantages and disadvantages. The primary concerns are relative cost, ease of application, and effect on strength of the fabric.

All resins used to impart durable press result in diminished tear strength and abrasion resistance, but there is a slight difference in degree. According to one authority,

> There is a direct (positive) correlation between wrinkle recovery or durable-press appearance and the amount of resin applied—and there is also a direct correlation between the wrinkle recovery or durable-press behavior and the tear strength. Some systems give a little better strength than others, but they all lower the strength by 40 to 50 percent in order to give durable-press properties to cotton fabrics.[11]

One solution to the problem of reduced strength has been to combine man-made fibers with cellulosic fibers in blended yarns.

Urea and melamine formaldehyde are subject to yellowing after exposure to chlorine. Discoloration results when fabrics with these finishes are bleached with chlorine bleaches.

If fabrics are not thoroughly rinsed after the final curing, it is possible for the catalyst and free formaldehyde to be left in the fabric. These chemicals can then set up a reverse action and break the cross-links formed. Another problem arises when these resins gradually wash away, resulting in decreased self-smoothing characteristics. In addition, the free formaldehyde vaporizes, and the odor becomes obvious in poorly ventilated fabric stores and stockrooms.

---

[11]J. W. Weaver, in a personal communication with the author.

Cyclic reactants used in durable-press processing include dimethylol ethylene urea (DMEU), dimethylol dihydroxy ethylene urea (DMDHEU), dimethylol dimethyl hydantoin, imidazolidones, and triazones. These reactants give good results in producing self-smoothing fabrics, and they are not visibly damaged by chlorine. However, the fabrics may be weakened by chlorine and, in some instances, a slight discoloration may occur.

All of these cyclic reactants, as well as the urea and melamine formaldehydes, contain nitrogen, and it is this element that reacts with free chlorine in a series of steps. The final reaction is the formation of hydrochloric acid, which causes the tendering of the fabric. Triazones may show less damage than other substances because they are highly basic, and the basicity nullifies the action of the acid formed. However, if not adequately rinsed after curing, the triazones tend to develop "fishy" odors that make them undesirable.

Carbamates are applied to white fabrics, especially white shirt fabrics, because they do not yellow during the high-temperature cure. They impart good self-smoothing characteristics and resist change from chlorine bleaches or acid "sour" rinses used in commercial laundries. It seems superfluous to send durable-press shirts or other items to commercial laundries, but families without access to proper laundering equipment may have no alternative.

Finishes can be fixed either by wet fixation—applying and finishing the fabric in a wet bath or in a vapor (steam) phase process—or by gamma or low-energy beta radiation.

Durable-press and minimum-care finishes are effective only on cellulosic fibers. In blends, the man-made synthetic fibers, such as polyester or nylon, do not accept the finishing chemicals. Blends are popular because the strong thermoplastic fibers contribute strength and make it possible to construct lightweight, durable fabrics. This offsets any reduction in strength of durable-press cellulose fibers. Consequently, blends in common use today provide both durability and comfort, as well as ease of care.

Despite their tremendous acceptance, durable-press fabrics pose problems to the consumer. Deferred-cure fabrics are almost impossible to alter. Precured fabrics are difficult to press into new shapes. However, most people are so pleased with the easy-care properties of durable press that problems are overlooked.

There are still some fabrics on the market with a wrinkle-resistant finish that is not durable press. Some linen, some all-rayon items, and a few all-cotton fabrics are in this category. Most such fabrics are labeled to indicate that they resist or recover from wrinkling. A few trade names are Everglaze, Disciplined cotton, Regulated cotton, Wrinkle-shed, Minicare, Belfast, and Unidure.

Durable-press fabrics appear in all types of apparel and in a wide selection of home furnishings. Consumers request the finish on men's and boys' trousers, shirts, and undershorts; on women's and girls' skirts, slacks, shorts, blouses, dresses, and jackets; and on household items, such as sheets, bedspreads, tablecloths, and draperies. Trade names for durable press include Koratron, Coneprest, Dan-press, Never-press, Primitized, Reeve-set, Grid-press, and Russ-press.

A growing trend in finishing techniques today is to incorporate several finishes into one fabric. One of the combinations includes durable press, soil release, stain resistance, water repellents, and softeners. Durable press, soil release, and softeners are often combined to produce a fabric that is easy to maintain and has a good hand. Most fabrics with durable-press finishes have dimensional stability, but the size may change as the finish gradually disappears. Manufacturers have had to cope with the problem of shrinkage. After application of the finish, the fabric or textile item shrinks to a high degree in the final cure. This has necessitated a different sizing system in order to produce products that fit the consumer. It partially accounts for irregularities in size that occur.

The finishing industry never stands still. Research is constantly underway to find new functional finishes that will make fabrics more adaptable and more satisfying for life today. It is safe to assume that the finishing industry will continue to expand, for without modern finishes there would be no modern fabrics.

Evaluation
of Finishes

Finishes add many desirable qualities to a fabric. Unfortunately, they also add to the cost of production, and this increase is passed on to the consumer. In some instances, the durability and the required method of care of a fabric are affected. However, it must be said that for the most part the advantages quite distinctly outweigh the disadvantages.

Standard test procedures for evaluating finishes are prepared by several textile organizations and can be obtained by anyone who wants them. The American Society for Testing and Materials (ASTM) publishes yearly a series of volumes that include a variety of standard test methods. Committee D-13, which deals with standards for textile materials, is responsible for tests related to fibers, yarns, and fabrics. This committee proposes new tests as they are needed and determines when old methods should be discontinued; it specifies suggested tolerances for items; and it maintains an up-to-date glossary of definitions of important textile terms.

Included in the standards are some test methods for qualitative analysis of selected finishes and a number of tests that can be used to determine specific fabric properties that are the result of finishing materials. Examples of the latter group include

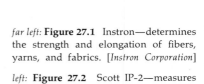

far left: **Figure 27.1** Instron—determines the strength and elongation of fibers, yarns, and fabrics. [*Instron Corporation*]

left: **Figure 27.2** Scott IP-2—measures the strength and elongation of fibers and yarns. [*Scott Testers, Inc.*]

above: **Figure 27.3** Scott Tester, Model J—tests the strength and elongation of fabrics; can be modified to test yarns. [*Scott Testers, Inc.*]

center: **Figure 27.4** Taber Abrader—determines abrasion resistance of fabrics. [*Taber Instrument Corporation*]

right: **Figure 27.5** Elmendorf Tearing Tester—measures the tearing strength of fabrics. [*Thwing-Albert Instrument Company*]

1.	D 583-58	Tests for Water Resistance of Textile Fabrics
2.	D 862-57	Tests for Evaluating Treated Textiles for Permanence of Resistance to Microorganisms
3.	D 1230-61	Tests for Flammability of Clothing Textiles
4.	D 1296-60T	Tests for Recovery of Woven Textile Fabrics from Wrinkling
5.	D 1905-61T	Tests for Dimensional Change in Woven or Knitted Fabrics

The American Association of Textile Chemists and Colorists (AATCC) has developed several methods of analysis that can be applied to evaluate specific fabric properties. These properties result from an interrelation of several factors: fiber content, yarn and fabric structure, and finish. The tests are used in controlled situations to determine the effectiveness of a finish upon certain fabric properties. They include tests for flammability, water repellency, shrinkage, wrinkle recovery, durable-press appearance, resistance to biological damage, and resistance to damage by abrasion. These are also tests to identify the finishes that have been applied to fabrics.

The American National Standards Institute (ANSI)—formerly the United States of America Standards Institute—has also developed and published tests for analysis of specific fabric properties. These can be used in the same manner as the AATCC test methods. In many cases the ASTM, AATCC, and ANSI procedures are identical and have been devised by interorganizational committees.

In addition, both Federal and state agencies have established specifications that serve as the basis for bidding by manufacturers who sell to the government. These are detailed descriptions of properties and tolerances that an item must possess to be considered for purchase. In order to verify that the item submitted does meet the specifications, the government tests it, and the tests are published as Federal Specification Test Methods CCC-t-191 b. Among the test methods available are those related to fabric behavior that may be affected by finishes, such as water repellency, abrasion resistance, shrinkage, mildew resistance, and flame resistance. State agencies, including hospitals and prisons, require adherence to similar criteria.

The average person does not have access to the testing equipment used for laboratory analysis (Figs. 27.1–27.5). However, many simple procedures can be followed in the home to provide helpful information about fabric finishes. Before describing them, it is necessary to emphasize several points:

1. At no time should a student or consumer who uses the following tests substitute the results for those from regular research or testing laboratories.
2. The tests give limited information about fabric and finish behavior. However, the consumer may gain valuable insight into care problems, so that he can more readily formulate procedures for maintaining fabric appearance and behavior.
3. Teachers and students can use some of the following tests when there is no regular testing equipment. By conducting a series of tests in an identical manner, comparisons can be made among fabrics tested, but generalizations about fabrics or fabric types should be avoided.

HOME TESTS FOR FABRIC PROPERTIES

Percent of Shrinkage

The possibility of shrinkage in a fabric is very important to the consumer. Even if a maximum percent of shrinkage is stated on the label, he may wish to verify this or to check for himself. It may be interesting to see if this technique results in more or less shrinkage than cited on a label.

If a person has access to a 22-inch square, he will obtain superior results by following the method in ASTM test D 1905. However, the sample available may be small, and for such occasions the following methods have proved informative.

Test #1   If a 12-inch sample can be obtained,

1. Mark a 10-inch square on the sample fabric, with the warp clearly identified.
2. Launder the sample with a regular wash load.
3. Dry and press, then measure to determine the shrinkage or stretch.

If the consumer is interested in learning the effect of a dryer upon dimensional stability of the fabric, he may wish to prepare two samples. One will be laundered and air dried, and the second sample will be laundered and dryer dried. By simple mathematical calculations the percent of shrinkage can be determined.

$$\% \text{ shrinkage} = \frac{\text{original measurement} - \text{final measurement}}{\text{original measurement}} \times 100$$

Calculate for warp and filling separately. A shrinkage of more than 3 percent is said to cause a full size change in wearing apparel. A 2-percent shrinkage is usually considered a maximum for consumer satisfaction.

This method will give more reliable and realistic results than the following test. However, if there is inadequate material available, a consumer can obtain some idea of potential fabric shrinkage with method #2.

Test #2

1. Trim a sample so the cut edges are parallel to warp and filling in woven fabrics and to wales and courses in knitted fabrics. Clearly indicate the warp. For practical results the scrap of fabric should be approximately 4 to 5 inches square.
2. Place the test sample on paper and draw around it so there is a clear record of the original sample.
3. Wash the sample by hand, using water temperature and detergent recommended for laundering.
4. Rinse thoroughly, dry, press if needed, and compare with the original size.

This method does not provide for change in size caused by raveling, and a fabric in which raveling of yarns does occur will give misleading results. This might be controlled to a degree by edge stitching the sample before the test.

Shrinkage Resistance to Dry Cleaning

Shrinkage in dry cleaning is usually caused by water added to the cleaning solvent to remove water-borne soil and stains, by agitation in the presence of water, or by steam pressing. Fabrics that are to be dry cleaned can be tested to determine shrinkage resulting from moisture and pressure by marking a sample as for laundry method #1 and thoroughly pressing with steam and pressure.

The consumer can obtain some idea of the potential shrinkage of a fabric in cleaning by subjecting a marked sample to the following steps:

1. Wet the sample in water and remove all excess moisture.
2. Place the sample in cleaning solvent and agitate frequently by stirring, shaking, or rubbing. Continue this procedure for eight or ten minutes.
3. Remove from the solvent, squeeze out excess fluid, and dry.
4. Measure the percent of shrinkage.

It is important for the individual making the test to observe all precautions that are required in the use of the cleaning solvent selected for the test.

The tester can also note the degree of retention of such finishes as glazing and stiffening. Subjective observation after laundering or soaking in cleaning solvents will give an idea of the durability of appearance finishes.

Durable Press

Test samples similar to those described for shrinkage tests are used to evaluate the quality of durable-press finishes.

1. Launder the sample.
2. Dry.
3. Evaluate by visual inspection, before pressing, to determine the need, if any, for touch-up ironing.

Unless standard test equipment is available, this method of evaluation is completely subjective. However, since the final decision on ironing is made by the consumer, the information gained by this home test can be helpful in the care of the finished garment.

Wrinkle Recovery

1. Fold a 2-inch-square fabric sample twice so as to make a 1-inch square.
2. Place a pound weight (perhaps a packaged item weighing one pound) on top of the folded sample, and let the weight remain for five minutes.
3. Remove the weight and unfold sample, but do not smooth.
4. Observe to see if wrinkles spring out quickly.
5. After five minutes, examine the sample a second time.

This is not a scientific test, but it will give some indication of the wrinkle recovery of the fabric. Fabrics that appear to smooth out easily will recover from wrinkling more quickly than those that retain clear creases.

Water Repellency

With a clothes sprinkler or eye dropper, drop or shake water onto fabric that has been laid on a smooth surface or placed in an embroidery hoop and held at a 45-degree angle. If the water forms tiny beads and rolls off without penetrating, the fabric is considered water repellent.

Oil Repellency

Place a drop of salad oil on a scrap of fabric. If the drop of oil forms a bead and can be removed easily with a blotter or a piece of absorbent tissue leaving no stain, the fabric has repellency to oil-borne stains and oily liquids.

Flame Resistance

1. Hold a piece of fabric in a pair of tongs.
2. Place a lighted match at the lower edge until the fabric appears to flame.
3. Remove the source of flame, and observe the behavior of the fabric.

If the flame extinguishes itself, the fabric is flame resistant to some degree. It is essential to observe extreme caution in doing this test to prevent damage to the individual and the surrounding area.

As with the other methods, this cannot be interpreted as equivalent to government tests. A small swatch will burn much more easily than the large samples that are used to determine flammability as related to legislation and standards.

SUMMARY

Some finishes are destroyed by the first laundering or dry cleaning; others last for more than fifty washings or cleanings. When purchasing fabrics with various finishes, the consumer should make careful note of labeling information and observe any recommended care procedures. The purchase of any textile fabric or product is a matter of personal choice, but to be an intelligent consumer, one must know how to evaluate a product.

The testing techniques given in this chapter require fabric scraps. They are, therefore, difficult, often impossible, to use on ready-made articles. For these the consumer must depend on label information, which is often meager, or must wait and observe what happens at the first laundering. If the item does not perform satisfactorily, it should be returned. Reliable manufacturers want to know about defective merchandise. The consumer with an unsatisfactory item should return it to the place of purchase and insist that the store return it to the manufacturer. Otherwise, the manufacturer has no idea of the source of dissatisfaction. Comments or reports from consumers are important in the development of quality merchandise.

Stretch Fabrics

The concept of stretch originated in 1589, when the first knitted fabrics were introduced. Since that time knits have been prized for their comfort and appearance, and many other methods of imparting stretch to fabric have been attempted. In the early 1960s textile manufacturers began using the principle in new and dramatic ways, and the stretch revolution was launched.

The word *stretch* has now acquired a more specific meaning. In modern terminology, a true stretch fabric has the ability to extend or stretch under tension, *plus* the equally important capacity to return to its original size after release of strain. The degree of potential stretch, sometimes referred to as *elongation*, varies from as little as 5 percent to as much as 500 percent. Most fabrics fall within either the 10- to 25-percent category or the 30- to 50-percent range. In stretch fabrics this amount of elongation occurs at low loads. Suggested standards from the Good Housekeeping Institute specify a load of 4 pounds per 2-inch-wide strip as the force desirable in determining the percent of stretch.[1] (Other organizations suggest testing a 1-inch-wide strip with 2 pounds of pull. The same proportion is obtained with the use of

---

[1] "Institute Report," *Good Housekeeping* (May 1965), p. 6.

a narrower sample of fabric.) The Institute believes that any 2-inch-wide test specimen that fails to stretch 20 percent or more under the 4-pound load should not be called a stretch fabric.

American Fabrics classifies stretch fabrics into two categories: comfort stretch and power stretch.

> *Comfort stretch* describes stretch fabrics that go into clothing for everday use with a stretch factor of up to 30 percent. *Power stretch* or *action stretch* describes stretch fabrics that have more snap and muscle power, more extensibility, and quick recovery. The stretch factor generally ranges from 30 to 50 percent and is best adapted to ski wear, foundation garments, swimwear, athletic clothing, and the more professional types of active sportswear.[2]

There is a nebulous area between these types where the amount of stretch may reach 40 percent and still provide comfort, or a viable action stretch may extend below 30 percent. Comfort stretch is designed for use under low loads, such as 2 pounds per inch, while power stretch usually encounters considerably higher loads.

Stretch fabrics as a group defy classification. The property of stretch can, under certain conditions, be introduced at any stage of fabric manufacture. For convenience, the following discussion has been divided into four categories:

1. fiber stretch—elastomeric fibers and fiber modification
2. yarn stretch—including heat setting and other techniques
3. fabric stretch—knitting and similar processes
4. finishing stretch—the application of special finishes to impart stretch after the fabric has been completed

FIBER STRETCH

Stretch has been produced through the use of elastomeric rubber fibers since about 1920. Fine filaments of rubber were covered with cotton, rayon, or silk, and then woven, knitted, or braided into fabrics with combinations of other yarns. Narrow elastic fabric is a good example of one type of construction. Rubber core yarns were early examples of elasticity and stretch. These yarns are still used in such items as elastic banding, elastic bandages, and some fabrics. However, spandex fibers have now replaced rubber in a large percentage of fabric construction.

Spandex, described fully in Chapter 15, owes its stretchability to its chemical molecular configuration, not to a mechanically imparted property. Spandex fibers can be used in several ways in stretch fabrics. Uncovered or bare, spandex filaments are combined with other fibers,

[2]"Stretch," *American Fabrics,* No. 63, p. xvi.

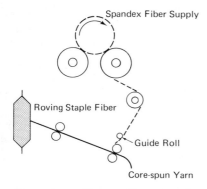

Spandex Fiber Supply

Roving Staple Fiber

Guide Roll

Core-spun Yarn

Figure 28.1 Diagram illustrating the core-spinning technique. Spandex fibers feed from one source into the zone of staple fibers that are to surround the spandex core. (See also Fig. 15.5.)

staple or filament. Such power stretch fabrics appear in items like foundation garments, swimsuits, and surgical hose. In small amounts, bare spandex may be included in fabrics for comfort stretch, but the trend in these fabrics is to core-spun yarns.

Core-spinning is a procedure by which an elastic spandex filament is fed directly into the twisting zone of the spinning frame (Fig. 28.1). The tension under which the elastomeric filament is held is carefully controlled and predetermined in relation to the end-use of the fabric. The resulting yarn has variable degrees of stretch, depending upon manufacturing controls. Spandex filament forms the core of the yarn and is surrounded by staple fibers of one or more types, which provide a flexible outer sheath. Core-spun yarns have two significant advantages. Only 3 to 10 percent spandex is required to produce a high-quality stretch yarn. Furthermore, the percent of stretch can be controlled from 10 percent or less to as much as 200 percent. Most core-spun yarns are used in comfort-stretch products and do not exceed 30 percent stretch at low loads, such as 2 pounds per inch. Core-spun yarn acquires the hand and appearance of the fibers used to form the outer sheath. Depending upon the fiber used as covering, the fabric may have good moisture absorbency. The care of core-spun yarns is, basically, the same as that required of the covering fibers. However, the presence of spandex, even in small amounts, makes it advisable to know those procedures that will prolong the life of the product.

Spandex yarns are also employed in *intimate-blend spinning*. Presently, the technique is new and might still be considered experimental. However, it shows promise of excellent success. This procedure involves the cutting of spandex into staple lengths to match the length of one or more types of rigid fibers. These are spun to produce a true blended yarn, with the amount of stretch dependent upon the amount of spandex fiber used. It has been suggested that such blends be formulated at the fiber-producer level and then sold to the yarn and fabric manufacturer.

Stretch through modification of fiber structure is relatively new. Probably the best-known product of this type on the consumer market is Cantrece® nylon. This fiber is found in women's hosiery and in certain other apparel items. Cantrece is a bicomponent nylon and is discussed in Chapter 10. Other bicomponent nylons and modified polyesters exist which may impart some stretch because of the fiber structure.

Bicomponent stretch occurs when the filaments are subjected to processing: one component shrinks and causes the other component to coil. The resulting filaments and yarns have a high degree of coil and, as a result, stretch. The stretch will vary according to the amount of coiling and "give" in the fibers.

Yarn Stretch

Of the new techniques for producing stretch fabrics, heat-set stretch yarns are considered the original. The first stretch yarn of this type was developed by the Swiss firm of Heberlein and Company, A.G. The original goal was to impart a crimp to man-made fibers so that they would resemble wool. After initial work on viscose fibers, Heberlein adapted the process to nylon and developed the first nylon stretch yarns, which were introduced in 1947 as *Helanca*. (See p. 208 for a description of this process.) Helanca was used in hosiery, leotards, and other figure conforming apparel.

The current techniques for producing stretch yarns by heat setting include the false-twist method, the twist–heat-set–untwist method, and the knife-edge method. These are discussed in some detail in Chapter 19. Though the stuffer-box method is not generally known for its stretch-producing ability, some authorities include it in the list.[3] Depending on the method employed, stretch can be obtained from either continuous filament or staple fibers.

Two methods similar to those used in producing stretch yarns from thermoplastic fibers were developed by the research laboratories of the United States Department of Agriculture and were called *back-twisting* and *crimping*.

In the back-twisting method cotton yarn is treated with cellulose cross-linking chemicals or resins such as DMEU (see p. 300), which are used in making minimum-care fabrics. The yarn is twisted, the twist is cured, and then the yarn is untwisted and retwisted in the opposite direction. The resultant yarns are kinky and springy and have good stretch properties.

In making crimped cotton the cotton yarn is treated with one of several chemicals that react with the cellulose to form a cellulose ester or cellulose ether that is thermoplastic. The modified cotton is then processed by one of the methods used for other thermoplastic fibers. This process is not in commercial use.

Stretch yarns made by the heat-set technique can be used satisfactorily in knitted or woven fabric construction.

FABRIC STRETCH

The most common method for introducing stretch in the fabric-construction stage is by knitting. Hand-knit fabrics have been popular since the sixteenth century, and the products of modern knitting machines are now equally sought after for every conceivable article of wearing apparel. Knits are discussed in detail in Chapter 21.

[3]Berkeley L. Hathorne, *Woven, Stretch and Textured Fabrics* (New York: John Wiley & Sons, Inc., 1964), p. 17.

FINISHING STRETCH

The process of imparting stretch to a fabric after it has been constructed is called variously *piece-goods stretch, mechanical stretch,* and *chemical stretch.* The first of these terms is probably the most convenient. The actual finishing procedure involves the use of chemicals and results in some physical change, so, theoretically, the term "chemical" is appropriate. However, chemical change of the molecular structure of the fiber does not occur at all times. Instead, a physical or "mechanical" change takes place. On the other hand, due to the fact that chemicals are required to bring this change about, the word "mechanical" can be somewhat misleading. Thus, of the currently proposed terms, *piece-goods stretch* is adequate.

As this phrase indicates, stretch is introduced into the fabric after it has been woven. Cotton, cotton blend, and wool fabrics are treated in this manner. The procedure used on cotton and cotton blends is called *slack mercerization.* For wool, special processes have been developed by leading fabric producers.

Slack mercerization utilizes the same principles as those applied in standard yarn or fabric mercerization, except that the fabric is not held under tension; hence the term *slack.* For horizontal or "filling" stretch, the fabric is held under lengthwise tension, or if no tension is used at this stage, the fabric is restretched and set for length at a later time. If both horizontal and longitudinal stretch are desired, the fabric is treated without tension.

The amount of stretch is determined by the concentration of sodium hydroxide solution and by the density of the original fabric. Production of filling stretch fabrics by this method has proved relatively successful, but warp stretch has posed several problems, the most critical being the tendency for warp "stretch" to be lost during washing and for the fabric to stretch out to its original pretreated length. According to recently published information, filling stretch is easier to dye and finish than warp or two-way stretch and appears to offer more durable fabrics.[4]

Production of stretch fabrics by slack mercerization requires carefully controlled conditions. Reliable manufacturers have tried to maintain quality in goods entering the market, and research workers have spent considerable time and effort in developing techniques that will stabilize the "stretch" performance of cotton and cotton-blend fabrics. One successful treatment employs cross-linking agents such as those used to produce wrinkle-resistance and minimum-care cottons. These cross-linking agents set and maintain a crimp in cotton that helps to hold stretch. Stretch fabrics produced by the slack mercerization process are less expensive than others; however, to date, durability of the stretch is unsatisfactory.

[4]"Stretch Surges Ahead," *American Fabrics,* No. 64, p. xi.

Special processes to make stretch fabrics of wool have been developed by two leading mills. These techniques, patented under the trade names Plus X and Restora, introduce a crimp into the wool yarns.

STRETCH FABRICS IN USE

Stretch can be produced in the horizontal or filling direction of fabrics, in the vertical or warp direction, or in both directions (two-way stretch). The latter is used primarily in action stretch, while horizontal or vertical stretch fabrics are common in comfort stretch items.

Considerable research has been conducted to determine how much stretch is required in fabrics and where and how it should be used (Figs. 28.2, 28.3). Studies to determine the amount of stretch required have been conducted by the DuPont Company and the J. P. Stevens Company. Both studies analyzed how much the body's skin stretches in various locations. Results of these studies indicate that the elbow, when bent, flexes from 35 to 40 percent in the vertical or lengthwise direction and from 15 to 22 percent in the horizontal or circumference direction; the knee flex is 35 to 45 percent in the lengthwise direction and from 12 to 14 percent in the circumference; the hip area stretches 4 to 6 percent in the horizontal direction when seated and as much as 60 percent in the vertical direction; the amount of stretch across the shoulders is from 13 to 16 percent.

Further studies by the J. P. Stevens Company recommended that the following amounts of stretch be used:[5]

Tailored clothing: 15 percent to 25 percent unrecovered stretch (see below).

Spectator sportswear: 20 to 35 percent stretch; no more than 5 percent unrecovered stretch.

Form-fit garments: 30 to 40 percent stretch; no more than 5 percent unrecovered stretch.

[5] *American Fabrics,* No. 64, p. xi.

left: **Figure 28.2** Stretch fabric in the relaxed state. Arrow indicates fabric at 0 percent stretch.

right: **Figure 28.3** Stretched fabric. Arrow indicates that the fabric has stretched approximately 35 percent.

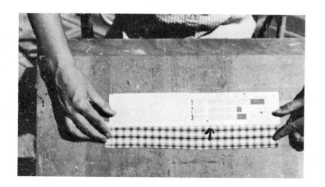

Active ski wear: 33 to 50 percent stretch; no more than 6 percent unrecovered stretch.

The optimum percentages suggested by another writer are somewhat different:[6]

Vertical stretch pants: elongation (stretch) potential of 40 to 60 percent with a "growth" of less than 6 percent.

Tailored clothing: elongation," of 25 to 35 percent with a growth of less than $2\frac{1}{2}$ percent.

Sportswear and dresses: elongation of 25 to 40 percent with a growth of less than 5 percent.

Standards developed by the Good Housekeeping Institute for stretch fabrics that may bear their seal require at least 35 percent stretch for active sportswear, 25 percent stretch for casual apparel, and 20 percent stretch for tailored clothes.[7] The Good Housekeeping Institute study, as well as a technical interpretation of existing standards for fabric stability, indicates that more than 2 percent growth in any fabric or item is unacceptable.

Most stretch fabrics do not completely return to their original measurements after elongation. The difference between the original size and that measured after elongation is referred to as *growth* or *nonrecovered stretch.* This growth is frequently eliminated in laundering, but it can lead to temporary (or in some cases, permanent) deformation. A growth or increase in size of $2\frac{1}{2}$ to 3 percent would result in an unattractive garment. The maximum growth of 2 percent would appear to be a good standard.

The problem of where and how stretch should be used has caused considerable controversy. Stretch in foundation garments, swimwear, and other items where considerable holding ability is involved is power stretch. This type of garment requires figure shaping and conformity. The amount of potential stretch should be high, but it should stretch only under greater force than comfort stretch. While a small growth will probably not alter size noticeably, it will reduce the holding power.

Comfort stretch garments should be sized in the same way as "rigid" fabric garments. Here fabric growth could lead to size change. For this group of garments—which is of considerable importance to manufacturers, retailers, and consumers—stretch becomes the factor that puts comfort into movement for various activities.

In the early application of the stretch principle to a variety of fibers and fabrics, and before adequate testing was done, some manufacturers tried to capitalize on the idea through premature and rapid promotion. The result was consumer confusion about the care and use of stretch fabrics and some inferior products on the market. Gradually, this gap

[6]Marylou Luther, "Stretchology," *McCall's Piece Goods* (Summer 1964), p. 76.
[7]*Good Housekeeping* (May 1965), p. 6.

is being closed, and stretch fabrics are used wisely by reputable manufacturers. However, education in the use of stretch is needed on all levels.

The consumer must decide if comfort stretch is adequate and worth the cost. He must understand the value of comfort and buy a garment in the same size as rigid fabrics. Stretch is not meant to take the place of proper garment sizing.

The consumer must also learn to read manufacturers' labels regarding proper care of the product. The type of stretch, fiber content, and finishes or explicit directions for care should be indicated on labels. Some stretch needs careful handling, while other kinds respond well to machine laundering and drying.

Sales staffs in retail establishments should be trained in "stretchology," so they will be able to talk intelligently to consumers and make informed suggestions concerning purchases. Sales people should know the values of stretch and how to care for it; they should have adequate opportunity to put the purchaser into a fitting room to try a garment.

Manufacturers must select desirable products for stretch and determine how to use the fabrics best. They must be well informed so as not to repeat previous mistakes such as lining stretch garments with rigid fabrics or constructing seams in the stretch direction so that either the stitching breaks or it prevents stretch.

The current concept of stretch is relatively new. However, the following advantages ensure continued market growth. Depending upon the methods used in manufacturing the stretch fabric and the techniques used in product construction, the consumer should experience the following:

- comfort
- good fit
- shape retention
- design flexibility
- psychological appeal
- wrinkle resistance
- appearance appeal
- longer wear
- reduced seam puckering

In addition to apparel, stretch has considerable value in home-furnishing fabrics for such items as fitted sheets and slipcovers.

To date, stretch created by means of knits, elastomeric fibers, and heat-set thermoplastic fibers has proved most satisfactory. Slack mercerization provides low-cost stretch products, but appearance retention and durability are not good.

Stretch fabrics are experiencing a surge in popularity, and their potential seems unlimited.

Dyestuffs
and Their
Application

The importance of color in textile products cannot be overemphasized. Color speaks louder than words. Its appeal is universal, and it repeatedly serves as a common language. In the modern marketplace, consumers are usually more concerned with selecting the "just right" color than they are with consideration for other fiber and fabric characteristics.

The textile industry is well aware of the consumer's desire for color. Manufacturers know that the consumer who selected an item because of its color will be extremely annoyed if the color is not maintained for the desired or anticipated life of the product. Consequently, research by dyestuff manufacturers has produced coloring agents that will satisfy the esthetic desires of consumers and provide lasting pleasure with the fabric if the color is properly applied and the fabric carefully maintained. This type of research never ceases, so each year new and better dyestuffs become available for the myriad of textile fibers.

Because of the tremendous importance of color, no textbook concerning textiles could be adequate without a discussion of dyestuffs and their application.

HISTORICAL REVIEW

Dyestuffs and dyeing are as old as textiles themselves and predate written history. Fabrics dating from 3500 B.C. have been found in Thebes that still possess the remains of blue indigo dye. Other fabrics, discovered in the ancient tombs of Egypt, were colored yellow with dye obtained from the safflower plant. Beautifully colored fabrics dating back several thousand years have been unearthed in China, Asia Minor, and some sections of Europe.

Prior to A.D. 1856 all dyestuffs were made from natural materials, mainly animal and vegetable matter. A bright red was obtained from a tiny insect native to Mexico. This insect was used by the Aztecs to color their fabrics, and when the Spaniards invaded Mexico in 1518, they called the insect and the dyestuff *cochineal.*

A tiny mollusk (*murex Brandaris* and *murex trunculus*) found on the Phoenician coast near the city of Tyre produces a beautiful purple color. By 1500 B.C. Tyre became the center for the trading and manufacture of this purple dye. Some historians believe this was the first example of a "city" industry. The dye was tremendously expensive to produce, because approximately 12,000 animals were needed to obtain but one gram of dyestuff. Thus, the expressions "royal purple" or "born to the purple" came into being as an indication that only the wealthy could afford the dye.

Other ancient dyes included Madder, a red dye from the roots of the *rubia tinctorium* plant; blue Indigo from the leaves of the *indigofera tinctoria* plant; and yellow from the stigmata of the saffron plant, a member of the crocus family. Logwood, extracted from the pulp of the Logwood tree, produced excellent dark colors and is still employed to some extent in modern dyehouses.

Many of the early natural dyestuffs were prohibitively expensive. It is quite probable that the growth and development of the dye industry came about, in part, in an effort to find less costly methods and provide colorful fabrics for people of all economic levels.

Early efforts at coloring fabrics were hampered by the fact that few of the natural dyes formed colorfast combinations with fibers. Eventually, scientists found that this defect could be partially overcome by the use of *mordants*—compounds that render colors insoluble on the fiber. This knowledge aided the dye industry tremendously. Further advancement accompanied the discovery of additional products that yield attractive colors.

During the Dark Ages there was little advancement in fabric coloring, and most dyeing was done in the home. The beginning of the Italian Renaissance saw the art of dyeing revived, and the first books on the subject were published in Italy during the fifteenth century. Other books soon appeared in France, Germany, and England. Dyeing techniques were constantly being improved during the ensuing centuries, and written materials kept pace with these developments. Several interesting publications appeared in the United States during the late eighteenth and early nineteenth centuries. All of these were devoted to the use of dyes from natural resources.

As long as man was dependent upon animals and vegetables for his dyestuffs, and for minerals as the mordanting agents, his progress was limited by the skill of the operator in mixing the natural dyes and in perfecting the techniques used. As with many of nature's products, the quality of materials varied considerably, and, consequently, the results were rather unpredictable.

The year 1856 marked the turning point in the history of dyes. Sir William H. Perkin, while trying to make artificial quinine from coal tar, accidentally produced the first synthetic dyestuff, a purple color called *mauve*. This discovery launched the modern dyestuff industry. Today, nearly all dyes are synthetically compounded and, in most cases, are superior in every way to natural dyes. The dyestuff industry is one of the leading chemical fields in the Western world.

The development of artificial dyes has been rapid, and the list of dyestuffs currently available numbers in the thousands. More than 1500 dyes are presently available from manufacturers in the United States alone. However, the dye industry cannot relax. Each new textile fiber requires either new dyes or new methods of dye application, and the chemist must meet these demands. In addition, modern consumers prefer colorfast dyes, and dye manufacturers attempt to please.

The history of dyes is fascinating. It is impossible to discuss adequately the development of coloring agents in a text of this scope. Much legend is included in the history of dyestuffs, and it is sometimes difficult to separate fact from fiction. The background does, however, provide insight into the chronicle of man's unceasing desire for a better and more colorful future.

SEEING COLOR

Color is a visual sensation. It results from the reflectance of certain visible light rays that strike the retina and stimulate cells in the nerves of the eye. The nerves send a message to the brain, which, in turn, produces the sensation of a specific hue. Thus, we "see" color. When all the visible light rays are reflected, an object appears white; if none of the rays are reflected, it appears black. When one or more rays are reflected, the viewer senses the color produced by the specific reflected ray or combination of rays.

The purpose of a dye is to absorb light rays on a selective basis, causing the substrate (fabric) to reflect the rays that are not absorbed. In other words, if all the rays except those producing blue are absorbed, the viewer sees blue. The ability of an organic compound to create this desired color derives from the presence of chemical groups called *chromophores*. Substances that include chromophores in various arrangements will produce the sensation of different color hues. These chromophores include such molecular arrangements as the azo group, $-N=N-$; the thio group, $=C=S$; the nitroso group, $-N=O$; the azoxy group,

$$-N=N-$$
$$\mid$$
$$O-$$

and the *p*-quinoid group,

All chromophores contain some type of double bond or some type of unsaturation point.

While chromophores confer color upon a substance, the intensity or brilliance of the color depends on the presence of one or more substances called *auxochromes*. Furthermore, the auxochromes give water solubility to the dye and provide the groups that form associative bonds with the fiber. These include $-SO_3H$, $-N(CH_2)_2$, $-NHCH_3$, $-NH_2$, $-OH$, and $-OCH_3$. Dyestuffs themselves or combinations of other chemicals with dyes in the dye bath contribute both chromophores and auxochromes.

The technical definition for a *dye* is a compound that can be fixed on a substance in a more or less permanent state and that evokes the visual sensation of a specific color.

TYPES OF DYES

Dyes can be classified in several ways: according to hue produced, according to chemical class, and according to the method of application and types of fibers to which they are successfully applied. The last method is preferred by many textile authorities. The important groups that will be discussed in some detail include substantive or direct dyes, naphthol or azoic dyes, vat dyes and sulfur dyes, acid dyes and mordant or metallized dyes, basic or cationic dyes, disperse dyes, fiber reactive dyes, and pigments.

Substantive or Direct Dyes

Substantive or direct dyes comprise the largest and most commercially significant group of dyestuffs. Direct dyes are water soluble; they are applied primarily to cellulosic fibers and occasionally to protein fibers and polyamides. When the dyestuff is dissolved in water, a salt is added to control the absorption rate of the dye by the fiber. The fabric is then immersed in the dye bath. The amount of dyestuff absorbed by the fiber depends upon two main factors—the size of the dye molecule and the size of the pore opening in the outer surface of the fiber. Direct colors are easy to apply and are considered the most inexpensive dyes on the market. They require no special equipment.

Direct dyestuffs exhibit relatively good colorfastness to sunlight, and some are considered to have excellent lightfastness. However, the colorfastness to washing may be poor. Because they are soluble in water, the laundry process tends to dissolve some of the dye and remove it from the fabric. In cases where a fabric requires good lightfastness but is not laundered as frequently as apparel (such as draperies), direct dyes are adequate. The problem of poor washfastness can be solved to a degree by the application of selected finishing compounds, including formaldehyde and the resins used for wash-and-wear and wrinkle-resistant finishing. These treatments produce fabrics that will wash quite satisfactorily. There is no chemical *reaction* between direct dyestuffs and the textile fibers to which they are applied, but there is a chemical *attraction*. Nevertheless, at no time can it be assumed that the colors are completely fast to all outside influences.

Direct dyes can be developed after application to the fabric. The developers are usually naphtholic compounds, and the end product is similar to the azoic dyes. To develop a direct dye, the dye molecule must have a radical, usually an amino group, that will react with the naphtholic developer. The resultant dye is insoluble, or possesses a molecule of such size that it will not pass through the fiber surface, or has improved attraction to the fiber, or actually forms a chemical bond with the fiber. Developing direct dyes frequently changes the hue.

Developed direct dyes have superior washfastness to nondeveloped direct dyestuffs; however, many of them also have decreased lightfastness. As often happens in textiles, improvement of one quality appears to reduce or eliminate some other property. Both direct and developed direct dyes are used for bright shades and for coloring fabrics to be sold at medium to low prices.

Azoic or Naphthol Dyes

Azoic dyes are used on cellulosic fibers and to a limited extent on man-made synthetic fibers such as polypropylene, nylon, acrylic, and polyester. Man-made fibers require special application methods.

Azoic dyes produce color as a result of a chemical action in the fiber between a diazotized amine and a coupling agent—a basic naphthol. The diazo compound is often applied to the fabric before the naphthol, but for certain procedures the two are applied simultaneously. The starting compounds are colorless; it is the coupling of the two that produces the color.

These dyes are sometimes referred to as "ice" colors, because they are applied from a low-temperature bath. However, after the dye application, the fabric is treated in a hot detergent solution to attain the desired result.

Azoic dyestuffs produce brilliant colors at relatively low cost. They exhibit good colorfastness to laundering, bleaching, alkalies, and light. The major problem is a tendency for dyes to *crock* or rub off onto other fabric.

Acid Dyes and Mordant or Metallized Dyes

Acid dyes are primarily organic acids, usually available to the dyer in the form of salts. They are generally applied to the fiber from solutions containing sulfuric, formic, or acetic acids. Most acid dyestuffs acquire their acidity from the presence of sulfonic acid groups ($-SO_3H$) in the molecule.

These dyes are used on protein, acrylic, nylon, and certain modified polyester fibers. They function particularly well on fibers containing nitrogenous basic radicals that attract the acid dye and between which associative bonds or forces are established. In recent times, these dyes have been applied to spandex and selected polypropylene olefin fibers.

Acid dyes have no affinity for cellulosic fibers, and they cannot be used safely on fibers that are sensitive to weak acid solutions. They exhibit varying degrees of colorfastness. Some are fugitive (quickly faded) to light, while others have excellent lightfastness; some withstand laundering well, others do not. Their colorfastness to dry cleaning and perspiration varies considerably. The selection of acid dyes, therefore, should be based upon the planned use of the fabric and the anticipated method of maintenance.

Mordant or chrome dyes, often referred to as *metallized* or *pre-metallized dyes,* have many properties in common with acid dyes. They are effective on the same groups of fibers. The major difference is the fact that a metal is added to the dye molecule. This metal reacts with the dye to form relatively insoluble dyestuffs with improved wet colorfastness and lightfastness. The metal most often used is chromium, which produces the "chrome" dyes. Other metals include cobalt, aluminum, and nickel. Copper serves as the metal mordant in some acid dyeing of acrylic fibers.

Metallized dyes usually have good colorfastness to dry cleaning, which is of considerable importance in wool and silk fabrics that are more frequently dry cleaned than laundered.

Cationic or Basic Dyes

Cationic dyes, frequently referred to by their older name, *basic dyes,* are salts of colored organic bases. This class includes the oldest synthetic dyestuff, *mauviene,* as well as some of the natural dyes. The important factor in cationic dyes is that the colored portion of the dye molecule (the *cation*) is positively charged.

Cationic dyes are excellent for coloring acrylic fibers. There appears to be a chemical reaction between a cationic dye and the SO_3 radicals at each end of the polyacrylic molecule. These radicals result from the reaction between the acrylic molecules and the substance used to initiate and terminate the polymerizing action. The dyes, therefore, produce relatively colorfast products. Cationic dyes are successful on modified nylon and modified polyester fibers, and a variety of color effects can be obtained by using different fiber modifications with various types of dyestuffs. Research continues on ways to successfully apply cationic dyes to other synthetic fibers.

Basic dyes were formerly used on cellulosic and protein fibers, but because they produce fugitive colors on these fibers, their popularity has decreased. Basic dyes occasionally serve as "topping" colors to increase the brilliance or brightness of a fabric. The property of producing brilliant colors has resulted in their application to paper products, inks, and any other item where durability is not a factor but bright color is important.

Disperse Dyes

Disperse dyes, formerly called *acetate dyes*, were originally developed for acetate fibers. They are used now for coloring acetate, polyester, acrylic, and polyamide (nylon) fibers. These dyes are relatively small molecules that contain amino groups, $-NH_2-$. The dye is only slightly soluble in water but is easily dispersed throughout a solution. The colored particles attach themselves to the fiber surface and then dissolve into the fiber.[1] This reaction occurs between disperse dyes and thermoplastic fibers such as acetate, nylon, and polyester. Some disperse dyes work best on one fiber and some on another. For this reason the choice of dye is very important.

The dyestuffs frequently exhibit color fading and actual hue change when exposed to nitrogen oxides in atmospheric fumes. This behavior is termed *fume fading*. It can be a serious disadvantage when the dyes are used on fabrics for coat and suit linings, umbrellas, and similar end-uses where they are exposed to the air for long periods of time.

Colorfastness to light, laundering, and dry cleaning is good. The reaction with nitrogenous oxides does not occur as quickly or as seriously when the dyes are used on nylon or polyester fibers. If disperse dyes are applied to polyesters or nylons at low temperatures, carriers are required to bring about penetration of the dye into the fiber. The combination of high temperatures with high pressure elimi-

---

[1] Kartaschoff, cited in William Postman, "Dyeing Theory and Practice," *Dyestuffs*, No. 42.

right: **Plate 12** Fabric of fiber-dyed yarns.

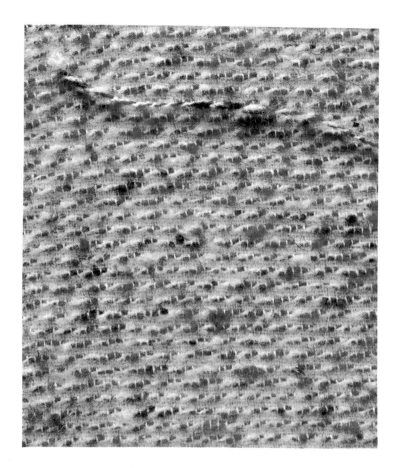

below: **Plate 13** Dope dyeing. [*Rhône-Poulenc-Textiles*]

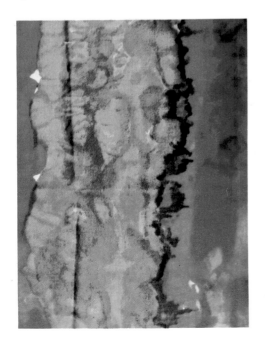

above: **Plate 14** Fabric of space-dyed yarns.
below: **Plate 15** Polychromatic dyed fabric.
right: **Plate 16** Tie-dyed fabric. [*Bernice Coleman*]

above: **Plate 17** Modern batik fabric. [*Bernice Coleman*]

left: **Plate 18** Silk-screen print. [*Bernice Coleman*]

below: **Plate 19** Screen printing with flat-bed machine. [*Allied Chemical*]

left: **Plate 20**   Hand screen printing. [*Allied Chemical*]

below left: **Plate 21**   Rotary screen printing. [*Stork-Boxmeer*]

below: **Plate 22**   Hand block printing. [*Jack Lenor Larsen, Inc.*]

nates the need for carriers; the *thermosol* process reduces the need. This process involves padding the dye into the fabric and passing the fabric into a heat zone, where the dye is fixed in the fibers. Thermosoling can be achieved by heated dry air, contact with hot surfaces, or exposure to infrared heating zones. The time of exposure to heat varies from ten seconds for the infrared technique to thirty to sixty seconds for heated dry air. The most common at present is the heated dry air method, and many dyes are fixed at approximately 205°C (400°F) in thirty to sixty seconds.

Pigment Colors

Technically, pigments are not dyes, but they are used to color some fabrics, so it is essential to include them in this discussion. Pigments are not considered dyes because they are completely insoluble in water or in the solvents used in dyeing and must be applied by some other mechanism.

Pigment colors are mainly organic coloring compounds, and, in addition to being insoluble in water, they have no affinity for fibers. To fix them on fibers some type of adhesive, resin, or bonding agent must be employed. The colors thus produced are relatively permanent, except that as the resin or bonding agent wears away, the color will also disappear. Even mild abrasion may diminish the color.

It is also possible to mix pigments thoroughly with the fiber solution, and thus the fiber is already colored as it is extruded. When pigment colors are added to the fiber solution or molten polymer, the terms *dope* or *solution dye* are used. Fibers colored in this manner exhibit good colorfastness to laundering, dry cleaning, light, perspiration, and crocking.

Pigment colors are used in solution dyeing of acetate, rayon, nylon, polyester, and other man-made fibers that accept color during the solution state. They can be used on cellulose fibers, cellulose/polyester blends, and other blends when they are attached by a bonding agent. Glass fabrics are frequently dyed with pigment colors.

Vat Dyes

The dyestuffs frequently publicized as having the best colorfastness are the vat dyes. In general, vat dyes do exhibit excellent colorfastness, although their lightfastness is inferior to pigments used in solution-dyed fibers. Furthermore, the dyes in this category exhibit some variation in colorfastness because of differences in chemical structure. If the dyeing procedure is not carefully controlled, the color may fade quickly.

Vat dyes were originally developed in Europe about 1910 and derived their name from the equipment used in applying the dye, which included a large vessel or vat. Today vat dyes can be applied in vats or in continuous-feed methods.

These dyes are insoluble in water and have no affinity for fibers until they are converted to a product that is soluble in an alkaline solution. The solution of dye and alkaline substance is called a *leuco* bath. According to one authority, all vat dyes are characterized by the presence of a $=C=O$ group, which may be reduced to

$$a \quad \overset{\diagdown}{\underset{\diagup}{C}}-O-H \text{ group}$$

that is soluble in sodium hydroxide. In the NaOH it changes to form

$$a \quad \overset{\diagdown}{\underset{\diagup}{C}}-O-Na \text{ group}$$

that is substantive to fibers.[2] This indicates that the dye will have a great attraction to the fiber. After application in the leuco state, the dye is reoxidized to its original form, at which time the final color develops. The reoxidized color is insoluble and is, thus, highly resistant to water.

There are soluble vat dyes on the market, but these are sodium salts of sulfuric-acid esters and, in reality, are already reduced. They still require oxidation for color development.

Vat dyes are adaptable to all cellulosic fibers and some of the newer man-made fibers, but they are rarely used on protein substance because of the alkaline bath, which would damage the fibers. Recent technological developments in processing have enabled the dyer to apply vat dyes to a wider selection of material. This system, called the *Pad-Steam* process, shortens the time of contact between the fiber and the alkaline bath and decreases the possible damage to alkali-sensitive fibers. There is a wide choice of colors available in vat dyes, and they withstand hard wear. Because of the excellent colorfastness of many of these dyes, the term *vat dye* has become almost synonymous with *fast color*.

Sulfur Dyes

Sulfur dyes are used primarily on cellulosic fibers. They produce fair to good colorfast dark shades—browns, blacks, and navies. Unless they are properly applied, they may eventually break down, causing tendering or weakening of the fabric and eventual disintegration. Sulfur dyes are also dissolved in an alkaline solution. The fabric is passed through the alkaline dye bath and absorbs the color. When the fabric is washed, the color remains on the fiber in an insoluble form. Since the dyes are not soluble in water, they give good washfastness. The cost of sulfur dyes is relatively low.

[2]Postman, "Dyeing Theory and Practice."

Reactive Dyes

The first practical reactive dyes were introduced in 1956. These dyes actually react chemically with the fiber molecule. They are used primarily on cotton, but some types are suitable for other cellulosics, nylon, wool, silk, and acrylics, as well as blends.

The reactive dye unites with the fiber molecule by addition or substitution. In the addition method the dye adds to the hydroxyl group to form a cellulose ether compound; in the substitution method the hydrogen of the hydroxyl group and chlorine in the dye split off to form hydrochloric acid, and the remaining dye molecule attaches to the fiber to form a cellulose ether. The acid is neutralized in the alkaline bath used for dyeing.[3]

Bright colors with excellent washfastness are available by means of reactive dyes, and the cost is decreasing, so they are gaining popularity. Colorfastness to light is good to excellent depending upon the dye base. One of the major problems of reactive dyes is their susceptibility to damage from chlorine. Colorfastness to crocking, perspiration, dry cleaning, and fume fading is good to excellent. Among the more common trade names are Procion, Cibacron, Remazol, Reactone, and Cavalite.

Development of each type of man-made synthesized fiber has necessitated, in most instances, the creation of special dyestuffs for the fiber. The majority of dye manufacturers develop these dyes simultaneously with the new fiber, and most give special trade names to their products. A few of these dyes are of a reactive nature in that they form chemical bonds with the fibers. Most of the dyes are held on the fiber by attraction.

Colorfastness of all dyestuffs will vary to some degree. It is, therefore, desirable when purchasing textile items to find label information that will provide data relative to colorfastness and recommendations for proper use and care.

APPLICATION OF COLOR

Color can be applied to textile products at various stages in their manufacture—to fibers, yarns, or fabrics.

Fiber Dye

Dyeing is often done to the raw fiber stock (Pl. 12, following p. 324). This provides the possibility of deep penetration of the dye into the fiber, which gives uniform color and a tendency to greater colorfastness. This method is referred to as *fiber dye* or *stock dye*. In the case of wool it may be called *top dyeing* when done to the fiber sliver.

[3] V. S. Ryan, "Recent Advances in Fiber Reactive Dyes," *American Dyestuff Reporter* (April 12, 1965), p. 73.

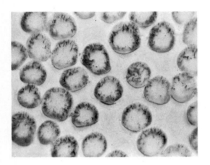

Figure 29.1 Photomicrograph of solution-dyed fabric, cross section.

Figure 29.2 Yarn-dyed fabric.

Figure 29.3 Yarn dyeing. [*Service Yarn Dyeing Corp.*]

Solution or Dope Dye

In the manufacture of man-made fibers, color can be added to the chemical solution before it is forced through the spinnerettes (Pl. 13, following p. 324). If the pigment is good, this method ensures not only even dyeing but colors that are an integral part of the fiber (Fig. 29.1) and, therefore, fast to most outside influences. Solution or dope dyeing has been especially important in coloring acetate and olefin fibers.

Yarn Dye

One of the oldest systems of dyeing textiles is to color the yarns (Fig. 29.2). This is still widely practiced. Yarns can be dyed either in a skein form (Fig. 29.3) or rolled on tubes. The latter is frequently called *package dyeing*. Recently, a system of dyeing yarns after they are rolled onto the warp beam has been introduced. This is called *beam dyeing* and has proved economical and successful. Yarn dyeing provides good color absorption and adequate penetration. It permits the use of various colored yarns in one fabric and gives the fabric designer wide latitude in designing plaids, checks, stripes, muted color arrangements, and iridescent effects. In a variation of this process, called *space dyeing*, selected areas of the yarns are dyed different colors to produce pattern in the fabric (Pl. 14, following p. 324).

Piece Dye

Most solid-color fabrics are dyed after the fabric has been completed. This is the easiest and cheapest method for adding color. Manufac-

turers can color fabrics as ordered and need not maintain a large stock of dyed fabrics that might become obsolete if the fashion changes. Piece dye does not always provide a thorough penetration of dyestuff, but for many uses it is quite satisfactory. Normally, piece-dyed fabrics are a single color, but when a fabric contains more than one fiber, a pattern can result from the different absorption rates of the fibers. The most important variations are union-dye and cross-dye techniques.

Union Dyeing   The term *union dye* is used to indicate that a fabric containing two or more fibers has been dyed a single uniform color. Various dyestuffs are applicable only on certain fiber types, so when a fabric containing two or more types of fibers is to be colored a solid color, the dyes must be carefully selected and properly applied in order to ensure color uniformity. If possible, dyes should be chosen that can be applied simultaneously to the fabric. However, in some cases dyers have found it preferable to introduce the dyes individually, each by its recommended procedure. These two methods are often called *one-bath* and *two-bath* processes.

Cross Dyeing   Fabrics of two or more fibers can be dyed so that each fiber accepts a different dyestuff amd becomes a different color. In some cases the dye bath is planned so that certain fibers will accept no color at all and remain white. The end product depends upon the fiber arrangement in the fabric. It may be a check, a plaid, a tweed, a stripe, a muted color, or some other design. Recent developments in fiber chemistry have produced fibers of the same general chemical type that are dye selective, so that fabrics of one fiber come from the dyebath checked or striped. This is especially successful with acrylic and polyester fibers.

Polychromatic Dyeing   Polychromatic dyeing is a new technique that provides for pattern effects in an almost infinite variety (Pl. 15, following p. 324). It is truly a dye method, however, not a printing process in the accepted sense of the word. There are two basic techniques: the *flow-form* method and the *dye-weave* method. These differ mainly in the point at which the dye is actually applied.

The basic principle consists of running streams of different colored dye solutions onto a moving substrate (fabric) and then crossing these with other colors. Saturated by the first set of solutions, the fabric resists the second set long enough for the pad-mangle to fix the pattern in place and remove excess solution. Timing, amount of color extruded, arrangement of color applicators, and movement of the substrate all determine the resulting pattern.

Many designs on the market are of the dye weave variety. These patterns are developed by running streams of dye solution down an inclined plane onto the moving fabric. The dyes are fixed into the fabric by the method most appropriate to the type of fiber and the type of dyestuff involved.

It is possible to make a complete product—for example, a dress—and then dye it. Some couture designers employ this technique, but it is not considered practical for general use by apparel manufacturers. Other methods of dyeing, such as tie-dye and batik, are discussed in relation to applied design in the following chapter.

30

Applied Design

This chapter considers the application of color to create surface designs or prints. *Printed fabrics* are defined as those that have been decorated by a motif, pattern, or design applied to the fabric after it has already been constructed.

PRINTING

The art of printing color onto fabrics originated thousands of years before Christ. Primitive peoples decorated garments with paints as they did their bodies. However, since these colors were not "fixed," they were of temporary value. Pictures printed on the walls of the tomb of Beni Hassan (c. 2100 B.C.) in Egypt depict figures costumed in fabrics with small conventional printed motifs. Actual remnants of printed fabric found near Thebes have been dated about 1594 B.C. The stiffness of the fabric, plus the presence of loops at the top, give evidence that it was meant for wall decoration, not for apparel. Printed fabrics are known to have existed in India and China as early as 500 B.C.

The oldest textile prints found in Europe date from about A.D. 600. In the Americas, examples of printed fabrics have been found in remains of the Inca civilization. However, it was during the latter part of the sixteenth century that textile printing attained general acceptance. Printing became a fine art in France during the eighteenth century, when Christophe Phillippe Oberkampf opened his textile printing factory at Jouy and began production of the famous "Toiles de Jouy," considered by many to be the finest patterned fabrics in the world. Other quality printed fabrics were manufactured at Orange.

The peoples of India, Java, and certain sections of South America and Mexico developed native techniques for decorating fabrics.

Resist Printing

Some historians consider resist printing one of the oldest methods of applying surface design. Early Javanese batiks and Japanese stencil prints are examples of resist printing, as is Plangi tie-dye, which was developed in Asia, probably China or India.

The basic principle of resist printing is the protection of certain areas of the fabric by some device to prevent color (dye) penetration.

Plangi Tie-dye   Plangi resist methods were used by primitive peoples in the Far East, and early designs were extremely delicate. In this technique, tiny puffs of fabric were pulled over a pointed object, and waxed thread was tied tightly below the small puff. Wherever the fabric was to resist the color, it was tied securely with the waxed thread. After tying, the fabric was dipped into a dye bath. If two, three, or more colors were desired, the thread was removed and the fabric retied.

In addition to tying with thread, plangi designs can be made by folding the fabric so certain areas are protected from the dye or by knotting or plaiting the actual fabric to prevent dye penetration. A third method involves hand or machine stitching. Called *tritik*, the stitching is put in according to special designs, and the thread is pulled so as to draw the fabric together. The fabric resists dye penetration in the specified areas.

Modern tie-dye prints are usually characterized by large blotchy areas of color (Pl. 16, following p. 324). The principle is the same for coarse designs, but the thread used for tying is not necessarily waxed. In the United States tie-dye resist is confined primarily to handcraft prints. They are quite popular in apparel and in home decoration fabrics. Many artists use tie-dye as a medium of expression.

Batik   A resist method perfected by the Javanese, batik has recently been revived (Fig. 30.1; Pl. 17, following p. 324). In batik prints wax serves as the resist substance. Using a *Tjanting*—copper cup attached to a reed handle—the batik printer applies wax to the fabric where it is to resist the dye (Fig. 30.2). The intricacy of the design determines the degree of skill required by the worker. Apprentices do simple designs or large background areas, while the experienced craftsman applies wax to delicate, fine-line designs.

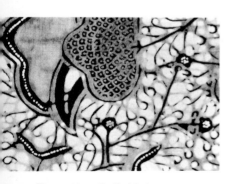

Figure 30.1   Batik fabric.

An important part of producing batik prints is the preparation of the fabric. It is thoroughly washed to remove starch or sizing, and oiled to facilitate dye penetration. The fabric is then washed again to remove any impurities acquired during oiling and stiffened with a special tapioca starch. The starch produces a smooth surface on which the design can be drawn easily and prevents spread or excessive penetration of the wax.

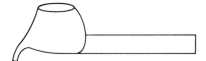

Figure 30.2   The Tjanting used to apply wax resist.

The batik design is produced by the following steps:

1. Melted wax is applied with the Tjanting to all areas that are to remain unaffected by the color.
2. The cloth is dipped in dye. All waxed areas repel the dyestuff, so only the untreated areas absorb the color.
3. The wax is removed by boiling water. As the water cools, the wax solidifies, floats on the surface, and is recovered for reuse.
4. Wax is again applied in preparation for the second color. It covers previously dyed areas and any other areas not to be colored by the second dye.
5. The cloth is dipped in the second dye color.

The process is repeated for as many additional colors as are required to complete the pattern.

Sometimes the wax cracks or is deliberately cracked so that fine lines of color appear in the background areas.

In order to speed up the batik process, early batikmakers developed the *Tjap*. This is a block with the design executed in copper wires and bands. The block is dipped in melted wax and transferred to the cloth. The wax coats the fabric with the pattern embedded in the block.

Using the Tjanting, Javanese women require from thirty to fifty days to complete a 2-yard length of fabric. The Tjap shortens the time considerably; a 2-yard length can be made in less than fifteen days. The Tjap did not replace the Tjanting, because fine designs are of better quality when done with the latter.

Ikat (Kasuri)   *Ikat,* also called *Kasuri,* is a resist method in which only the warp yarns are printed. The areas not to receive color are tied before dyeing, the yarns are dipped in the dye, and the final warp has yarns with some areas colored and others white. Patterns can be created by artistic handling of the yarns. This technique was the forerunner of modern warp prints and vigoureux printing. However, the two latter are now done by direct methods. Resist-warp printing is not practiced in the United States. Ikat is still found in some Asian cultures and is an important craft in Japan and Guadalcanal.

Stencil Printing

Stencil printing was first developed by the Japanese and was the precursor of modern screen printing. In modern times, it is considered a handcraft, and it is a method from which beautiful designs can be executed. In stencil printing design areas are cut from sheets of paper coated with oil, wax, or varnish, or from thin sheets of metal. Usually a separate stencil is cut for each color. The stencils must be planned so they register or fit together properly to result in a perfect print. A difficulty with stencil print is that the design areas must be connected to prevent parts of the stencil from falling out. To offset this problem, Japanese stencil artists developed a method of tying the various sections together with silk filament or human hair.

In producing stencil prints today, the color can be applied by hand brush, air brush, or spray gun. The method is used for limited yardage, one-of-a-kind scarves, and similar products.

Silk-screen Printing

Screen printing is considered by many textile authorities to be the newest technique for decorating fabrics (Pl. 18, following p. 324). It developed from stencil printing and is essentially a stencil process. A screen resist is made by covering a frame with bolting cloth of silk, metal, nylon, or polyester filament yarns. The fabric is covered with a film, and the design areas are cut out of the film, leaving the fine mesh fabric open for the dyestuff to pass through and print the fabric.

In the early years of screen printing, the design was cut from the film before it was sealed to the screen fabric. For contemporary handcraft screen prints, the fabric may be painted with a film-producing substance to cover the areas where color is not to penetrate. In modern screen-printing mills, screens are prepared by photochemical processes, chemical-resist reactions, or engraving techniques. After the screen is completed, it is fastened to a frame. In printing, the frame is laid on the fabric to be printed, the dye is placed at one edge of the frame, and a rubber knife or *squeegee* moves the dye across the screen and forces the color through the open areas and onto the fabric.

One screen is prepared for each color, and the size is large enough to include at least one repeat pattern (Pl. 19, following p. 324). All screens used for a specific design are arranged to register or fit together accurately for reproduction. This can be done mechanically by attaching fixed stops to the printing table, so each screen can be located in relation to each stop. The machines are equipped with electronic controls to accomplish this automatically.

Screen printing is desirable for the reproduction of large patterns and for fabrics that require considerable dye, such as terry cloth. The process is versatile: It can be used economically for sample runs as well as for limited yardage. Unfortunately, screen printing is slower than roller printing, and some types of prints are difficult to reproduce. New developments in machines and controls may overcome these problems.

Before the introduction of automatic screen printing, the amount of yardage that could be printed was limited by the length of the printing table, the speed of the operators, and the number of colors used. The fabric was spread flat on a padded table and fastened securely in position. The operator laid the screen in place, a color at a time, and forced the dye through the screen with a squeegee (Pl. 20, following p. 324). This was repeated as many times as necessary to completely print the fabric. It was then moved to overhead racks to complete drying.

Since 1950 most screen-printing organizations have installed automatic equipment. During the early 1960s the number of automatic flat-bed screen-printing machines in the United States more than doubled. Since that time many screen printers have also installed rotary equipment. Some rotary machines are particularly flexible: they can be used with either roll-type screens or flat-bed screens. A few models combine both in one operation. Automatic screen printing is five to seven times faster than hand screen methods and enables two men to do the work formerly done by fifteen.

Automatic screen printing is a continuous process. The cloth moves slowly along the flat surface of the printing table, and the screens and color are applied to the fabric by electronic devices. After printing, the fabric is fed through drying equipment. There are several types of screen printing machines that can print between 60 and 75 yards per minute, with an unlimited number of colors and screens.

Screen-print equipment that employs rolls instead of flat screens has been introduced (Pl. 21, following p. 324). While the roll reduces the flexibility of the repeat size for designs, it increases printing speed. Furthermore, it is easier to add screens for additional colors than in flat-bed machines. In rotary screen printing the screen is rather rigid to retain its circularity. Color is fed into the center of the roll, where a squeegee spreads the dye as the roll turns.

Despite the increased use of automatic machines, screen printing is slower than roller printing (not to be confused with rotary screen printing, cited above). However, it offers several advantages. Designs can be larger; even a small amount of yardage with one design can be economically profitable for the manufacturer; fabrics that might be distorted in roller printing can be screen printed successfully; dyes or pigments can be laid on in thick layers to produce a handcraft effect; machines are available to print fabrics as wide as 105 inches. The future of automatic screen printing appears limitless.

Discharge Prints

Discharge printing is used when print designs are to be applied to fabric that has previously been dyed a solid color. A design roll is coated with a reducing bleach that removes the base dye and leaves a white pattern on a colored ground. The printing machine may replace the color removed with a different color. This is done by a second design roll that applies dye to the discharged areas. More than one color can be added by means of additional rolls. It is also possible to discharge and print in one operation. The chemical to discharge removes the color while a dye in the solution is printed on in place of the original color.

A difficulty that may possibly be encountered with this process of discharge printing is that the chemical used for discharging the dye background may weaken the fabric. However, if the proper procedures are followed, no damage should occur. Dark fabrics with white designs, such as small floral prints and polka dots, are frequently the products of discharge printing.

Direct Printing

Direct printing is the most common method of applying designs to fabric. Block printing, a type of direct color application, seems to have been the oldest printing technique.

Block Printing   Actual samples of fabric stamped with block prints and dated as early as 1600 B.C. have been discovered by archeologists. Wall paintings indicate a possible application of pattern by block stamping as early as 2100 B.C. These fabrics appear to have been printed with small blocks no larger than 1 to 2 inches in diameter. By the fifteenth century blocks had grown to between 12 and 18 inches in diameter and $2\frac{1}{2}$ to $3\frac{1}{2}$ inches thick. Block printing is still practiced today as a handcraft (Pl. 22, following p. 324). The blocks can be of carved wood or of metal shapes attached securely to a wood base. Each block prints only one color, so if a design of several colors is desired, a block must be made for each. In preparing a block, the design area remains raised while the background is carved away.

In producing a block print, the procedure is more or less a standard one. The fabric is first laid flat on a table that has been covered with protective padding. The block is either dipped in the printing paste so only the raised portion picks up the dye, or a roller is coated with paste and then rolled over the block, depositing a layer of color on the raised areas. The block is then pressed on the fabric, forcing color into the surface (Fig. 30.3).

For hand printing a mallet is used to pound the block and to produce clear color on the fabric. Although there is no mechanical block printing, there are a few firms which have semimechanized presses for block printing. Since the latter part of the 1940s, most of the designs suitable for block printing have been reproduced on screen-printing equipment.

Figure 30.3   A wood block (*above*) and the fabric printed from it (*below*).

Roller Printing   Most printed fabrics are produced by direct roller printing (Fig. 30.4). The equipment can be completely automatic with few men needed to operate the machines. The process is fast and economical.

Roller printing had its origin in the use of flat plates or *blocks* on a flat printing press. These were natural developments from block printing. The flat-bed press was used by Oberkampf at Jouy.

Thomas Bell, a Scotsman, invented the roller-print machine in 1783. Other men have made improvements and developed more efficient and more complicated equipment, but the basic idea incorporated in Bell's roller-print apparatus is still used in modern machines. Bell's process combined developments in metal engraving and color printing.

In roller printing the design, which can never have a repeat greater than the circumference of the printing roll, is reproduced to conform to the rolls, and the colors are clearly indicated (Fig. 30.5). A roll is made for each color.

The design is transferred to the metal rolls by means of a pantograph or other guiding mechanism. This device controls an instrument that outlines the design on the roll. The area of the roll that is to print the specific color involved remains clear, while other areas of the roll are coated with a chemical-resistant paint. The roll is dipped in acid, which etches or burns away the surface layer of metal wherever there is no chemical-resist coating. After etching, the resist coating is removed, and the roll is polished. A recent development in producing the metal design roll involves photoengraving techniques. This permits shading and more subtle effects.

Figure 30.4   Direct printed fabric.

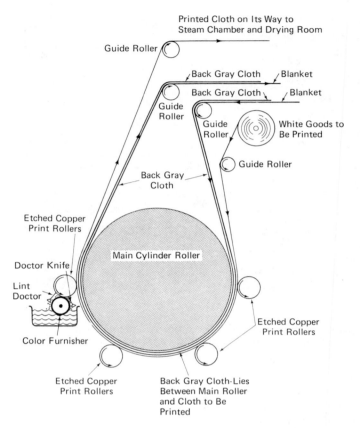

Printed Cloth on Its Way to
Steam Chamber and Drying Room

Guide Roller

Back Gray Cloth / Blanket

Back Gray Cloth / Blanket

Guide
Roller

Guide
Roller

White Goods to
Be Printed

Guide Roller

Back Gray
Cloth

Etched Copper
Print Rollers

Main Cylinder Roller

Doctor Knife

Lint
Doctor

Etched Copper
Print Rollers

Color Furnisher

Etched Copper
Print Rollers

Back Gray Cloth-Lies
Between Main Roller
and Cloth to Be
Printed

Figure 30.5 Diagram of a roller print-ing machine. There must be a print roller for each color used in the pattern.

The engraved rolls are arranged around the main cylinder and locked into place on the printing machine. As many as fourteen or sixteen rolls may be used. Each has a trough for its particular dye and a "doctor" blade, which scrapes the dye away from the smooth raised surface, leaving it in the engraved areas. Modern equipment has electronic devices that register each roller into accurate position so the final design is clear, even, and well-matched.

The large cylinder is covered with padding blanket. A printing blanket is arranged to contact the padded roll, and a "back gray" cloth is sometimes used on top of the printing blanket. The cloth to be printed is on the outer surface. The layers move together from the supply rolls around the machine roll and the printing occurs. After leaving the last roll, the printed cloth goes into a drying oven, which sets the color, and the base fabrics return to the supply rolls.

Roller machines can print up to sixteen colors at one time at a speed of approximately 40 to 50 yards per minute. The end result depends on the accuracy of roll preparation and efficient machine operation and control. Colored backgrounds can be printed onto fabrics from etched rollers at the same time as designs are applied.

The quality of print fabrics has steadily increased during the past decade. Automatic controls assure accurate printing, while a willingness to experiment with unusual patterns and designs has resulted in beautiful and original print fabrics.

Duplex Prints Duplex prints are produced by modified direct roller-print equipment. The machinery is set up so the design is printed on both sides of the fabric. The completed fabric is identical on face and back. Duplex prints resemble fiber-dyed fabrics, yarn-dyed fabrics, or decorative-weave fabrics.

Photographic Prints Photographic prints are produced in a manner very similar to that used in making photographs. The fabric is treated with a light-reactive dye. A negative is placed on the fabric, light is transmitted to it, and the color is developed. After stabilization, the fabric is thoroughly washed, and the print is as permanent as any good photo.

Sublistatic Printing

A new process for printing fabrics, especially knits, sublistatic printing is a modern adaptation of the old decalomania technique. It relies on the use of preprinted release papers that can transfer their prints to fabrics. The release papers are printed with disperse dyestuffs, which, under heat, sublimate and transfer their colors to the fabric.

The preprinted paper is currently manufactured only in Europe. Rolls of fabric can be printed at a rate of up to 500 yards per hour; cut garment sections are processed at about 150 units per hour.

The heat printing, a dry process that simultaneously thermosets the fabric, takes fifteen to twenty seconds to complete. The variety of design is limited only by the artistic talents of the designer, the number of colors available, and the size of the design roll or garment unit.

EMBROIDERY

The application of yarn, thread, or floss is a very old method of decorating fabric (Fig. 30.6). Exquisite hand embroideries were made in Europe during the fourteenth, fifteenth, and sixteenth centuries, and some of these are now treasured in museums. Embroidery is still practiced as a craft, and many years of training may be required to attain the skill necessary for intricate designs.

Today, most of the embroidered fabrics on the consumer market are produced by machines, much like those used in making lace, which duplicate the fine stitches in one or several colors. In fact, embroidery techniques can also be adapted to lacemaking. For example, Schiffli

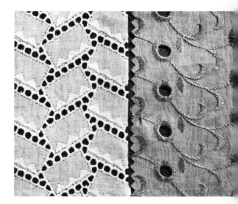

Figure 30.6 Fabric decorated with embroidered designs.

embroidery is sometimes applied to a very thin fabric composed of fibers that are soluble in certain solutions. The base fabric is dissolved leaving the embroidery, which forms a lace pattern. Depending upon the fiber to be removed, the chemical used in dissolving the base fabric might be dilute sodium hydroxide, dilute acids, or, in some cases, water. The embroidery must be done with fibers that are not affected by the solvent. An example would be cotton embroidery on a sheer silk base. The silk could be removed by dilute caustic soda, which would not harm the cotton yarns.

In addition to regular embroidery, fabrics can be decorated by quilting and by appliqué. In quilting, the design is produced by hand or machine stitching through two layers of fabrio with padding between. Simple geometric patterns or elaborate designs are possible.

Appliqué involves the "application" of designs cut from other fabrics. The extra fabric is fastened in place by means of visible (decorative) or invisible stitching. Except for some laces, appliqué is rarely used for fabrics sold on the commercial market. It is primarily a home method of decoration.

DESIGNS BY FINISHING PROCESSES

Attractive patterns can be added to fabrics by various finishing processes (see Chap. 26). These include plissé, moiré, embossing, frosting, flocking, glazing, and "burned-out" patterns.

Evaluation
of Color

Standard test methods for evaluating dyestuffs on fabric are available from the organizations listed in Chapter 27. The following discussion will include home tests for colorfastness.

Dyes still cause problems for consumers, despite the tremendous improvement of dyestuffs and printing processes in the past twenty-five to thirty years. Dissatisfaction may be the result of improper fabric care (when the consumer ignores label information) or of inadequate directions provided by the manufacturer. Problems with color can also be caused by improper dye selection by the converter for the fiber or fabric involved or by the choice of inferior quality dyes in an attempt to reduce manufacturing costs.

The consumer cannot always assume that fabric manufacturers have selected optimum methods of dye application or the best dyestuff. The manufacturer must adjust his costs to the prevailing selling price of the fabric, and to do this he may be forced to adopt dyes and dyeing methods that do not produce the most desirable product.

Label data related to care, fiber content, and color or washfastness are important and should be requested by the consumer. In places where label information seems inadequate, the consumer, in buying textile products (including yardage), should consider it possible to undertake home testing to determine color behavior in various environmental conditions.

Several nontechnical tests are included here as guides. If used for home analysis, they can save numerous consumers disappointment or dissatisfaction.

HOME TESTS FOR COLORFASTNESS OF FABRICS

Dry Cleaning

Test #1   Sponge a sample of fabric (or the seam allowance at a point where it is hidden) with a dry cleaner or spot remover, using a clean white cloth. If the color is not fast to cleaning, it will run onto the cloth. Dry cleaning solvents and spot removers should be used in well-ventilated areas, and some must never be used near flame. It is important to avoid inhaling the vapor of many of these substances.

Test #2   When a sample of sufficient size, approximately 2 inches by 4 inches, is available, it can be immersed in cleaning solution for a period of ten to twenty minutes. Observe to determine if any color has "bled" into the cleaning solution, then dry and compare the sample with an original piece to determine color change.

Laundering

Test #3   A small fabric sample is required. This can be clipped from the seam allowance or hem if a fabric scrap is not available. When yardage is used, a 2-by-2-inch swatch is adequate.

1. Use a pint jar and water at the temperature suggested for the fiber content.
2. Place one cup of water and one teaspoon of soap or synthetic detergent into the jar.
3. Add the sample.
4. Shake the jar frequently and allow the fabric to remain in solution for ten minutes.
5. Observe the color of the wash water.
6. Rinse the sample in warm water at least twice, and observe any loss of color into the rinse water.
7. Dry the sample and compare it with the original fabric to determine if any color change occurred.

Laundering with Other Garments

Test #4   When dye is evident in the wash water, the consumer may wish to see if it will discolor other fabrics laundered at the same time.

1. Repeat test #3 with a new sample of fabric and include small samples of white fabrics of cotton, nylon, polyester, rayon, and acetate.
2. Observe the white samples to determine if they have picked up any color from the sample under investigation.

Colorfastness to laundering, frequently called *washfastness,* is desirable in any textile item that requires frequent cleaning by water and detergents. When color is picked up by other fibers or fabrics, the consumer must launder the problem item alone or with fabrics that are not affected.

Fabrics may lose color without giving visible evidence of color change. This occurs when excess dye is in the product. In such instances, the product can be laundered with other items after the excess color has been completely removed and when no additional color loss occurs.

Label information can be misleading. Although a label states that a fabric is colorfast, it may not be colorfast to laundering. The term *washfast* is preferred and, when used properly, indicates products that can be laundered without loss of color.

Sunlight

Test #5   Textile fabrics that will be exposed to sunlight for many hours each day, such as curtains and draperies, should be colorfast to sunlight or *sunfast.* It is difficult for a consumer to execute accurate tests at home. However, it is possible to assess some degree of sunfastness.

1. Expose the fabric (directly or through glass) to sunlight between 10:00 A.M. and 4:00 P.M. Standard Time, during the period between May and September.
2. Keep a record of the number of hours of exposure.
3. Compare the sample with an original at frequent intervals.

Fabric cannot be considered satisfactory for use at windows or in patios unless it will resist fading for a minimum of 120 to 140 hours. A consumer wishing to pretest a fabric sample before final purchase has the problem of time. Frequently, by the time sunfastness is checked, the fabric is no longer available. Thus, the consumer, in reality, must depend upon label information and past experience. When the word *sunfast* appears on a label, or if the manufacturer guarantees the color, the consumer should expect satisfaction.

Other conditions that can affect color include atmospheric fumes, perspiration, dry heat, and wet heat. Mechanical rubbing of fabric by other fabrics or foreign substances can also cause color loss. Some of these factors can be checked easily and quickly by the consumer.

However, it is quite difficult to test for colorfastness to fumes or perspiration without laboratory equipment, chemicals, or actually subjecting the fabric to the same conditions for the same length of time as the end-use product.

Knowledge of fiber properties can help in preventing disappointment, especially where atmospheric fumes or perspiration are involved. For example, silk fibers and colored silk fabrics are often damaged by perspiration; acetate fabrics may fade from exposure to atmospheric fumes.

Ironing

Test #6   Color may be altered by ironing or pressing with either dry or wet heat. When dry heat is used, the fabric will usually return to its normal color after cooling. The consumer can check this satisfactorily with an iron on either a sample of fabric or an inside seam allowance.

1. Press the fabric with an iron set at the temperature recommended for the fiber. Observe any color change.
2. If color does change, observe the fabric as it cools to determine if it returns to its original shade.
3. If not, try a lower temperature or hand smoothing. When all else fails, contact the retailer to determine if the item should be returned.

Pressing with steam or with a damp press cloth affects some dyes. Fabrics can be checked as for dry heat, using steam or a wet press cloth. If colors change with this test and not with the dry heat or vice versa, the consumer knows which ironing method is required. Dyes affected by wet heat may stain the ironing board cover, the press cloth, or, under certain circumstances, other fabrics they encounter under warmth and pressure.

Rubbing and Crocking

Test #7

1. Place a small square of white cotton fabric—preferably muslin or percale—over the forefinger.
2. With even pressure rub the white fabric at least ten times over a colored item.
3. Observe to see if the color rubs off onto the white square of fabric.
4. Repeat with a white square of fabric that has been moistened thoroughly.

Either by test, using home methods or laboratory equipment (Figs. 31.1–31.5), or by label information a consumer has the ability to

top left: **Figure 31.1**   Launder-Ometer—determines colorfastness to laundering or dry cleaning. [*Atlas Electric Devices Company*]

top right: **Figure 31.2**   Fade-Ometer—tests colorfastness to light. [*Atlas Electric Devices Company*]

center left: **Figure 31.3**   Crockmeter—determines colorfastness to rubbing or crocking. [*Atlas Electric Devices Company*]

above left: **Figure 31.4**   Perspiration Tester—measures colorfastness to perspiration. [*Atlas Electric Devices Company*]

above right: **Figure 31.5**   Gardner Color Difference Meter—tests color change and whiteness retention. [*Gardner Laboratory, Inc.*]

determine, to some degree at least, the care a product will require. As mentioned above, not all manufacturers use the best type of dye for a fabric. The influence of fashion upon color choice sometimes results in the dyer applying preferred colors without considering which is the most desirable dye for the particular fiber. Of course, most reputable dyers attempt to use the dye best suited to the intended application. Unfortunately, some consumers and some product manufacturers do not restrict themselves to the recommended end-use. When this occurs, the dye may not live up to its promise.

The consumer is often faced with buying merchandise that has no label information regarding the fastness of dyes to various agents. If these consumers would request such information and then follow directions, retailers and manufacturers would tend to provide the desired data. However, a major problem is apathy on the part of the consumer in regard to labeling. Furthermore, some consumers do not follow care directions when they are available. They mistreat the product, and then return it for adjustment when it fails to meet expectations. Such inconsiderate people sometimes cause retailers or manufacturers to refuse to guarantee the fastness of selected dyes that actually would resist damage from sun, laundering, or other effective conditions.

The consumer who desires a special color or print for limited usage does not need to give substantial attention to the durability of the dyestuffs used. However, the long life and easy maintenance of a textile product require careful attention to color and label information.

Fabric End-Use

Fibers, yarns, fabrics, finishes, dyes—all are directed at one goal: the end-use of a product. With rare exceptions, textiles are not meant to be hung on a wall and admired. They are to be worn, sat upon, and walked on. They protect man as well as enhance his person and his environment.

The concluding chapters of this text discuss the geometry of fibers, yarns, and fabrics; the factors to be borne in mind in selecting fabrics for a particular end-use; and the standards that a fabric must or should live up to and how they are enforced.

Fiber, Yarn, and Fabric Geometry

Fiber properties are governed by the inherent characteristics of the base polymer and such geometric features as fiber length, cross-section shape and area, crimp, stiffness, and surface contour.

Yarns obtain their properties from fiber characteristics and from such geometric yarn features as amount of twist, shape and diameter, and compactness of fibers within the yarn, as well as from the basic yarn structure.

Fabric properties are the result of fiber and yarn characteristics, and the special geometry of the fabric itself, which includes fabric structure, number of yarns per inch, and tension on yarn components.[1]

Fabric geometry can be defined as the relationship of fibers and yarns to their ultimate shape and arrangement in finished fabric. According to several authorities, geometric factors influence the following properties important to consumers:[2]

---

[1] E. R. Kaswell, *Textile Fibers, Yarns and Fabrics* (New York: Reinhold Publishing Corporation, 1953), p. 175.

[2] Kaswell, *Textile Fibers, Yarns and Fabrics,* p. 177; E. R. Kaswell, *Handbook of Industrial Textiles* (New York: Wellington Sears, 1964), p. 449; F. T. Peirce, "The Geometry of Cloth Structure," *Journal of the Textile Institute,* Vol. 28 p. T45; and Samuel MacFarlane, ed., *Technology of Synthetic Fibers* (New York: Fairchild Publications, Inc., 1953), p. 251.

- transmission of heat through fabric, or its prevention
- air permeability or the ability of a fabric to permit air flow
- transmission of moisture and moisture absorbency, except that caused by inherent fiber properties—an aspect of which is water repellency
- dimensional stability—the absence of undesirable shrinkage or stretch of the textile
- abrasion resistance—the ability of a fabric to resist damage by rubbing or flexing
- fabric soiling and cleaning problems and the production of fabrics that are soil resistant
- the hand and drape of fabric, which determine product appearance and manipulativeness

Knowledge of geometric features provides clues to possible fabric behavior and can be used by the informed consumer in fabric selection and care. It is important to emphasize, however, that many finishing techniques in current use will alter fabric properties despite known inherent features and geometric factors. On the other hand, some finishes show optimum effectiveness when applied to fabrics with special geometric features.

THERMAL PROPERTIES AND FABRIC GEOMETRY

Fabric for use in selected items of apparel or home furnishings must have properties that make it adaptable for various climatic conditions. The end-use may require that the fabric hold heat against the body, prevent heat from penetrating to the body, or conduct heat away from or toward the body.

For clarity in understanding thermal properties and fabric geometry, a brief discussion of the mechanisms involved in heat loss is included. Heat can be lost from the body by

1. conduction
2. convection
3. radiation
4. evaporation of perspiration

Heat is conducted away when materials known to be good conductors are placed next to the body. The conductor carries the body heat to the surrounding atmosphere, where it is dissipated into the air.

Heat loss by convection occurs when air currents move over and around the body and carry the heat away. This chill factor is one reason why a person feels colder on a windy day than on a day when no wind is apparent, even when air temperatures are identical.

Heat loss by radiation means that heat is given off from the body in the form of rays. This occurs when the air temperature is below body temperature.

Heat loss by evaporation of perspiration is a mechanism to control body temperature. When the body becomes too warm, noticeable amounts of fluid are given off through the skin. This fluid, perspiration, evaporates and the body is cooled. The cooling takes place because the water requires heat of vaporization to evaporate, and this heat comes from the body.

Fabrics used in cold environments should entrap and hold body heat. They should prevent heat loss by conduction, convection, or radiation. (Loss by perspiration evaporation is seldom a problem in cold weather.) To reduce heat loss by conduction, desirable fabrics should possess the following geometric features: fibers should be combined into yarns in a manner that leaves air spaces to serve as insulating areas; the yarns might be soft, fluffy and thick, resulting in fabrics that are bulky but not necessarily compact. Fabrics with rough or irregular surface textures would be good choices.

To prevent heat loss by convection and, to some degree, by radiation, the following features would be desirable: a fabric with smooth yarns that are packed densely; a high thread count with very tiny interstices between yarns and interlacing of yarns to reduce air permeability.

Some items of clothing for cold climates are constructed of two layers of fabric. The inner layer is often a napped or pile fabric made of fibers that are poor conductors, while the outer layer is a tightly woven taffeta, poplin, or similar fabric. In addition to preventing heat loss by convection, this outer fabric reduces heat loss by radiation, because it will stop the heat and may, in some cases, actually reflect heat back to the body.

In warm areas clothing should accelerate or at least aid in heat loss. Where a minimum of insulation is desired, fabrics of loose weave, smooth yarns, fine diameter yarns, and large interstices will be wise selections. These fabrics contribute to heat loss by convection, for air currents can circulate freely; by radiation, since the open spaces permit the rays to move to the surrounding air; and by evaporation, because the moisture vapor can move easily into the air. If the fibers are good conductors, and if the fabrics have few if any spaces to entrap air, heat loss by conduction is aided.

In regions where air temperatures exceed 98°F, particularly desert and tropical areas, there may be occasions when fabrics are chosen that will prevent passage of heat toward the body. These fabrics will be similar to those recommended for cold climates.

AIR PERMEABILITY AND FABRIC GEOMETRY

The air permeability of a fabric is often closely related to thermal properties and frequently is a major factor in body comfort and in protection against moisture. Fabrics with good air permeability provide for heat loss by convection, while fabrics with low permeability prevent heat loss by air movement.

The air permeability of a fabric is influenced by several factors: the type of fabric structure; the design of a woven structure, such as lacy fabrics versus compact weaves; the number of warp and filling yarns per inch; the gage in knitted fabrics; the amount of twist in yarns; the size of the yarns, and the resulting size of the interstices in the fabric. Fabrics with low thread counts and fine yarns usually have good air permeability, while compact fabrics with high thread counts and very tiny interstices between yarn interlacings will prevent the passage of air.

In addition to influencing heat loss by convection, air permeability is sometimes related to the movement of moisture or water vapor. Fabrics that are geometrically constructed so they provide for good air permeability will allow the diffusion of water vapor. To make comfortable hot-weather apparel from hydrophobic fibers, fabrics should be of a porous construction, so they permit both air permeability and moisture-vapor diffusion.

MOISTURE RELATIONSHIP AND FABRIC GEOMETRY

The movement of moisture into fibers or along the surface of fibers, yarns, and fabrics is the result of wicking, moisture adsorption, and moisture regain. Hydrophilic fibers will absorb moisture and permit passage of moisture through geometric openings in the fabric. Hydrophobic fibers do not absorb moisture, but the fabrics provide for moisture transmission by geometric openings and by adsorption. Both hydrophilic and hydrophobic fiber fabrics allow moisture transmission through the wicking characteristics of fibers and yarns.

The movement of liquid moisture and vapor through fabric is dependent upon the compactness or looseness of the weave, the yarn structure, the size of the interstices, and the wicking and adsorption characteristics. When fabrics are composed of hydrophilic fibers vapor moves through both the fibers and the fabric interstices. Hydrophobic fibers permit the passage of moisture vapor only through the interstices, and, therefore, this movement is enhanced by loose fabric structure with large interstices. The transmission of liquid moisture may be enhanced by loose structures; however, liquid moisture moves comparatively well through a porous but more compact structure if wicking or fiber adsorption occurs. Wearer comfort or discomfort appears to be closely allied to the ability of a fabric to pick up water by means of wicking, adsorption, or absorption. In addition, the fabric should provide for some water-vapor transmission. Fabrics constructed of hydrophobic fibers may be uncomfortable if the fabric's geometric features are not properly planned and if wicking is not present or is hindered in some way.

Blends that include cotton are especially good in water transmission, because cotton fibers have a high degree of wickability. Other cellulosic fibers also improve moisture passage. They are extremely efficient in picking up water from the skin surface and carrying it into the fabric. The geometric relationship of the fibers in blended yarns is of extreme importance to the moisture movement. The wicking fibers must be evenly distributed throughout the yarn.

The type of fiber, filament or staple, is an important aspect of comfort. Studies indicate that filament-yarn fabrics tend to plaster themselves to skin when liquid moisture is present, while spun-yarn fabrics are held away by the minute fiber ends extending from the surface.

Water repellency is determined to some degree by the geometry involved in relation to moisture characteristics. Fabrics with large interstices will permit rapid transfer of water, but fabrics so constructed that the interstices are extremely minute will tend to repel water. This behavior can be enhanced by fibers that are hydrophobic, do not wick, or do not absorb moisture. For fabrics made of hydrophilic fibers, good water repellency can be produced by applying special finishes to structures with high thread or fabric count. Fabrics made of soft cotton yarns can provide good water repellency by virtue of their ability to absorb considerable moisture and swell. As the yarns swell, they close the interstices and reduce the transmission or passage of water.

In contrast, there are situations when a high degree of water absorption is desired. The geometric features for such fabrics are altered to provide a maximum surface of soft yarns with good absorptive properties. For example, terry cloth is frequently selected for its absorbency. The geometry of this fabric includes relatively dense construction with yarns that are soft and loosely twisted. Furthermore, the loops of yarn tend to hold moisture within the pile until it moves into the yarns and fabrics.

WRINKLE RECOVERY AND FABRIC GEOMETRY

Fabric geometry is important in relation to wrinkle recovery. Several writers have reported on research in this area, but there is need for additional investigation. The following discussion includes selected geometric features and their relation to wrinkle recovery.[3]

1. Very short fibers tend to be displaced easily when yarns are folded and will, therefore, retain permanent deformation and wrinkles. Medium-length staple fibers, very long-staple fibers, and filament fibers produce good wrinkle recovery if properly combined into yarns and fabrics.

[3] As postulated by Kaswell, *Textile Fibers, Yarns and Fabrics*, p. 285; includes new data provided by N. J. Abbott, "The Relationship between Fabric Structure and Ease-of-Care Performance of Cotton Fabrics," *Textile Research Journal*, Vol. 34, No. 12 (December 1964), p. 1049.

2. Actual fiber diameter and shape influence wrinkle recovery. Fibers that are round in cross section resist bending more than flat fibers. Fibers that resist bending usually recover from light- to medium-weight folding rather quickly. However, if a sharp, heavy wrinkle is formed in the round fiber, it recovers more slowly than in flat fibers, because the factor of strain in this case tends to hold the crease in place. Strain refers to a change of form or size caused by the application of some external force.

3. Degree of yarn twist contributes to wrinkle recovery. Medium-twist yarns provide no opportunity for fiber displacement, so they return to their original position. Yarns with very low twist permit fiber displacement when bent, and, upon release of strain, these fibers do not return to their original position. Thus permanent folds result. High-twist yarns do not recover from wrinkling as quickly as medium-twist yarns, because the high-twist yarns retain folds as a result of fiber and yarn strain, particularly if the wrinkling force has been severe.[4] Crepe yarns present a problem in this theory. Crepe yarns are highly twisted, but crepe fabrics tend to recover from wrinkling more easily than many fabrics of low- or medium-twist yarns.

4. Coarse yarns composed of fine staple fibers or filaments resist wrinkling and, if wrinkled, recover more effectively than either fine yarns of fine fibers or coarse monofilament yarns.

5. Woven fabrics of basket-, twill-, or satin-weave constructions recover more easily from wrinkling than plain-weave fabrics if the fabrics are of equal compactness and yarn size. The basket, twill, and satin weaves have higher yarn mobility than the plain weave because of the fewer yarn interlacings for the former. In general, the lower the number of yarns per inch, the better the wrinkle recovery. A compact twill-weave fabric with a relatively high thread count made of fine yarns will probably have a lower wrinkle recovery than a coarse plain weave made from larger yarns.

   Apparently, any type of fabric, regardless of the weight or tightness of weave, would give better ease-of-care performance if constructed from basket, twill, or satin weaves rather than plain or Oxford weave. Wrinkle recovery in cotton fabrics can be improved by opening up a tightly woven construction—for example, using a twill or sateen instead of a plain weave—but the behavior of more open fabrics is not affected by changing the fabric construction.[5]

6. Thick fabrics, the result of coarse yarns, usually have good wrinkle recovery. Rough surfaces are less likely to show wrinkles than smooth fabrics. This is due, partly, to visual effects. When fabric

---

[4] Buck and McCord, cited by Kaswell, *Textile Fibers, Yarns and Fabrics*, p. 287.
[5] Abbott, p. 1060.

surfaces have fibers napped or inserted at right angles to the base fabric, as in a pile construction, they tend to obscure wrinkles and create the effect of a smooth surface.

DIMENSIONAL STABILITY AND FABRIC GEOMETRY

The dimensional stability of fabric is its ability to resist shrinking or stretching. While fiber content has some influence on this property, geometric factors are extremely important. One of the most significant elements in dimensional stability is the degree of tension under which yarns are held during fabric construction. Yarns are held taut during weaving, and after removal from the loom they relax. This relaxation is accelerated when the fabric is first subjected to moisture. As the yarns relax, they return to their original length and pull closer together, so that fabric shrinkage results. Extremely compact fabrics with firm yarns and a high fabric or thread count are less subject to size change than those with loose, soft yarns and low thread count. Firm yarns spaced to give a low thread count will shrink as much as soft yarns.

It is well known that wool fabrics shrink by reason of the felting property of wool fibers. In addition, wool is also subject to relaxation shrinkage.

Heat-sensitive fibers shrink when exposed to temperatures above that at which the fabric or fiber was heat-set. Many of these fibers show very little relaxation shrinkage if properly heat-set, and the fabrics have good dimensional stability if properly processed. Relaxation shrinkage may occur on these fabrics if the yarns were not permitted to relax sufficiently during the heat-setting process.

ABRASION AND FABRIC GEOMETRY

Fabrics can be damaged by flat abrasion or rubbing, by flex abrasion, or by edge abrasion. Damage from abrasion is the result of certain inherent fiber properties and selected geometric factors. Smooth fabrics constructed of firm yarns with optimum yarn interlacing and relatively compact yarn arrangement are less subject to damage by flat abrasion than fabrics with irregular surfaces, low yarn count, and minimal yarn interlacing. The latter are easily roughened and snagged by rubbing. Loop yarns and other complex yarns with similar characteristics, as well as pile fabrics, are subject to abrasion damage.

The size of yarns also influences abrasion resistance. Thick yarns resist damage from abradants, while fine yarns may abrade easily. Yarn uniformity is important, for irregular yarns may show wear very quickly if abraded. Although the surface of satins and sateens can be abraded because of the large floating yarns, this behavior may be used to advantage in protecting the yarns beneath from damage. This could prolong durability, but it does not preserve appearance, and consumers usually become dissatisfied with the product.

Despite the abrasion resistance of firm fabrics, the consumer should be aware of at least one disadvantage of extremely compact fabrics. With use, these fabrics lose some of their flexibility, and if they become too rigid, abradants can do considerable damage very quickly and easily.

Flex abrasion damage occurs when fabric is flexed or folded upon itself or other fabrics. In this situation the fabric surface is of less importance, and fiber properties—for example, stiffness and yarn mobility—are much more significant. Edge abrasion is effected by the same factors applicable to flex damage and rubbing of the surface.

SOILING, CARE, AND FABRIC GEOMETRY

Geometric factors play an important part in soiling, soil resistance, and maintenance of fabrics. Fibers with a smooth surface and comparatively large diameter, made into smooth yarns and relatively firm fabrics, tend to resist soiling, because they do not provide as many places for soil to become lodged in the fabric structure. Normal cleaning procedures remove dry surface or occluded soil rather easily. However, oil- or water-borne soil or stains are not always easy to remove.

Fabrics of a rather loose structure, regardless of fiber shape and cross section, tend to permit penetration of soil into the interstices. This may hinder cleaning. Fibers with irregular cross section retain soil, but the fabrics will not show soil as readily as those made from round cross-section fibers. Thus, they will not look as dirty when the same amount of soil is present.

Loosely twisted yarns that are somewhat coarse are readily penetrated by soil, as are yarns composed of short staple fibers. The latter may be somewhat difficult to clean, for the soil may be lodged tightly in the yarns and resist efforts at removal.

HAND, DRAPE, AND FABRIC GEOMETRY

Geometric factors are important to the hand and drape of a fabric. The hand of fabrics is influenced by flexibility (pliable to stiff), compressibility (soft to hard), extensibility (stretchy to nonstretchy), resilience (springy to limp), density (compact to open), surface contour (rough to smooth), surface friction (harsh to slippery), and thermal character (cool to warm). Several of these characteristics can be measured objectively by standard test procedures. However, in describing the overall property of hand, the consumer depends primarily upon a subjective evaluation. He may describe a fabric as soft or firm, smooth or rough, springy or limp, stretchy or rigid, compact or loose, pliable or stiff. The geometric factors involved in the creation of the subjectively described fabric and its hand and drape include contour, cross section, and length of fiber; arrangement of fibers in yarns; and arrangement of yarns in fabrics.

The way a fabric will hang and drape depends on the same general factors that influence hand. In one study consumers were unable to make adequate distinction between the terms *hand* and *drape*.[6] It is probable that the two expressions are almost synonymous to the average person. They are descriptive and must be explained subjectively. To do this, the consumer must actually fold, bend, and handle the fabric to understand and express his impression of the draping qualities.

It is the purpose of this chapter to alert the student and the consumer to the fact that fabric is characterized by geometric factors as well as by inherent fiber properties. Knowledge of the geometric factors, coupled with personal observation and an understanding of other fiber and fabric properties discussed in this book, should help develop consumer judgment in the selection of fabric for desired end-uses. Further information on selection, use, and care of fabrics can be gained through laboratory tests discussed in Chapter 35 and throughout the book. Actual laboratory testing, included as an integral part of a course in textiles, will aid in comprehension of fabric behavior.

[6]Kaswell, *Handbook of Industrial Textiles*, p. 450.

Factors
in the Selection
and Care of Fabrics

Fiber content, yarn structure, fabric construction, color, and finish—all of these are important in the selection and care of fabrics. These factors combine to determine the appearance, durability, maintenance, and comfort of a fabric. Each merits different degrees of emphasis, depending upon the desired end-use, and end-use depends upon the consumer.

FABRIC APPEARANCE

The *appearance* of a fabric can be described as its visual effect upon the consumer, who may use his sense of touch for greater comprehension of what he sees.

Fiber luster and texture influence the appearance of fabric. Yarn structure can produce different effects, such as a smooth or rough surface. Fabric structure is important. Color is an obvious property of appearance, and finish may alter the visual impact of a fabric. These have all been discussed in preceding chapters.

Depending upon the interaction of the various factors, fabrics may appear to be soft or stiff, rough or smooth, delicate or coarse, lustrous or dull, bulky or sheer, bright or gray, light or dark. A continuum can be established for each of these comparisons, and fabrics will rank at any one of the infinite number of positions on the scale. After fabric appearance has been established, the preservation of the appearance must be considered.

FABRIC DURABILITY

Durability is defined as the ability to last or endure; the power of lasting or continuing in any given state without perishing. Consumers, in general, do not wish to have fabrics last a lifetime. Thus, for purposes of this discussion, fabric durability is considered to be the ability to retain properties and characteristics for a *reasonable* period of time.

The five factors examined below are related to appearance as well as to durability. However, the importance of durability varies with the predetermined end-use of a product. For example, a consumer expects much longer life from a bath towel than from an evening gown. Other variables include comfort and maintenance. Improper care may cause fiber deterioration, while proper care aids in preserving fiber life.

Generalizations concerning fabric durability might include the following:

1. Fiber content influences durability. Some fibers, such as acetate, are valued for their beauty of hand and drape more than for their wearing qualities. Others, such as nylon, are selected for their strength, abrasion resistance, and other properties that contribute to durability.
2. Yarn structure determines durability to a degree. For example, complex yarns with loops and similar surfaces are easily snagged, producing damage and reducing fabric usefulness. Simple-ply yarns with medium twist will generally give good wear. Yarns with even twist throughout their length are less likely to show wear than those with slubs or irregular areas of very low twist, which may pull apart. Other aspects of yarn durability can be found in Part III.
3. Fabric structure is an important factor in durability. Plain- and twill-weave fabrics are more durable than satin weave, for the floating surface yarns in satins are subject to snagging, breaking, and damage from abrasion. Decorative weaves that include long floats are easily damaged by rough or sharp objects, which may snag the floating yarn. Decorative weaves with short floats, high thread counts, and strong yarns and fibers may give more satisfaction than inferior plain-weave fabrics. Filling- or weft-knit fabrics of plain design are subject to the formation of runs when the yarn is broken. The run spreads more rapidly when fibers and yarns are smooth in surface, such as monofilament nylon found in women's hose. Yarns of rough fibers, such as wool, tend to adhere to each other, and damage from runs is reduced. Mesh con-

structions made on weft-knitting machines often do not form runs. Holes will form, however, if the yarns are broken. Warp-knit fabrics, such as tricot, are comparatively run resistant. They also have sufficient fabric give to resist damage from bending or extending.

Felt fabrics usually have good durability unless they are subjected to considerable pulling force or abrasion. Nonwovens may be durable or disposable depending on planned end-use. Nonwoven interfacing should be able to withstand various methods of maintenance to give satisfactory life in the final product.

Decorative fabrics, such as lace, are not usually purchased with durability in mind. However, with proper handling, lace fabrics will last for many years.

In general, basic-weave constructions and plain-knit fabrics are more durable than complicated fabric constructions, which could show signs of wear quite easily as a result of surface distortion or damage. Fabrics with medium to high yarn count are usually considered more durable than fabrics with low count; however, this is greatly influenced by the type of yarn structure and fiber involved.

4. Color selection and method of application influence durability. Choosing the proper dyestuff for each fiber type is essential if the fabric color is to give good service. The consumer, however, should be aware of scientific, economic, and fashion effects on color. It is impossible to find all colors in all classes of dyestuff, and this may mean selection of a second-choice dye in order to obtain a currently stylish color. Moreover, not all dyes within any one class are equally good, and it is difficult to find dyes in all colors that prove to be colorfast to all degrading environmental conditions. Dyestuffs differ in cost and in the expense of application; thus, it is necessary to choose the dye that will be consistent with a reasonable wholesale price. Furthermore, dye selection should be based on the planned end-uses of the fabric. This, too, may mean a compromise.

The consumer should be willing to accept advice and select fabric for the end-uses for which it was planned. The use of a dress fabric for draperies, for example, may lead to trouble. It is very unlikely that the dress-fabric color will resist sunlight as a regular drapery fabric should.

Loss of color during maintenance of any item generally results in consumer dissatisfaction with the product and premature discard. The durability of color is of considerable importance to appearance. In fact, color behavior emphasizes the importance of the interrelationship among appearance, care, and durability.

5. Finishes, whether applied for appearance or to alter certain behavioral properties, often affect fabric durability. Surface finishes such

as glazing can be removed by improper care, so that the useful life of the product is reduced. Most wrinkle-recovery, minimum-care, or permanent-press finishes reduce fabric tear strength, breaking load and elongation, and abrasion resistance. However, some finishes provide fabric characteristics that are more desirable to the consumer than long life. Minimum-care finishes, despite reduced durability, are frequently more attractive than comparable fabrics that do not have such finishes, and this may more than compensate for the shorter wear-life. In addition, these finishes usually reduce the time spent in care.

It should be evident that durability is determined by several factors, including fabric geometry (see Chap. 32). It is the responsibility of the consumer to decide the relative importance of durability in terms of product end-use and to make his selection accordingly.

FABRIC COMFORT

The importance of fabric comfort will vary according to predetermined end-use of the product and personal preferences of the consumer. Geometric factors discussed in the previous chapter are quite important in assessing comfort. Such considerations as air permeability, thermal retention, texture and tactile characteristics, moisture absorbency, wicking properties, weight-per-unit area, and rigidness versus stretch appear to be related to human comfort. They can be measured accurately by standard test procedures, *but* the actual relationships between the properties and human comfort have not been measured satisfactorily at the present time.

Comfort is affected by various psychological considerations, which are not easily discerned. For example, loss of color would hardly influence physical comfort, but it might help create a psychological atmosphere leading to an illusion of physical discomfort.

Analysis of various geometric principles and other characteristics of fiber—fiber diameter, yarn structure, fabric structure, surface texture, color, and finish—will enable a consumer to make decisions concerning comfort. Since no two people are identical, fabric comfort will be related to individual differences and will require careful decision-making on the part of each consumer based on his or her ideas and attitudes.

FABRIC MAINTENANCE

The care given a textile product is dependent upon the fiber or fibers used, yarn structure, fabric construction, method of imparting color, type of dyestuff, finishes, and geometric factors. Considerable attention has been given to problems of maintenance in previous chapters. A few of the more important points to consider are cited here in an effort to aid the student and consumer in determining the most desirable fabric care methods.

Inherent and geometric fiber properties influence care. Thought should be given to the reactions various fibers exhibit when laundered, dry cleaned, or otherwise processed. Part II indicates care procedures for each type of fiber.

Yarn structure may affect maintenance. Such characteristics as the number of turns per inch and the type of construction—simple, complex, single, ply, or cord—all play a part. For example, a yarn with a very high number of turns per inch, such as a crepe, may relax when subjected to moisture and undergo excessive shrinkage. A complex yarn can be damaged by abrasion from equipment or other fabrics during the maintenance process. It is, therefore, important to consider the yarn in arriving at desirable methods of product care.

Blended yarns should receive the type of care recommended for the more easily damaged fiber. However, proper blending may result in easier care for *both* fibers. For example, a blend of a thermoplastic fiber with cotton produces a fabric that is easily laundered, dries quickly, and requires lower ironing temperatures than pure cotton. A blend of wool and acrylic can be home laundered, while wool itself must usually be dry cleaned.

Fabric construction is related to maintenance in several ways. For example:

- Fabrics with long floating yarns, such as satins, can be snagged during cleaning. This results in unattractive surfaces, destruction of appearance, and weakening of fabric durability.
- Knitted fabrics may need only a minimum of ironing, but some require reblocking and reshaping to retain size and appearance after maintenance procedures.
- Sheer fabrics of leno-weave construction might require careful handling, but they are easier to maintain than sheer fabrics of a plain weave. The interlocking of the warp yarns reduces yarn slippage. However, other factors, such as fiber content, alter the care of sheer fabrics.
- Nonwoven fabrics demand careful maintenance, for incorrect care may result in loss of adhesive or in fiber separation.
- Fragile fabrics, such as lace, may require special handling. This is influenced by fiber content; fragile, sheer fabrics of strong, tough fibers are easier to maintain than firm fabrics of weak fibers.
- Firm, compact, and smooth fabrics usually respond satisfactorily to care procedures suitable to the fiber used.
- Pile fabrics may require brushing to remove lint from the surface. Those that can be laundered look better if tumble dried, because the dryer action tends to raise the pile and fluff the fabric.
- Tufted fabrics can generally be laundered and tumble dried. If

the product is too large for home equipment, many dry cleaners and commercial laundries will do the job satisfactorily. It is important to tumble dry such products in order to maintain the appearance of the tufts.

Proper laundering techniques should be observed for all washable fabrics. Excessively long wash cycles may cause redeposition of soil upon fabrics, which produces grayed or dull garments instead of clear and clean ones.

Color is an important aspect of care. Whether fabrics are dyed or printed, color may be lost during laundering or dry cleaning. Tests for evaluation of color have been mentioned (see Chap. 31). When convenient, it is a good idea to determine colorfastness of dyestuffs before using specific procedures. Proper maintenance in relation to color results in continued good appearance.

The presence of finishes is of considerable importance in fabric care. Finishing substances may improve the behavior of fabric, but they may also cause new problems. Some finishes respond well to laundering but are destroyed by dry cleaning; others thrive if dry cleaned but are damaged by laundry methods. A review of the chapters on finishes will provide some guidelines to maintenance problems.

The selection, use, and care of fabrics depends upon many factors. The importance of appearance, durability, comfort, and maintenance are relative. It is the responsibility of each consumer to evaluate the qualities of a fabric in terms of its ultimate end-use and then make the decisions required concerning its use and care.

34

Textile Performance: Standards and Legislation

When shopping for textile items, consumers often ask

- Will the product be durable?
- Will the product be easy to maintain?
- Will the product be comfortable?
- Will the product be attractive for the desirable wear-life?
- Does the product include any guarantee?
- What performance should be expected for the cost?
- What care procedures are recommended, if any?
- What label information is required?
- What is the fiber content?

Other questions might be related to such esthetic concerns as color schemes or the best selection for a specific occasion. Discussion of the latter problems will not be included in this text, for they relate more specifically to clothing styles or interior decoration. Federal legislation, suggested label symbols from retail merchant organizations, existing standards of performance such as L-22, and some general consumer problems will be examined.

The Textile Fiber Products Identification Act of 1960 and the Wool Labeling Act of 1939 require that the fiber content of a textile product be clearly stated on an accompanying label. Labels may be sewn into the product or attached to it by some other means, or the information may be included on an adjacent sign or on the package or container.

The Wool Products Labeling Act, described in Chapter 8, requires that wool or part wool products be labeled to indicate whether the wool content is new wool (labeled *wool*, *Virgin wool*, or *lamb's wool*), reprocessed wool, or reused wool.

The Textile Fiber Products Identification Act of 1960 (see p. 24) provides that each textile fiber product—with certain items exempted, such as already installed upholstery fabric and coated fabrics—shall be labeled or invoiced in conformity with the requirements of the act, and that advertising of textile fiber products shall conform to specified requirements of the act. It is important to mention certain exemptions. The absence of labeling requirements for upholstery fabric already securely attached to a frame is regrettable. This area is one in which fiber labeling would be of value to the consumer in making decisions about buying furniture.

Except for natural fibers, which are labeled by commonly accepted names, fibers must be identified by generic term (see Chap. 3). Selection of the original sixteen terms by the government committee responsible was based on previously accepted words, those having identity with the chemicals used, and the development or "coining" of terms that would be recognized and learned easily. Most generic terms are accompanied by a trademark name when used for advertising identification with the producing company. The law gives specific regulations about labeling procedures and lists those items exempted. Copies of this law and regulations governing its operation can be obtained from the Federal Trade Commission, the group responsible for its enforcement.

The Textile Fiber Products Identification Act as amended classifies most man-made fibers into one of the seventeen generic groups. While there are seventeen groups, there are actually nineteen generic terms, since two groups—acetate and rubber—provide for specific generic names on fibers that qualify. The use of a generic term reduces the confusion of trade names. The law provides for amendments to add new generic terms whenever new fibers are developed that do not conform to definitions of existing groups.

The fiber labeling legislation requires that correct fiber content data be transmitted by fiber producers to product manufacturers, who, in turn, must provide adequate and accurate information to the retailer. The retailer is then responsible for proper labeling at the consumer level.

If a fabric is composed of two or more fibers, the label must state the actual percentage of each fiber in descending order of the amount present. For example, a fabric of 65 percent polyester and 35 percent cotton should have the fibers listed thus:

65 percent polyester
35 percent cotton

All fibers present in amounts of 5 percent or more must be identified on the label. When less than 5 percent of a fiber is included, it may be identified in one of the following ways:

1. If the fiber serves a clearly established and definite function, it should be identified by generic name, percent, and significant property.
2. If the fiber is included for appearance, or for other nonfunctional reasons, it is to be identified as "other fibers," with the percent indicated.

The real responsibility for enforcing the law rests with the consumer. Many people have little or no knowledge of legal requirements; they may have no interest in the fiber content of a textile product and are probably not yet convinced that, in order to give fabrics proper care, they should have accurate fiber content information. Consumer apathy, or, more likely, ignorance of legal requirements for fiber identification, results in very few complaints and limited legal action relative to labeling procedures.

Retailers are sometimes lax in maintaining properly labeled textile merchandise, especially when their customers seem disinterested in having the information. Such retailers cannot be excused. However, the situation emphasizes the importance of consumer education, the need for consumers to demand proper label information, and most important, the need to provide information about how fiber content data can be used to advantage. Consumers who encounter merchandise that is not labeled or is labeled incorrectly should inform the nearest office of the Federal Trade Commission.

In addition to the two acts just discussed, there are some legislative controls of advertising. The Wheeler-Lea Act of 1938 is directed against false or misleading advertising other than labeling and includes controls of printed advertisements.[1] This law prohibits advertisements that are misleading not only because of what they *do* say, but also because of what they *do not* say. The regulations written into the act are particularly valuable in preventing statements in advertising that imply that fabrics will respond in a certain manner when, in actuality, they will not. The act also forbids unfair or deceptive practices in commerce.

Despite the laws, there are always examples to be found of misleading advertisements, deceptive methods of selling, and unfair com-

---

[1]C. M. Edwards and R. A. Brown, *Retail Advertising and Sales Promotion*, 3d ed. (Englewood Cliffs, N.J.: Prentice-Hall, Inc., 1959), p. 642.

petition. This results, in part, from lack of adequate controls for enforcing the legislation, in part from "ignorance" on the part of the retailer or seller, the advertising media, or the local public.

Quality controls of consumer products, including textiles, are often established by manufacturers. Such companies maintain outstanding reputations for quality by means of testing and research in company-owned and -operated laboratories. Retail companies operating laboratories for quality control include the J. C. Penney Company and Sears, Roebuck. Most large textile mills maintain research and development laboratories, and companies without their own facilities may contact special testing groups for quality control. Any organization in the textile industry is entitled to help from independent research groups such as Fabric Research Laboratory and Harris Research Laboratory. Quality control and other types of testing will be performed for requesting organizations by U.S. Testing and the Better Fabrics Testing Bureau. Some groups of manufacturers conform to established standards of quality by self-policing practices in various stages of merchandising.

Many consumers who read labels and who know something about fiber content prefer to have helpful information concerning methods of care. Labels with good directions for maintenance can prevent dissatisfaction. When these labels are sewn into the item, they provide a permanent record of care procedures. Hang tags that include care techniques are valuable only as long as the consumer keeps them where they are easily referred to when laundering or preparing the item for cleaning. A good filing system for such labels is very helpful.

Consumers frequently complain, justifiably, that hang tags are easily lost, destroyed, removed from the garment before purchase, or put away never to be found again. To help alleviate this criticism, the National Retail Merchants Association has developed a series of "sure care" symbols for use on labels either sewn into the garment or attached temporarily. The sewn-in label is preferred.

Such symbols on labels would have many advantages:

1. A quick glance could interpret the sketches.
2. The symbols could become universal in their use and application.
3. They require little space, yet provide valuable information.
4. They could be learned easily by a vast majority of the consuming public.

Delay in the complete development, promotion, and general use of sure care symbols by manufacturers is partially the result of consumer apathy about textile care information. Only the informed, shrewd, interested consumer is always aware that the tremendous variety of products available makes it literally impossible to know, by appearance and feel, what care procedures are required. Therefore, if at the time of purchase the consumer is more concerned with appearance or fashion than with information about fabric behavior and care, he may be dissatisfied when the item is laundered or dry cleaned.

In 1966 an industry advisory committee on textile information developed a set of recommendations concerning textile labeling. Their report was submitted to the President's Committee on consumer interests, and their opinions and recommendations were published. The recommendations included the following:

1. Textile consumer goods should be permanently labeled with appropriate maintenance information whenever proper care requires special techniques and whenever it is not obvious to the consumer that the item can be successfully cared for by conventional procedures.
2. Care labels should be phrased in common language to describe conventional laundry and cleaning methods, except where special symbols or nomenclature and their explanations are made available to consumers.

The committee further suggested that the group continue to function as a unit to aid in implementing the labeling recommendations. There is increased emphasis on the importance of permanently attached care labels.

The industry is concerned, also, about performance, and a set of minimum performance standards for fabrics in various end-uses has been developed and published. This set of ground rules is called *American Standards Performance Requirements for Textile Fabrics, L-22,* and products that conform to these standards can be purchased with the knowledge that they will give service if the label information is observed.

In 1956 the American Standards Association, now the American National Standards Institute, and the Textile Distributors Institute pooled their resources to permit development of one set of standards. The American Standards Association, under the sponsorship of the National Retail Merchants Association, assumed leadership of the proposed project and became the coordinating group. The resulting standards and test methods used in evaluating performance were the result of the close cooperation of such groups as the American Association of Textile Chemists and Colorists (AATCC), the American Association of University Women (AAUW), the American Federation of Labor–Congress of Industrial Organizations (AFL–CIO), the American Home Economics Association (AHEA), the American Hospital Association (AHA), the American Institute of Laundering (AIL), the American Society for Testing and Materials (ASTM), Consumers Research, Inc., Consumers Union of the United States, National Institute of Drycleaning (NID), the Textile Distributors Institute, Inc., (TDI), and many other organizations interested in textile performance.

New L-22 standards were approved and published in 1960 and revised in 1968. The 1968 revision consolidated 75 performance standards into 67 and added one new one. Performance requirements for durable-press fabrics and stretch fabrics were added to applicable standards. The performance standards apply to wearing apparel and home furnishings; they refer to fabric behavior and seam strength, not to product design and construction. Each standard specifies minimum acceptable results for selected standard textile tests. Manufacturers who adhere to the standards usually label products clearly, both to indicate results of some performance testing and to provide consumers with information about maintenance.

Test methods used to evaluate L-22 performance standards are those developed and published by AATCC, ASTM, USDC–CS Standards, American National Standards Institute, and cooperating research organizations and manufacturing concerns.

Many manufacturers *are* using the standards for their textile products. However, these same manufacturers prefer to use their own label, rather than the type recommended by L-22. Furthermore, these manufacturers do not publicize the fact that they are adhering to the standards.

The Apparel Research Foundation (ARF) of the American Apparel Manufacturers Association (AAMA) is conducting research on the feasibility of the establishment of voluntary industry standards for apparel. The ARF has published a testing manual for member organizations and is presently conducting a series of workshops on standards and testing programs at apparel manufacturing centers throughout the United States. This group is seriously concerned with maintaining quality in consumer products and achieving reputations for its member manufacturers.

Information concerning care is not a legal requirement. Therefore, it becomes the responsibility of the consumer to request such information if it is not provided as a service by the manufacturer or retailer. Persistent requests by consumers might stimulate retailers and manufacturers to provide more and better care instructions on all merchandise.

The consumer has a basic responsibility that many do not accept. When merchandise proves to be unsatisfactory, the consumer should return it. Every time he does not do so, the production of inferior merchandise is encouraged.

All manufacturers are in business to make money. Many manufacturers of apparel and some fiber and fabric producers are marginal concerns. They sell whatever they can at whatever price they can get for as long as they can get away with it. If merchandise is exceptionally poor, these concerns go out of business quickly, but not quickly enough to prevent the sale of inferior products. Other manufacturers in various areas of the textile business do not identify themselves, and, thus, do not attempt to build a reputation. The reliable manufacturers with publicized names do care about repeat purchases and try to market merchandise that is worthy of their name and the price asked. To encourage these firms, the consumer should ask for their products. We should reject inferior merchandise or, if purchased, return it as soon as the fault becomes evident.

In our affluent society it is possible to sell many items that will give poor service. Because the consumer may not care if the item does not prove to be worth the price, and, in fact, may actually appreciate the excuse to purchase a new item, inferior merchandise is not returned and retailers may not be aware of such products.

The development of guides or standards for the production and sale of quality merchandise at fair prices is the responsibility of manufacturers, retailers, and consumers. Legislation may help by requiring information that will aid the consumer, but it is not and should not be the sole solution. Legislation can never replace intelligent buying.

Wearing apparel is often chosen for its color or fashion value and, where this is the only consideration, care and comfort may be unimportant. A bride, for example, will choose her dress for its lovely appearance; other factors are secondary or of no importance. On the other hand, though sheets may be selected to blend with the bedroom color scheme, maintenance, durability, and comfort are usually of primary importance.

Decisions a consumer makes regarding such items as this will be the result of a personal storehouse of textile knowledge and the label information provided. Fiber content will give clues about basic principles of fabric care. Some technical knowledge of fabric structure, color, finish, and geometric factors will aid in the decision-making process. Information gained from courses, textbooks, periodical articles, or practical experience will help the consumer to anticipate performance or behavior of any textile product.

In using basic information as criteria in selecting textile products, it should be emphasized that the consumer must also consider such factors as

- length of time the product is expected to last
- maintenance procedures that should be or would be used
- relative importance of durability
- relative importance of comfort
- relative importance of appearance

No one can make decisions for someone else. Therefore, the behavior of textiles in relation to appearance, comfort, durability, and maintenance becomes specific to the consumer and the anticipated end-use. The informed consumer has the background information available to apply to the decision-making process and arrive at the most desirable selection for the end-use under consideration.

Consumers should remember that

- durability may or may not be important
- the price may not indicate quality
- maintenance techniques will vary according to the product and to available care facilities

Factors that influence the behavior of a textile and, in turn, determine its appearance, durability, maintenance, and comfort include the following:

- fiber content—inherent fiber properties and fiber geometry
- yarn structure—types of yarn, geometrical arrangement of fibers in the yarn, yarn size
- fabric structure—type of construction, geometry of the fabric, thread count
- color—proper selection of dye for fibers used, method of application, resistance of color to various environmental factors
- finish—type of finish, method of application, effect on fiber and fabric behavior, durability to environmental conditions

35

Fabric
Performance

TESTS FOR PERFORMANCE EVALUATION

Standard test procedures administered in the laboratory provide reliable data that can be used for evaluation and, in some instances, to predict fabric behavior. Several basic tests will be outlined in this final chapter. Informal experiments that can be done in the home or in laboratories that are not equipped with testing machines have been discussed in other chapters (see Chaps. 27 and 31).

Though scientific studies of textiles yield quantities of information, they also raise questions that cannot be answered by a simple "yes" or "no," "right" or "wrong." What is adequate for one end-use may be inadequate for another. What pleases one consumer could displease another. However, test results do provide guidelines for evaluating textile performance and as such serve a valuable function in the classroom or laboratory.

The American Society for Testing and Materials, the American Association of Textile Chemists and Colorists, the American National Standards Institute, and a number of Federal Government agencies, have established tests, and in some cases, recommended specifications for textile behavior.

The government has developed test methods that are used by such agencies as the Natick Labs of the U.S. Army in testing textile items submitted for approval by an armed forces purchasing unit. Many of these tests are the same as those published by AATCC and ASTM. The various testing organizations and government agencies work together in developing techniques for evaluating fabric performance, and many test procedures are the result of intergroup activity. These are identified by a number specific to each group involved.

For general laboratory analysis of fabrics the student will find adequate test procedures in the Book of Standards published each year by the American Society for Testing and Materials. Chapters 24 and 25 of this book are concerned with textile materials and are the work of committee D-13. An equally important volume for the student interested in textile evaluation is the Technical Manual released annually by the American Association of Textile Chemists and Colorists.

Various laboratory manuals designed for use in the college classroom will contribute to a better understanding of fabric and fiber analysis. The 1970 yearbooks of the two organizations just mentioned have served as resources for this chapter.

The following discussion includes those tests used for general textile evaluation and those that apply to consumer use and care of textile products. The purpose and scope of the tests are mentioned, but actual test procedures are not given.

The reader is reminded that it is important to refer to the most recent manual when selecting test methods; he is cautioned against using test data as proof of good or poor textile performance. These data can help predict *possible* performance, but when a product is put into actual use, other factors play important parts in behavior. Accidental damage is an obvious variable; human differences in handling are less evident in their effect on performance. Nonetheless, the following tests should be valuable tools for fabric assessment.

Fabric Description (ASTM D1910-64)

A textile fabric is described by citing its average width, length, number of yarns per inch in warp and filling, yarn size, and weight in ounces per square or linear yard. This information is important to product manufacturers who are billed for the fabric. If fiber content is known and constant, it provides a basis for price comparison and for determination of suitable end-uses.

The width suggests how patterns must be laid for efficient cutting; the length and width indicate how many items can be cut. Width is also important in determining how much yardage must be purchased by the consumer for home sewing.

The number of yarns per inch (thread count) and the yarn size give some idea of the compactness and density of the fabric. *Balanced thread counts*—in which the number of yarns per inch in warp and filling are similar—frequently are considered to offer better wearing qualities than unbalanced counts. However, this is also influenced by yarn size or yarn number (ASTM D1059-69T and D1907-69). If yarns in one direction have a very tiny diameter and yarns in the opposite direction are thicker, the number of threads per inch may vary considerably and still provide a satisfactory fabric. Furthermore, fabrics such as broadcloth and poplin have uneven thread counts and yarns of similar size in both warp and fill, but they exhibit good wearing properties.

The weight of the fabric determines to a great extent the weight of the end product, and it is important to fabric hand, appearance, and comfort. If a fabric is extremely heavy, it may be uncomfortable in wearing apparel, or, depending upon the garment, the weight may make a product hang more attractively. Lightweight fabrics can be comfortable, and they often drape attractively, but, for certain end-uses, they are too filmy, soft, and light.

Knitted fabrics (ASTM D231-62) are tested for weight, width, and length. Instead of measuring yarns per inch, the number of wales and courses per inch are counted. This provides information concerning the density of the fabric in relation to the closeness of yarns. The fineness of knitted fabrics is expressed as the gage, which indicates the number of needles per unit width across the wales, frequently cited as the number per $1\frac{1}{2}$ inches.

Fabric Thickness (ASTM D1777-64)

The thickness of a fabric is one of its basic characteristics. It is usually expressed in thousandths of an inch. Warmth in textile products can be estimated from the thickness of fabric. Tests that measure change in thickness resulting from rubbing or shrinkage are helpful in evaluating performance.

Air Permeability (ASTM D737-46)

The amount of air flowing through a fabric is measured by an apparatus that conforms to the basic requirements established by the test methods. Air flow is expressed as cubic feet of air per square foot of fabric at a given pressure drop across the fabric. The ability of a fabric to permit flow of air and water vapor through it is one aspect of comfort. Air permeability may also indicate which fabrics are suitable for protective apparel and coverings.

Thermal Properties (ASTM D1518-64)

Thermal transmission tests also provide data on fabric comfort. They determine the amount of heat that passes through a fabric and thus the relative insulative value. A low degree of heat transfer would indicate a good fabric for insulation and protection from undesirable temperatures. Weight and thickness of fabric are calculations used in evaluating thermal properties and, hence, serve as factors in end-use performance. A fabric with good resistance to heat transfer could be a poor choice if too heavy and too thick for comfort.

Abrasion Resistance (ASTM D1175-64T)

The abrasion resistance of a fabric can be determined by several testing instruments. Results from different instruments should not be expected to agree and, thus, may give misleading data if not used properly. These results are influenced by the machine, the abradant used, the area abraded, the tension on the sample, and the type of abrading action. Reliability of the data is limited.

Resistance to abrasion is affected by such factors as inherent and geometric properties of fibers, yarns, and fabrics and the kind and amount of finishing substances.

In assessing fabric performance, the results of abrasion testing should be interpreted carefully. Test results obtained on fabrics intended for the same end-use can be compared for a general prediction of expected performance.

Strength—Breaking Load (ASTM D1682-64)

Woven fabric strength is usually expressed as the force in pounds per inch required to break or rupture the fabric sample. The American National Standards Institute, in the L-22 standards, maintains that a breaking load of 20 pounds is desirable for blouse and dress fabric, 50 pounds for bathing suit fabric, and 55 pounds for men's shirt fabrics. Standards are given for a variety of other items.

Knitted fabric strength is expressed as the force in pounds per square inch required to rupture the fabric. The two methods most frequently used are the following:

1. the force of glycerine or water pushing against a rubber diaphragm over which the test specimen is clamped
2. the force required to push a ball through fabric that is held in a spherically clamped disc

For most satisfactory comparisons of breaking strength, results obtained on identical instruments with identical procedures should be evaluated.

Breaking strength data can predict resistance to damage by pulling or similar forces. They should not, however, be considered as essential fabric criteria, for some fabrics with comparatively low breaking strengths can perform adequately in selected end-uses.

TESTS FOR FINISH PERFORMANCE

Resistance to Loss of Finish in Laundering or Dry Cleaning (AATCC 94-1969)

The AATCC test cites procedures for determining quantitatively the amount and kind of finish substances used on a fabric. By checking samples before and after laundering or dry cleaning, the loss of finish can be measured. In addition to this quantitative analysis, a subjective analysis will provide information of value to the consumer, such as change in appearance and hand. The quantitative analysis is a reliable objective test that supports the visual inspection.

Data concerning finish loss should be used with care. End-use, desired performance, and anticipated life of the item must be considered.

Resistance to Water: Rain Test (AATCC 42-1967 and ASTM D583-63)

The rain test is applicable to any fabric with or without water-repellent finishes. It measures the fabric's resistance to penetration of water by impact and predicts the behavior of the fabric in rain. Equipment used for this test can apply water under different amounts of pressure so behavior of fabric in various conditions can be determined. Water repellency depends upon the inherent and geometric properties of fibers, yarns, and fabrics, as well as the presence of water-repellent finishes.

Spray Test (AATCC 22-1967 and ASTM D583-63)

The spray test is used on fabrics that may or may not have been treated for water resistance or repellency. It measures the resistance of fabrics to wetting and provides visual standards to determine fabric ratings. Ratings range from 100 (no sticking or wetting of upper surface) to 0 (complete penetration of fabric). The method is recommended for measuring the water repellency of apparel fabrics. It is especially useful in evaluating the effectiveness of water-repellent finishes. Results depend primarily on the water resistance of fibers, yarns, and finishes, not on fabric construction.

Resistance to Insect Damage (AATCC 24-1963 and AATCC 28-1956)

Damage to fabrics by insects, such as moths and carpet beetles, costs the consumer millions of dollars each year. Besides the economic factor, a consumer may have to discard a well-liked and prized article. The purpose of these tests is to ascertain the susceptibility of a fiber to insects and the efficacy of finishes used in the prevention of such destruction. This information is important to most consumers, for knowledge that fabric is not resistant alerts them to the need of taking precautions in their own homes to prevent damage.

Resistance to Mildew and Rot (AATCC 30-1957)

Fabrics composed of fibers damaged by mildew or rot, or which contain substances attacked by mildew, can be treated to resist such destruction. Test procedures determine the susceptibility of a fabric to mildew damage, as well as the effectiveness of mildew preventatives available. Several methods are given for testing the response of fabric under varied conditions. Soil-contact tests are used for items such as tarpaulins and sleeping bags. Exposure on weathering racks is the method employed for evaluating fabrics in such products as awnings.

Resistance to Fire: Flammability of Clothing Textiles (AATCC 33-1962 and ASTM D1230-61)

Federal law regarding fire safety for textiles is concerned with reducing the *speed* of burning; it does not require that burning be eliminated altogether. The present legislation is based on U.S. Commercial Standard Test 191-53. Other test procedures have been developed, but no single test has been identified as the best.

Testing for flammability identifies fabrics that burn intensely and rapidly enough to be dangerous. Furthermore, the tests can determine the effectiveness of flame-retardant finishes. Fabrics are rated according to the speed and manner in which they burn. They are classed by weight, and the burning rate permitted depends on this weight classification.

Dimensional Change: Effect of Laundering (AATCC 96-1967, AATCC 99-1962, and ASTM D1905-68)

Perhaps one of the greatest disappointments for the consumer is to have a textile product shrink or stretch out of shape so that it no longer fits, hangs correctly, or looks attractive. The dimensional change of fabric in dry cleaning or laundering can be determined by standard test procedures. Methods approximate those used in home laundering, commercial laundering, or dry cleaning. The laundering test provides variables in order to approximate the treatment best suited to different fabrics. These include different washing temperatures, laundering agents, drying procedures, and restoration procedures.

Label information regarding shrinkage or stretch is desirable. If dimensional change equals or exceeds 2 percent, it will noticeably alter the fit of apparel.

Recovery from Wrinkling (AATCC 66-1968, AATCC 128-1969, and ASTM D1295-67)

Consumers naturally want fabrics that will resist the formation of undesirable wrinkles and creases. If creases do form, it is gratifying to have them spring out quickly. The purpose of this test is to determine the recovery of fabrics from creasing. Both warp and filling of woven fabrics are tested. The test is applicable to any fabric of any fiber content, with or without additional finishes.

Fabrics that have a high degree of recovery from creasing will tend to retain a satisfactory appearance and require less care.

Durable-press Properties (AATCC 88B-1969, AATCC 88C-1969, and AATCC 124-1967)

The purpose of durable-press tests is to evaluate the retention of original appearance after repeated care periods. The analysis is somewhat subjective. Plastic replicas and photographs of fabrics at various stages of wrinkling are used as standards. The item being tested is compared with the standards and rated according to the replica or photograph it most closely approximates. The test should be repeated periodically to determine the expected duration of the fabric appearance. It is flexible to provide for variations in laundering procedures, so the washing method most desirable for the particular fabric should be used. The replicas and the fabric are evaluated under selected conditions of light, distance, and angle. The fabric surface, the seam construction, and deliberate pleats and creases are all studied.

Soil-release Properties (AATCC 130-1969)

Tests are available to measure the ability of a fabric to release oily soil during laundering. Fabric is stained with oily substances in a specified manner; it is then washed according to standard procedures. The test fabric is compared with stain-release standards and evaluated. Many new finishes and some fibers tend to hold oily stains. This test yields data that will give the consumer, as well as the manufacturer, some idea of care requirements and appearance retention.

Resistance to Yarn Slippage (ASTM D434-42)

Some fabrics are prone to show pull or slippage at seams. The useful life of the item is therefore reduced because of undesirable appearance and because slippage is followed very quickly by the development of rips and holes. Slippage occurs more readily on fabrics made from filament yarns.

The test is particularly applicable to rayon, silk, and acetate woven fabrics, but it can be used on any fabric. Resistance to slippage is expressed as the number of pounds of pull per inch across a seam required to produce an elongation of $\frac{1}{4}$ inch in excess of the normal fabric stretch.

Colorfastness of Fabrics (AATCC 3-1962, 8-1969, 15-1967, 16-1964, 32-1952, 36-1969, 61-1968, 71-1956, 72-1969, 87-1965, 104-1969, 105-1967, 106-1968, 107-1968, 116-1969, 132-1969, and 133-1969)

Standard test procedures subject colored fabric to specific environmental conditions that may cause fading. These include light, laundering techniques, perspiration, sea water, chlorine, dry cleaning, rubbing, or crocking. The tests can be used for interlaboratory testing and establishment of specifications.

The importance of colorfast dyestuffs has been discussed in previous chapters. If consumers have some idea of the color behavior of a textile, they can make sensible choices of care techniques as well as reasonable selections of fabric for particular end-uses. It is the responsibility of the consumer to utilize any label information about color behavior.

Modern test procedures can provide a range of predictability that is a tremendous boon to the consumer. Ideally, any textile product could have a label affixed to it telling the purchaser exactly what the fabric can and cannot do, will or will not endure. No item, regardless of how beautiful, will give pleasure to its owner if it shrinks from a size 12 to a size 6 after the first laundering, fades from navy to powder blue, or looks as though the wearer has slept in it five minutes after he puts it on. Consumers who use their textile training can make intelligent selections of textile products and wise evaluations of anticipated textile performance.

Bibliography

Books

Alexander, Peter, Robert Hudson, and Christopher Earland. *Wool, Its Chemistry and Physics.* New York: Reinhold Publishing Corporation, 1963.

American Fabrics Encyclopedia of Textiles. New York: Doric Publications, 1960.

Analytic Methods for a Textile Laboratory, J. W. Weaver, ed. Research Triangle Park, N.C.: American Association of Textile Chemists and Colorists, 1968.

ASTM Standards (published annually), Part 24, *Textile Materials.* Philadelphia: American Society for Testing and Materials, 1970.

ASTM Standards (published annually), Part 25, *Textile Materials.* Philadelphia: American Society for Testing and Materials, 1970.

Baity, Elizabeth C. *Man Is a Weaver.* New York: The Viking Press, Inc., 1949.

Bell, J. W. *Practical Textile Chemistry.* New York: Chemical Publishing Company, Inc., 1955.

Bendure, Z., and G. Pfeiffer. *American Fabrics.* New York: The Macmillan Company, 1947.

Brewster, R. Q., and W. E. McEwen. *Organic Chemistry.* Englewood Cliffs, N.J.: Prentice-Hall, Inc., 1962.

Brown, H. B., and J. O. Ware. *Cotton.* New York: McGraw-Hill, Inc., 1958.

Buresh, Francis M. *Nonwoven Fabrics.* New York: Reinhold Publishing Corporation, 1962.

Carroll–Porczynski, C. Z. *Manual of Man-made Fibers.* New York: Chemical Publishing Company, Inc., 1961.

———. *Natural Polymer Man-made Fibers.* New York: Academic Press, Inc., 1959.

Cook, J. Gordon. *Handbook of Polyolefin Fibres.* London: Merrow Publishing Company, 1967.

———. *Handbook of Textile Fibres,* 2 Vols. London: Merrow Publishing Company, 1968.

Cotton from Field to Fabric, 5th ed. Memphis, Tenn.: National Cotton Council, 1951.

Cowan, Mary L., and Martha E. Jungerman. *Introduction To Textiles.* New York: Appleton-Century-Crofts, 1969.

Crawford, M. D. C. *The Heritage of Cotton.* New York: G. P. Putnam's Sons, 1924.

Dembeck, Adeline A. *Guidebook to Man-Made Textile Fibers & Textured Yarns of the World,* 3d ed. New York: United Price Dye Works, 1969.

Edwards, C. M., and R. A. Brown, *Retail Advertising*

and Sales Promotion, 3d ed. Englewood Cliffs, N.J.: Prentice-Hall, Inc., 1959.

Encyclopedia of Polymer Science and Technology, H. F. Mark, N. G. Gaylord, and N. M. Bikales, eds. 15 Vols. New York: Wiley-Interscience, 1964–1971.

Fiberglas Yarns for the Textile Industry. Toledo, Ohio: Owens-Corning Fiberglas Corporation, 1951.

Garner, W. *Textile Laboratory Manual,* 6 Vols., 3d ed. London: National Trade Press, 1967.

Grover, E. B., and D. S. Hamby, *Handbook of Textile Testing and Quality Control.* New York: Interscience Publishers, Inc., 1959.

Hall, A. J. *The Standard Handbook of Textiles.* New York: Chemical Publishing Company, Inc., 1965.

————. *Textile Finishing,* 3d ed. New York: Chemical Publishing Company, Inc., 1966.

Hamby, Dame S., ed. *The American Cotton Handbook,* 2 Vols., 3d ed. New York: Wiley-Interscience, 1965.

Handbook of Asbestos Textiles, 2d ed. Philadelphia: Asbestos Institute, 1961.

Harris, J. C. *Detergency Evaluation and Testing.* New York: Interscience Publishers, Inc., 1954.

Harris, Milton, ed. *Handbook of Textile Fibers.* New York: Textile Book Publishers, Inc., 1954.

Hathorne, Berkeley L. *Woven, Stretch, and Textured Fabrics.* New York: John Wiley & Sons, Inc., 1964.

Hearle, J. W. S., and R. H. Peters. *Fiber Structure.* New York: Butterworth & Co., 1963.

Hess, Katherine P. *Textile Fibers and Their Uses,* 6th ed. Philadelphia: J. B. Lippincott Co., 1958.

Heyn, A. N. J. *Fiber Microscopy.* New York: Interscience Publishers, Inc., 1954.

Hollen, M., and J. Saddler. *Modern Textiles.* New York: The Macmillan Company, 1968.

Hoye, John. *Staple Cotton Fabrics.* New York: McGraw-Hill, Inc., 1942.

Kaswell, E. R. *Handbook of Industrial Textiles.* New York: Wellington Sears, 1964.

————. *Textile Fibers, Yarns, and Fabrics.* New York: Reinhold Publishing Corporation, 1953.

Klapper, Marvin. *Fabric Almanac.* New York: Fairchild Publications, Inc., 1966.

Kornreich, E. *Introduction to Fibres and Fabrics.* New York: American Elsevier Publishing Company, Inc., 1966.

Krcma, Radko. *Nonwoven Textiles.* Manchester: Textile Trade Press, 1967.

LaBarthe, Jules. *Textiles: Origins to Usage.* New York: The Macmillan Company, 1964.

Lancashire, J. B. *Jacquard Design and Knitting.* New York: National Knitted Outerwear Association, 1969.

Lee, Henry, Donald Stoffey, and Kris Neville. *New Linear Polymers.* New York: McGraw-Hill, Inc., 1967.

Leggett, W. F. *Story of Linen.* New York: Chemical Publishing Company, Inc., 1945.

————. *Story of Wool.* New York: Chemical Publishing Company, Inc., 1947.

Linton, G. E. *Applied Textiles,* 6th ed. New York: Duell, Sloane & Pearce–Meredith Press, 1961.

————. *The Modern Textile Dictionary,* 2d ed. New York: Duell, Sloane & Pearce–Meredith Press, 1963.

————. *Natural and Manmade Textile Fibers.* New York: Duell, Sloane & Pearce–Meredith Press, 1966.

Linton, G. E., and H. Cohen. *Chemistry and Textiles for the Laundry Industry.* New York: Textile Book Publishers, 1959.

Lynn, J. E., and J. J. Press. *Advances in Textile Processing.* New York: Textile Book Publishers, 1961.

Lyons, John W. *The Chemistry and Uses of Fire Retardants.* New York: Wiley-Interscience, 1970.

Man-made Textile Encyclopedia, J. J. Press, ed. New York: Textile Book Publishers, Inc., 1959.

Mark, H. F., S. M. Atlas, and E. Cernia. *Man-made Fibers,* 3 Vols. New York: Interscience Publishers, Inc., 1968.

Marsh, J. T. *Textile Finishing,* 2d ed. Metuchen, N.J.: Textile Book Service, 1966.

Matthews, J. M., and H. R. Mauersberger. *Textile Fibers,* 6th ed. New York: John Wiley & Sons, Inc., 1954.

Moncrieff, R. W. *Man-made Fibers,* 5th ed. New York: John Wiley & Sons, Inc., 1970.

Morton, Maurice. *Introduction to Rubber Technology.* New York: Reinhold Publishing Corporation, 1959.

Morton, W. E. and J. W. S. Hearle. *Physical Properties of Textile Fibers.* London: Butterworth & Co., 1962.

Moss, A. J. Ernest. *Textiles and Fabrics.* New York: Chemical Publishing Company, Inc., 1961.

Padgett, R. W. *Textile Chemistry and Testing in the Laboratory.* Minneapolis: Burgess Publishing Co., 1966.

Peters, R. H. *Textile Chemistry,* Vol. I. New York: American Elsevier Publishing Company, Inc. 1963.

————. *Textile Chemistry.* Vol. 2. New York: American Elsevier Publishing Company, Inc., 1967.

Pizzuto, J. J. *101 Weaves in 101 Fabrics.* New York: Textile Press, 1961.

Pizzuto, J. J., and P. L. D'alessandro. *101 Fabrics.* New York: Textile Press, 1952.

Potter, M. D., and B. P. Corbman. *Textiles: Fiber to Fabric,* 4th ed. New York: McGraw-Hill, Inc., 1967.

Reichman, Charles. *Double Knit Fabric Manual.* New York: National Knitted Outerwear Association, 1961.

———. *Knitting Dictionary.* New York: National Knitted Outerwear Association, 1966.

———. *Knitted Stretch Technology.* New York: National Knitted Outerwear Association, 1965.

Reichman, Charles, J. B. Lancashire, and K. D. Darlington. *Knitted Fabric Primer.* New York: National Knitted Outerwear Association, 1967.

Reisfeld, A. *Warp Knit Engineering.* New York: National Knitted Outerwear Association, 1966.

Schwarts, A. M., J. W. Perry, and J. Berch. *Surface Active Agents and Detergents.* New York: Interscience Publishers, Inc., 1957.

Sienko, M. J., and R. A. Plane. *Chemistry.* New York: McGraw-Hill, Inc., 1966.

Silk. New York: Japan Silk Institute, 1962.

Skinkle, J. H. *Textile Testing, Physical, Chemical, Microscopial,* 2d ed. New York: Chemical Publishing Company, Inc., 1949.

Stout, Evelyn E. *Introduction to Textiles,* 3d ed. New York: John Wiley & Sons, Inc., 1970.

Stoves, J. L. *Fiber Microscopy.* Princeton, N.J.: D. Van Nostrand Company, Inc., 1958.

Swirles, Frank M. *Handbook of Basic Fabrics,* 2d ed. Los Angeles: Swirles & Co., 1962.

Technology of Synthetic Fibers, Samuel B. MacFarlane, ed. New York: Fairchild Publications, Inc., 1953.

Textile Chemicals and Auxiliaries, 2d ed., H. C. Speel and E. W. K. Schwarz, eds. New York: Reinhold Publishing Corporation, 1957.

Textile Fibers and Their Properties. Greensboro, N.C.: Burlington Industries, 1970.

Textile Handbook, 4th ed. Washington, D.C.: American Home Economics Association, 1970.

United States Department of Agriculture, Bureau of Mines. *Asbestos, Materials Survey.* Washington, D.C.: Government Printing Office, 1959.

Von Bergen, Werner. *Wool Handbook,* Vol. 1, 3d ed. New York: Wiley-Interscience, 1963.

———. *Wool Handbook,* Vol. 2, Part 1, 3d ed. New York: Wiley-Interscience, 1969.

———. *Wool Handbook,* Vol. 2, Part 2, 3d ed. New York: Wiley-Interscience, 1970.

Walton, Perry. *The Story of Textiles.* New York: Tudor Publishing Co., 1936.

Ward, D. T. *Tufting: An Introduction.* London: Textile Business Press, 1969.

Wingate, Isabel. *Dictionary of Textiles.* New York: Fairchild Publications, Inc., 1967.

———. *Textile Fabrics,* 6th ed. Englewood Cliffs, N.J.: Prentice-Hall, Inc., 1970.

Recommended Periodicals

This list is not comprehensive but includes those periodicals that are of particular interest to the textile student.

American Dyestuff Reporter
American Fabrics Magazine
America's Textile Reporter
Ciba Review
Journal of the Society of Dyers and Colourists
Journal of the Textile Institute
Knitting Times
Modern Textiles
Textile Bulletin
Textile Chemist and Colorist
Textile Industries
Textile Month
Textile Organon
Textile Research Journal
Textile World

Glossary
of Textile Terms

absorption The attraction and retention of gases or liquids within the pores of a fiber; also, the retention of moisture between fibers within yarns and between fibers or yarns within fabrics.

adsorption The retention of gases, liquids, or solids on the surface areas of fibers, yarns, or fabrics.

antique satin A satin-weave fabric made to resemble silk satin of an earlier century. It is used for home-furnishing fabrics.

art linen A heavy plain-weave fabric used for tablecloths and as the basis for many types of embroidered household items.

balanced yarns Yarns in which the twist is such that the yarn will hang in a loop without kinking, doubling, or twisting upon itself.

barathea A closely woven dobby-weave fabric with a characteristic pebbly surface. It is generally made from silk or rayon, often combined with cotton or worsted. The fabric is used for dresses, neckties, and lightweight suits.

batiste A fabric named for Jean Baptiste, a French linen weaver. (1) In cotton, a sheer, fine *muslin*,[1] woven of combed yarns and given a mercerized

[1]Italics denotes a cross reference within the Glossary.

finish. It is used for blouses, summer shirts, dresses, lingerie, infants' dresses, bonnets, and handkerchiefs. (2) A rayon, polyester, or cotton-blend fabric with the same characteristics. (3) A smooth, fine wool fabric that is lighter than *challis*, very similar to fine nuns' veiling. It is used for dresses and negligees. (4) A sheer silk fabric, either plain or figured, very similar to silk mull. It is often called *batiste de soie* and is made into summer dresses.

Bedford cord Lengthwise ribbed durable cloth for outer garments or sport clothes. The corded effect is secured by having two successive *warp* threads woven in plain-weave order. Heavier cords are created with wadding—a heavy, bulky yarn with very little twist—covered by *filling* threads.

beetling A finish primarily applied to linen whereby the cloth is beaten with large wooden blocks in order to flatten the yarns.

bengaline A ribbed fabric similar to *faille*, but heavier, with a coarser rib in the *filling* direction. It may be silk, wool, acetate, or rayon *warp*, with wool or cotton *filling*. The fabric was first made in Bengal, India. It is used for dresses, coats, trimmings, and draperies.

bicomponent fibers Fibers in which two *filaments* of different composition have been extruded simultaneously.

bouclé A fabric woven with bouclé yarns, which have a looped appearance on the surface. In some bouclés only one side of the fabric is nubby; in others, both are rough. Sometimes the bouclé yarn is used as a *warp* rather than a *filling.* Bouclé yarn is very popular in the knitting trade; there are many varieties and weights.

breaking load The minimum force required to rupture a fiber, expressed in grams or pounds.

brins The two adjacent silk *filaments* extruded by the silkworm.

broadcloth A term used to describe several dissimilar fabrics made with different fibers, weaves, and finishes. (1) Originally, a silk shirting fabric so named because it was woven in widths exceeding the usual 29 inches. (2) A tightly woven, high-count cotton cloth with a fine crosswise rib. Fine broadcloths are woven of *combed* yarns, usually with high thread counts, such as 136 × 60 or 144 × 76. They are usually mercerized, Sanforized, and given a soft, lustrous finish. (3) A closely woven wool cloth with a smooth nap, velvety feel, and lustrous appearance. Wool broadcloth can be made with a two-up-and-two-down twill weave or plain weave. In setting up a loom to make the fabric, the loom is threaded very wide to allow for great shrinkage during the fulling process. The fabric takes its name from this wide threading. In higher qualities the cloth is fine enough for garments that are to be closely molded to the figure or draped. Its high-luster finish makes it an elegant cloth. Wool broadcloth is 10 to 16 ounces per yard and is now being made in chiffon weights. (4) A fabric made from silk or man-made *filament fiber* yarns, woven in a plain weave with a fine crosswise rib obtained by using a heavier *filling* than *warp* yarn.

brocade Rich Jacquard-woven fabric with an allover interwoven design of raised figures or flowers. The name is derived from the French word meaning "to ornament." The brocade pattern is emphasized with contrasting surfaces or colors and often has gold or silver threads running through it. The background may be either satin or twill weave. It is used for dresses, draperies, and upholstery.

brocatelle Supposedly an imitation of Italian tooled leather, in which the background is pressed and the figures embossed. Both the background and the figures are tightly woven, generally with a *warp* effect in the figure and a *filling* effect in the background. Brocatelle is employed mainly in upholstery and draperies.

cambric A closely woven white cotton fabric finished with a slight gloss on one side.

Canton flannel A heavy, warm cotton material with a twilled surface and a long soft nap on the back produced by napping the heavy soft-twist yarn. It is named for Canton, China, where it was first made. The fabric is strong and absorbent; it is used for interlinings and sleeping garments.

card sliver A ropelike strand of fibers about ¾ inch to 1 inch in diameter; the form in which fibers emerge from the *carding* machine.

carding A process by which natural fibers are sorted, separated, and partially aligned.

cavalry twill A sturdy twill-weave fabric with a pronounced diagonal cord. It is used for sportswear, uniforms, and riding habits.

challis or challie One of the softest fabrics made, named from the Anglo-Indian term *shalee,* meaning soft. It is a fine, lightweight, plain-weave fabric, usually made of worsted yarns. Challis was formerly manufactured with a small flower design, but now it is made in darker tones of allover prints and solid colors, in the finest quality fabrics.

chambray (1) A plain-woven fabric with an almost square count (80 × 76), a colored *warp* and a white *filling,* which gives a mottled, colored surface. It is used for shirts, children's clothing, and dresses. The fabric is named for Cambrai, France, where it was first made for sunbonnets. (2) A similar but heavier carded-yarn fabric used for work clothes and children's play clothes.

cheesecloth A very loosely woven plain-weave cotton fabric. The yard width is called tobacco cloth. It is used for curtains, costumes, and cleaning cloths.

chiffon A term used to describe many light, gossamer, sheer, plain-weave fabrics. Chiffon can be made of silk, wool, or man-made fibers. It is an open weave with tightly twisted yarns.

china silk A lightweight, soft, plain-weave silk fabric used for lingerie, dress linings, and soft suits.

chino A type of army twill made of combed, two-ply, mercerized yarns in a vat-dyed khaki color. It is now available in a variety of colors.

chintz   A highly lustrous, plain-woven cotton with a bright, glazed surface, generally made by finishing a print cloth construction.

cohesiveness   The ability of fibers to adhere to one another in yarn-manufacturing processes.

combing   A process by which natural fibers are sorted and straightened; a more refined treatment than *carding.*

co-polymer   A *polymer* composed of two or more different *monomers.*

corduroy   A ribbed, high-luster, cut-pile fabric with extra filling threads that form lengthwise ribs or wales. The thread count varies from 46 × 116 to 70 × 250.

core yarn   A yarn in which a base or foundation yarn is completely wrapped by a second yarn.

cotton linters   Cotton fibers that are too short for yarn or fabric manufacturing.

course   A series of successive loops lying crosswise of a knitted fabric.

covert   Generally called covert cloth, a closely woven *warp*-face twill. Its characteristically flecked appearance is produced by using a two-ply yarn so that one dark thread alternates with a light thread. Covert is generally made of wool or cotton, but man-made fibers and blends can also be used.

crepe   A lightweight fabric of silk, rayon, cotton, wool, man-made, or blended fibers, characterized by a crinkled surface that is produced by hard-twist yarns, chemical treatment, weave, or embossing.

cretonne   A plain-weave fabric similar to unglazed *chintz,* usually printed with large designs.

crimp   The waviness of a fiber, usually visible only under magnification.

crystallinity   The degree to which fiber molecules are parallel to each other, though not necessarily to the longitudinal fiber axis.

damask   A firm-textured fabric with raised patterns, similar to *brocade,* but lighter and reversible. Table damasks are Jacquard woven in lustrous designs.

denier   A unit of yarn number equal to the weight in grams of 9000 meters of the yarn.

denim   A twilled fabric made of hard-twist yarns, with the *warp yarns* dyed blue and the *filling yarns* undyed. Sports denim is softer and lighter in weight. It is now available in many colors and in plaids and stripes.

dimensional stability   The degree to which a fiber, yarn, or fabric retains its shape and size after having been subjected to wear and maintenance.

dimity   Literally, double thread; a fine checked or corded cotton sheer made by bunching and weaving two or more threads together.

dotted swiss   A sheer, crisp cotton fabric with either clipped spot or swivel dots.

doupion   Silk yarns made from two cocoons that have been formed in an interlocked manner. The yarn is uneven, irregular, and larger than regular *filaments.* It is used in making *shantung* and doupioni.

drawing   The process by which slivers of natural fibers are pulled out or extended after *carding* or *combing.*

drill   A strong cotton material similar to *denim,* which has a diagonal 2 × 1 weave running up to the left *selvage.*

duck   A durable plain-weave, closely woven cotton, generally made of *ply yarns,* in a variety of weights and thread counts. Often called canvas, it is used for belting, awnings, tents, and sails.

duvetyn   A very high quality cloth that resembles a compact velvet. It has a velvety *hand* resulting from the short nap that covers its surface completely, concealing its twill weave. It is used for suits and coats.

elastic recovery   The ability of a fiber, yarn, or fabric to return to its original length after the tension that produced *elongation* has been released.

elongation   The amount of stretch or extension that a fiber, yarn, or fabric will accept.

faille   A soft, slightly glossy silk, rayon, or cotton fabric in a rib weave, with a light, flat, crossgrain rib or cord made by using heavier yarns in the *filling* than in the *warp.*

felt   A nonwoven fabric in which the fibers develop a tight bond and will not ravel. It is used for coats, hats, and many industrial purposes.

fiber morphology   The form and structure of a fiber, including its biological structure, shape, cross section, and microscopic appearance.

fibrils   Bundles of fiber cells.

filament fibers   Long, continuous fibers that can be measured in meters or yards, or in the case of man-made fibers, in kilometers or miles.

filling yarns   Yarns that run perpendicular to the longer dimension or *selvage* of a fabric.

flannel   A catch-all designation for a great many otherwise unnamed fabrics in the woolen industry.

Flannel is woven in various weights of worsted, woolen, or a mixture of both. It can even be made of man-made fibers. The surface is slightly napped in finish. A wide range of weights is available: an 11-ounce flannel is made for suits, and there are tissue-weight flannels for dresses.

flannelette A soft, plain- or twill-weave cotton fabric lightly napped on one side. The fabric can be dyed solid colors or printed. It is popular for lounging and sleeping garments.

fleece Wool sheared from a living lamb.

flexibility The property of bending without breaking.

foulard A lightweight silk, rayon, cotton, or wool fabric characterized by its twill weave. Foulard has a high luster on the face and is dull on the reverse side. It is usually printed, the patterns ranging from simple polka dots to elaborate designs. It is also made in plain or solid colors. Foulard has a characteristic *hand* that can be described as light, firm, and supple.

gabardine A hard-finished, clear-surfaced, twill-weave fabric made of either natural or synthetic fibers. The diagonal lines are fine, close, and steep and are more pronounced than in serge. The lines cannot be seen on the wrong side of the fabric.

gauze A plain-weave fabric with widely spaced yarns, used for such things as bandages. Some weights of gauze can be stiffened for curtains or other decorative or apparel purposes.

gingham A light- to medium-weight plain-weave cotton fabric. It is usually yarn-dyed and woven to create stripes, checks, or plaids. The fabric is mercerized to produce a soft, lustrous appearance; it is sized and calendered to a firm and lustrous finish. Gingham is used for dresses, shirts, robes, curtains, draperies, and bedspreads. The thread count varies from about 48 × 44 to 106 × 94.

greige The state of a fabric before a finish has been applied.

grenadine A tightly twisted *ply yarn* composed of two or three *singles*.

grosgrain A closely woven firm corded fabric often made with a cotton filling. The cords are heavier than in *poplin*, rounder than in *faille*.

habutai A soft, lightweight silk dress fabric originally woven in the gum on hand looms in Japan. It is sometimes confused with china silk, which is technically lighter in weight.

hackling A *combing* process that separates short fibers from long fibers.

hand The "feel" of a fabric; the qualities that can be ascertained by touching it.

herringbone A fabric in which the pattern of weave resembles the skeletal structure of the herring. It is made with a broken twill weave that produces a balanced, zigzag effect and is used for sportswear, suits, and coats.

homespun A coarse plain-weave fabric, loosely woven with irregular, tightly twisted, and unevenly spun yarns. It has a hand-woven appearance and is used for coats, suits, sportswear, draperies, and slipcovers.

homopolymer A *polymer* composed of one substance or one type of molecule.

honan Originally, a fabric of the best Chinese silk, sometimes woven with blue edges. It is now made to resemble a heavy *pongee*, with slub yarns in both *warp* and *filling*. Honan is manufactured from silk or from man-made fibers. It is used for women's dresses.

hopsacking An open-basket-weave *ply-yarn* fabric of cotton, linen, or rayon. The weave is similar to the sacking used to gather hops, hence the name. It is used for dresses, jackets, skirts, and blouses.

huck or huckaback A toweling fabric with a honeycombed surface made by using heavy filling yarns in a dobby weave. It has excellent absorbent qualities. Huck is made in linen, cotton, or a mixture of the two. In a mixture it is called a "union" fabric.

hydrophilic Water loving; having a high degree of moisture absorption or attraction.

hydrophobic Water repelling; having a low degree of moisture absorption or attraction.

jean A sturdy cotton fabric, softer and finer than *drill*, made in solid colors or stripes. It is used for sport blouses, work shirts, women's slacks, and children's playclothes.

jersey Elastic knitted fabric in a stockinette stitch. It was first made on the Island of Jersey off the English coast and used for fishermen's clothing. Jersey can be made from wool, cotton, rayon, nylon, other man-made fibers, or a combination of any of these. The term is frequently applied to tricot-knitted fabrics used for dresses.

kersey A thick, heavy, pure wool and cotton twill-weave similar to *melton*. It is well fulled, with a nap and a close-sheared surface. Kersey is used for uniforms and overcoats.

lace An open-work cloth with a design formed by

a network of threads made by hand or on special lace machinery, with bobbins, needles, or hooks.

lamb's wool Wool clipped from sheep less than eight months old.

lawn A lightweight, sheer, fine cotton or linen fabric, which can be given a soft or crisp finish. It is sized and calendered to produce a soft, lustrous appearance. Lawn is used for dresses, blouses, curtains, lingerie, and as a base for embroidered items.

linear polymer A *polymer* formed by end-to-end linking of molecular units. The resulting polymer is very long and narrow. It is typical of fibrous forms.

loft The springiness or fluffiness of a fiber.

longcloth A fine, soft cotton cloth woven of softly twisted yarns. It is similar to *nainsook* but slightly heavier, with a duller surface. Longcloth is so called because it was one of the first fabrics to be woven in long rolls. It is also a synonym for *muslin* sheeting of good quality. The fabric is used for underwear and linings.

luster The gloss, sheen, or shine of a fiber, yarn, or fabric.

macromolecule A large molecule formed by hooking together many small molecule units. The term can be used synonymously with polymolecule or *polymer.*

madras (1) A finely woven, soft, plain- or Jacquard-weave fabric with a stripe in the lengthwise direction and Jacquard or dobby patterns woven in the background. Some madras is made with woven checks and cords. It can be used for blouses, dresses, and shirts. (2) A fabric handwoven in India from cotton yarns dyed with native vegetable colorings. The designs are usually rather large, bold plaids that soften in color as the dyes fade and bleed.

marquisette A light, strong, sheer, open-textured curtain fabric, often with dots woven into the surface. The thread count varies from 48 × 22 to 60 × 40.

matelassé A soft double or compound fabric with a quilted appearance. The heavier type is used in draperies and upholstery, while crepe matelassé is popular in dresses, semiformal and formal suits and wraps, and trimmings.

melton A thick, heavily felted or fulled wool fabric in a twill or satin weave, with a smooth, lustrous, napped surface. In less expensive meltons the *warp yarn* may be cotton instead of wool.

micronaire fineness The weight in micrograms of 1 inch of fiber.

moisture regain The moisture in a material determined under prescribed conditions and expressed as a percentage of the weight of the moisture-free specimen.

molecular orientation The degree to which fiber molecules are parallel to each other and to the longitudinal axis of the fiber.

monk's cloth A heavy, loosely woven basket-weave fabric in solid colors or with stripes or plaids woven into the fabric. It is used chiefly for draperies and slipcovers.

monofilament yarn Yarn composed of only one fiber *filament.*

monomer A single unit or molecule from which *polymers* are formed.

mousseline de soie Literally, "muslin of silk"; silk organdy, a plain-weave silk chiffon-weight fabric with a slight stiffness.

multicomponent fabric A fabric in which at least two layers of material are sealed together by an adhesive.

multifilament yarn Yarn composed of several fiber *filaments.*

multilobal A fiber with a modified cross section exhibiting several lobes.

muslin A large group of plain-weave cotton fabrics ranging from lightweight to heavyweight. The sizing may also be light or heavy. Muslin can be solid colored or printed. It is used for dresses, shirts, sheets, and other domestic items.

nainsook A fine, soft cotton fabric in a plain weave. Better grades have a polished finish on one side. When it is highly polished, nainsook may be sold as polished cotton. In low-priced white goods *cambric, longcloth,* and nainsook are often identical before converting; the finishing process gives them their characteristic texture, but even so it is often difficult to distinguish one from the other. Nainsook is heavier and coarser than *lawn.* It is usually found in white, pastel colors, and prints and is used chiefly for infants' wear, lingerie, and blouses.

ninon A smooth, transparent, closely woven *voile,* with the *warp yarns* grouped in pairs. It is available in plain or novelty weaves. Man-made fibers are generally used for glass curtains and dress fabrics.

nonthermoplastic Not capable of being softened by heat.

oleophilic Tending to absorb and retain oily materials.

oleophobic Tending to repel oily materials.

organdy A thin, transparent, stiff, wiry cotton *muslin* used for dresses, neckwear, and trimmings. Organdy, when chemically treated, keeps its crispness through many launderings and does not require restarching. It crushes readily but is easily pressed. Shadow organdy has a faint printed design in self-color.

organzine A yarn of two or more plies with a medium twist.

orientation See *molecular orientation.*

Osnaburg Named for the town in Germany where it was first made, a coarse cotton or blended fiber fabric in a plain weave that resembles crash. It is finished for use in upholstery, slacks, and sportswear. It was originally used unbleached for grain and cement sacks.

ottoman A heavy corded silk or synthetic fabric with larger and rounder ribs than *faille,* used for coats, skirts, and trimmings. *Fillings* of the cloth are usually cotton or wool, and they should be completely covered by the silk or man-made fiber *warp.*

outing flannel A soft, lightweight, plain- or twill-weave fabric usually napped on both sides. Most outing flannels have colored yarn stripes. Outing flannel soils easily, and the nap washes and wears off. It is used chiefly for sleeping garments.

Oxford shirting A cotton or blended fabric in a basket weave first made in Oxford, England and used for shirts, blouses, and sportswear.

percale A medium-weight, plain-woven printed cotton, such as 80 × 80; a staple of dress goods. Percale sheets are high quality, with a count of at least 180 threads per square inch. Most percales are made of combed yarns with a count of 84 × 96 or 180 threads per inch. Some fine percale sheets count over 200 threads per inch, such as 96 × 104 or 96 × 108.

picks See *filling yarns.*

pilling The formation of tiny balls of fiber in the surface of a fabric.

piqué Strictly a ribbed or corded cotton with wales running across the fabric, formed by *warp* ends. The term is often used in the trade to refer to Bedford cord or warp piqué, in which the cords run lengthwise.

plissé Usually a print cloth treated with chemicals that cause parts of the cloth to shrink, creating a permanently crinkled surface.

ply yarn A yarn in which two or more single strands are twisted together.

polymer A large molecule produced by linking together many *monomers.*

polymerization The conversion of *monomers* into large molecules or *polymers.*

pongee (1) A thin, natural tan-colored silk fabric originally made of wild Chinese silk with a knotty rough weave, named for the Chinese *Pun-ki,* meaning "woven at home on one's own loom." It is used primarily for summer suits and dresses, and both plain fabrics and prints are used for decorative purposes. (2) A staple fine-combed cotton fabric finished with a high luster and used for underclothing. (3) A man-made fiber fabric simulating pongee.

poplin A tightly woven, high-count cotton with fine cross ribs formed by heavy *filling yarns* and fewer, finer *warp yarns.* Poplin has heavier ribs, heavier threads, and a slightly lower count than broadcloth, ranging from 80 × 40 to 116 × 56.

pulled wool Wool pulled from the hide of a slaughtered animal.

raw silk Silk that has not been degummed.

reeling The process of winding silk filaments onto a wheel.

reprocessed wool Wool fibers reclaimed from scraps of fabric that have never been used.

resiliency The ability of a fabric to return to its original shape after compressing, bending, or other deformation.

retting The removal, usually by soaking, of the outer woody portion of the flax plant to gain access to the fibers.

reused wool Wool fibers reclaimed from fabrics that have been worn or used.

roving The process by which a sliver of natural fiber is attenuated to between $\frac{1}{4}$ and $\frac{1}{8}$ of its original size; also, the product of this operation.

sailcloth A very heavy, strong, plain-weave fabric made of cotton, linen, or jute. There are many qualities and weights. Sailcloth can be used for sportswear, slipcovers, curtains, and other heavy-duty items.

sateen A cotton or spun-yarn fabric characterized by floats running in the *filling* direction. It is usually mercerized and used for linings, draperies, and comforters.

saturation regain   The moisture in a material at 95 or 100 percent relative humidity.

scroop   A characteristic rustling or crunching sound acquired by silk that has been immersed in solutions of acetic or tartaric acid and dried without rinsing. It is probably caused by acid microcrystals in the fiber rubbing across each other.

scutching   The separation of the outer covering of the flax stalk from the usable fibers.

seersucker   A lightweight cotton or cotton blend with crinkled stripes woven in by setting some of the *warp yarns* tight and others slack.

selvage   The long, finished edges of a bolt of fabric.

sericulture   The raising of silkworms and production of silk.

shantung   Originally, a hand-loomed plain-weave fabric made in China. Made of *wild silk,* the fabric had an irregular surface. Today the term shantung is applied to a plain-weave fabric with heavier, rougher yarns running in the crosswise direction of the fabric. These are single complex yarns of the slub type. The fabric can be made of cotton, silk, or man-made fibers.

sharkskin   (1) A cotton, linen, silk, or man-made fiber fabric with a sleek, hard-finished, crisp, and pebbly surface and a chalky luster. *Filament yarns,* when used, are twisted and woven tightly in either a plain-weave or a basket-weave construction, depending upon the effect desired. *Staple fiber yarns* are handled in the same manner, except for wool. (2) A wool fabric characterized by its twill weave. The yarns in both *warp* and *filling* are alternated, white with a color, such as black, brown, or blue. The diagonal lines of the twill weave run from left to right; the colored yarns from right to left.

shed   The opening between *warp yarns* through which *filling yarns* are passed.

shoddy   See *reused wool.*

silk noil   Short ends of silk fibers used in making rough, textured, spun yarns or in blends with cotton or wool; sometimes called *waste silk.*

singles   A strand of several filaments held together by twist.

specific gravity   The density of a fiber relative to that of water at 4°C.

spinning quality   The ease with which fibers lend themselves to yarn-manufacturing processes; cohesiveness.

spun silk   Yarns made from short fibers of pierced cocoons or from short ends at the outside and inside edges of the cocoons.

spun yarns   Yarns composed of *staple fibers.*

staple fibers   Short fibers that are measured in inches or fractions of inches.

suede fabric   A woven or knitted fabric of cotton, man-made fibers, wool, or blends, finished to resemble suede leather. It is used in sport coats, gloves, linings, and cleaning cloths.

surah   A soft, usually twilled fabric often of silk or man-made fibers, woven in plaids, stripes, or prints. It is used for ties, mufflers, blouses, and dresses.

swiss   See *dotted swiss.*

synthesized fibers   Fibers made from chemicals that were never fibrous in character.

synthetic fiber   A fiber made from chemicals that were never fibrous in form; more frequently referred to as "man-made synthesized fiber."

taffeta   A fine, plain-weave fabric, smooth on both sides, usually with a sheen on its surface. It is named for the Persian fabric "taftan." Taffeta may be a solid color or printed or woven so that the colors appear iridescent. It is often constructed with a fine rib; this fabric is correctly called *faille taffeta.*

tapestry   A fabric in which the pattern is woven with colored *weft* threads. It is used extensively for wall hangings and table covers.

tenacity   The tensile strength of a fiber, expressed as force per unit of linear density of an unstrained specimen. It is usually expressed in grams per *denier* or grams per *tex.*

tensile strength   The maximum tensile stress required to rupture a fiber, expressed as pounds per square inch or grams per square centimeter.

terry cloth   A heavy, absorbent cotton made with extra heavy *warp* threads woven into loops on one or both sides.

tex   A system of yarn numbering that measures the weight in grams of one kilometer of yarn.

textile   Any product made from fibers.

thermoplastic   Tending to become soft and/or moldable upon application of heat.

thermosetting   A procedure in which a substance is softened by heat, whereupon the substance undergoes chemical change, becomes firm, and assumes a completely different structure and different properties. The substance cannot be softened by reapplication of heat.

three-dimensional polymer   A *polymer* formed when molecules unite in both length and width, producing a relatively rigid structure. This is typical of polymers used in processing.

ticking   A heavy twill made with a colored yarn stripe in the *warp*. It is used for mattress covers, home-furnishings, and sportswear.

trademark   A word, letter, device, or symbol used in connection with merchandise and alluding distinctly to the origin or ownership of the product to which it is applied.

trade name   A name given by a manufacturer or merchant to a product to distinguish it as one produced or sold by him. It is called, more accurately, a trademark name and may be protected as a trademark.

tram silk   A low-twist ply silk yarn formed by combining two or three single strands.

trilobal   A fiber with a modified cross section having three lobes.

tropical suiting   A lightweight plain-weave suiting for men's and women's summer wear. It has various weaves and is made of a variety of fibers. If called *tropical worsted,* it must be an all wool worsted fabric.

tweed   A term derived from the river Tweed in Scotland, where the fabrics were first woven. It is now used to describe a wide range of light to heavy, rough-textured, sturdy fabrics characterized by their mixed color effect. Tweeds can be made of plain, twill, or herringbone weave, in practically any fiber or mixture of fibers.

Tussah silk   See *wild silk.*

unbalanced yarns   Yarns in which there is sufficient twist to set up a torque effect, so that the yarn will untwist and retwist in the opposite direction.

velour   A soft, closely woven, smooth fabric with a short, thick pile. It is named for the French word for velvet. Velour is often made of cotton, wool, or mohair.

velvet   A fabric with a short, soft, thick, warp-pile surface, usually made of silk or man-made pile fiber with a cotton back. It is sometimes made of all silk or all cotton. The fabric is often woven double, face to face, and then, while still on the loom, it is cut apart by a small shuttle knife. There are several varieties of velvet, which differ in weight, closeness of pile, and transparency.

virgin wool   New wool that is made into yarns and fabrics for the first time.

voile   A sheer, transparent, soft, lightweight plain-weave fabric made of highly twisted yarns. It can be composed of wool, cotton, silk, or a man-made fiber. Voile is used for blouses, dresses, curtains, and similar items.

waffle cloth   A fabric with a characteristic honeycomb weave. When made in cotton it is called waffle piqué. It is used for coatings, draperies, dresses, and toweling.

wale   A column of loops that are parallel to the loop axis and to the long measurement of a knit fabric.

warp yarns   Yarns that run parallel to the *selvage* or longer dimension of a fabric.

waste silk   See *silk noil.*

weft yarns   See *filling yarns.*

whipcord   A twill-weave fabric similar to *gabardine* but with a more pronounced diagonal rib on the right side. It is so named because it simulates the lash of a whip. Cotton whipcords are often four-harness warp-twill weaves.

wickability   The property of a fiber that allows moisture to move rapidly along the fiber surface and pass quickly through the fabric.

wild silk   Silk produced by moths of species other than *Bombyx mori.* It is tan to brown in color and is coarser and more uneven than ordinary silk. It is usually called *Tussah silk.*

woof yarns   See *filling yarns.*

zibeline   A fabric made of wool, cotton, camel hair, mohair, or man-made fibers. It is characterized by a long, sleek nap brushed, steamed, and pressed in one direction, thus hiding the underlying satin weave.

Index